Integrated Assessment of Climate Change Impacts and Urban Resilience

Integrated Assessment of Climate Change Impacts and Urban Resilience

From Climate and Hydrological Hazards to Risk Analysis and Measures

Editors

Beniamino Russo
Eduardo Martínez-Gomariz

MDPI • Basel • Beijing • Wuhan • Barcelona • Belgrade • Manchester • Tokyo • Cluj • Tianjin

Editors
Beniamino Russo
AQUATEC (SUEZ Group)
EUPLA (Universidad de Zaragoza)
Universitat Politècnica de Catalunya
Spain

Eduardo Martínez-Gomariz
Cetaqua · Water Technology Centre
Universitat Politècnica de Catalunya
Spain

Editorial Office
MDPI
St. Alban-Anlage 66
4052 Basel, Switzerland

This is a reprint of articles from the Special Issue published online in the open access journal *Sustainability* (ISSN 2071-1050) (available at: https://www.mdpi.com/journal/sustainability/special_issues/Flood_Damage_Assessment_Urban).

For citation purposes, cite each article independently as indicated on the article page online and as indicated below:

LastName, A.A.; LastName, B.B.; LastName, C.C. Article Title. *Journal Name* **Year**, *Volume Number*, Page Range.

ISBN 978-3-03943-162-5 (Hbk)
ISBN 978-3-03943-163-2 (PDF)

Cover image courtesy of Eduardo Martínez-Gomariz.

Contents

About the Editors

Beniamino Russo has a Ph.D. in Civil engineering and is a Full Professor of Hydraulics and Hydrology at the Technical College of La Almunia (University of Zaragoza, Spain) and Director of the GIHA (Group of Hydraulic and Environmental Engineering) Research Group. He is also an R&D project manager of AQUATEC (Suez Group) and an associate professor at the Technical University of Catalonia. He is the author of more than 150 papers published in JCR-indexed journals and international conferences proceedings in the field of flood risk management and urban drainage.

Eduardo Martínez-Gomariz has 15 years of experience in water-related projects and he currently holds the position of R&D Project Manager at Cetaqua (SUEZ group). He has been involved in the management and development of urban resilience-related EU-funded projects, such as PEARL (FP7), BINGO (H2020), and RESCCUE (H2020). Moreover, Eduardo teaches hydraulics to undergraduates at the Universitat Politècnica de Catalunya (UPC) since 2018. He is the author of 28 articles in peer-reviewed journals, and is a member of the panel of editors of the Journal of Flood Risk Management. He has presented 26 research works in national and international conferences. Furthermore, he has supervised 8 academic works (3 Bachelor final projects, 3 Master Thesis, and 2 Doctoral Thesis).

Preface to "Integrated Assessment of Climate Change Impacts and Urban Resilience"

Modern cities are complex, vulnerable, and continuously evolving systems. In these dynamic areas, the existence of interacting strategic services and interconnected critical infrastructures, as well as the involvement of a multiplicity of stakeholders, adds complexity to their management.

Furthermore, urban resilience refers to the ability of an urban system—and all its key constituent parts—to maintain or rapidly return to desired functions in the face of a disruption, to adapt to change, and to quickly transform systems that limit current or future adaptive capacity.

Besides, the significant impacts of climate dynamics (such as intense precipitation events, tidal effects, droughts, or heat waves) on the urban strategic services, people, natural environment, and economy, as well as the aggravation of current conditions and the emergence of new hazards, also need to be considered in their management.

In this context, this Special Issue will publish cutting-edge methods and tools for a comprehensive assessment of climate change impacts and urban resilience and, also, for the proposal and prioritization of adaptation measures and strategies to cope with climate-related hazards and risks.

Beniamino Russo, Eduardo Martínez-Gomariz
Editors

Editorial

Integrated Assessment of Climate Change Impacts and Urban Resilience: From Climate and Hydrological Hazards to Risk Analysis and Measures

Marc Velasco [1,*], Beniamino Russo [1,2] and Eduardo Martínez-Gomariz [3,4]

[1]   Aquatec, SUEZ Advanced Solutions-Urban Drainage and Resilience Direction, Paseo de la Zona Franca, 46–48, 08038 Barcelona, Spain; brusso@aquatec.es

[2]   Grupo de Ingeniería Hidráulica y Ambiental (GIHA) (Group of Hydraulic and Environmental Engineering), Escuela Politécnica de La Almunia (EUPLA, Universidad de Zaragoza) (Technical College of La Almunia, University of Zaragoza), Calle Mayor, 5, 50100 Zaragoza, Spain

[3]   Cetaqua, Water Technology Centre, Carretera d'Esplugues, 75, 08940 Barcelona, Spain; eduardo.martinez@cetaqua.com

[4]   Flumen Research Institute, Universitat Politècnica de Catalunya, Jordi Girona 1–3, 08034 Barcelona, Spain

*   Correspondence: marc.velasco@suez.com

Received: 6 August 2020; Accepted: 7 August 2020; Published: 10 August 2020

Abstract: This Special Issue brings together recent research findings related to urban resilience, in particular taking into account climate change impacts and hydrological hazards. Taking advantage of the work done in the H2020 RESCCUE project, 12 different papers dealing with several issues related to the resilience of urban areas have been published. Due to the complexity of cities, urban resilience management is one of the key challenges that our societies have to deal with in the near future. In addition, urban resilience is a transversal and multi-sectorial issue, affecting different urban services, several hazards, and all the steps of the risk management cycle. This is precisely why the papers contained in this Special Issue focus on varied subjects, such as impact assessments, urban resilience assessments, adaptation strategies, flood risk and urban services, always focusing on at least two of these topics.

Keywords: urban resilience; climate change; impact assessment; adaptation strategies; urban services

---

1. Introduction

We live in a world of cities, and that trend will continue in the future. Today, 54% of the world's population lives in urban areas, a proportion that is expected to increase to 66% by 2050 [1]. In addition, climate change is adding pressures and uncertainties that pose challenges to society, economy, and environment. In this case, focusing on the impacts to urban areas [2], climate change can affect basic urban services, such as water or energy supply, making the continuously functioning city capacity crucial for most parts of the world population. As the United Nations have stated, managing urban areas has become one of the most important development challenges of the 21st century [1].

According to UN-Habitat [3], urban resilience refers to the ability of human settlements to withstand and to recover quickly from any plausible hazards. Resilience against shocks and stresses not only refers to reducing risks and damage from disasters (i.e., loss of lives and assets), but also the ability to quickly bounce back to a stable state; the ability to adapt and transform towards sustainability. While typical risk reduction measures tend to focus on a specific hazard, leaving out risks and vulnerabilities due to other types of perils, resilience adopts a multiple hazards approach, considering all types of plausible climate-related threats.

Cities face a growing range of adversities and challenges in the 21st century, and increasing urban resilience is the only way to survive and adapt to the coming shocks and stresses that may occur [4]. Due to climate change, critical disruptions occur too often in cities around the world. On the other hand, urban areas are complex systems that cannot be understood by sectorial and disciplinary approaches alone [5]. In this context, the RESCCUE (RESilience to cope with Climate Change in Urban arEas—a multisectorial approach focusing on water) project aims to assess current and future resilience (related to future climate change scenarios) through a multisectorial approach, taking water sector as the focus. Climatic drivers and pressures affecting the urban water cycle, such as droughts or heavy rains, can produce critical direct impacts on strategic urban services (water supply, wastewater and stormwater drainage, wastewater treatment, solid waste, telecommunication, energy supply, transport, etc.), and cause cascading collateral impacts on other services. Given the interdependencies existent between the several city services, RESCCUE focuses on the cascading failures that involve several urban functions [6].

2. The RESCCUE Project

The RESCCUE project (www.resccue.eu) aims to help cities around the world to become more resilient to physical, social, and economic challenges [7]. During the last four years, RESCCUE has generated models and tools to bring this objective to practice, make these tools available to be deployed to different types of cities, with different climate change pressures. RESCCUE has also supported cities preparing their resilience action plans (RAP) by developing guidance materials and plan templates. The consortium is led by Aquatec—SUEZ Advanced Solutions, and consists of a total of 17 partners with the three city councils of the research sites (Barcelona, Bristol and Lisbon), the United Nations agency UN-Habitat, several urban services companies, research centers, universities and SMEs (small and medium enterprises), all of them with a key role on resilience management in the three research sites.

2.1. RESCCUE Goals

The main goal of RESCCUE is to help cities around the world to become more resilient. To achieve this overall goal, a set of specific objectives are pursued, including:

1. Compilation, generation, and analysis of different local climate simulations to set up future climate-related scenarios in a coherent way and suitable for users' needs.
2. Improve the understanding of the effects of selected climatic drivers on the urban water cycle in each research site, and identify vulnerabilities of each urban service that will lead to increased social security.
3. Assess the direct impacts of these drivers on all the urban services and the cascade collateral impacts on the ones connected to them for the current situation and future climate change scenarios. The impacts will be assessed in terms of hazard and risk for each analyzed urban service, for the whole set of selected scenarios at each research site.
4. Develop a methodology to assess urban resilience with respect to different climatic pressures, based on the interaction among different urban services. In addition, the improvements related to the operational performance of urban systems will be evaluated, as the urban resilience framework generated will be designed to be adapted to the operational platforms currently in use by urban operators.
5. Explore and assess the economic and societal impacts of multiple feasible mitigation and adaptation measures and technologies to reduce climate change effects on the urban services and their collateral impacts. Based on the impacts evaluated on key urban services and on the needs of end users enrolled in the RESCCUE project, an inventory of the most appropriate mitigation and adaptation options with a special focus on nature-based solutions will be established. The result will constitute a portfolio of validated and prioritized improving resilience strategies, based on multi-criteria analysis, integrating technological and non-technological alternatives, to better cope with challenges raised by climate change.

6. Elaborate a RAP for each of the case study cities, considering the inputs of all local partners and stakeholders of each site, and led by the three involved local resilience offices. The civil protection and emergency sectorial plans will be analyzed to improve coordination during shocks and stresses, as these plans can benefit from RAPs inputs and vice versa.

7. Build a shared awareness and perception of challenges and opportunities, to guide actions and future collaborative approaches, by engaging leading universities and research centers, local governments, large companies, SMEs, non-governmental organizations and citizens from the three research sites.

2.2. RESCCUE Methodology

The RESCCUE project is being implemented through a set of eight WPs (work packages) described below (WP1 to WP6 is where the technical work is focused, whereas WP7 deals with communication and exploitation and WP8 is related to project management). Figure 1 depicts the project structure adopted by RESCCUE, specifying the relations among WPs and the main outputs.

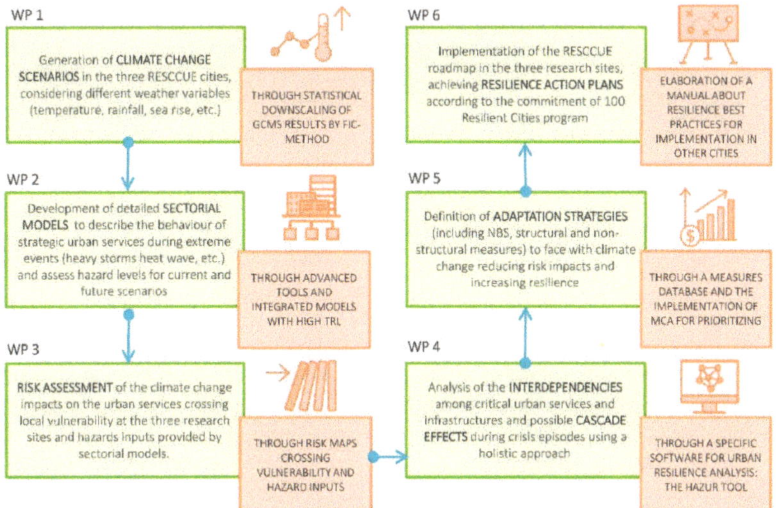

Figure 1. Resilience to Cope with Climate Change in Urban Areas (RESCCUE) Project structure and technical details. Source: [7].

The use of detailed models and software tools is essential to analyze the behavior and the response of strategic services and critical infrastructures with respect to specific pressures and drivers related to climate change. Moreover, the outputs of these sectorial models will be used to assess hazard, vulnerability, and risk levels for current and future scenarios, where a large set of measures and strategies will be simulated and evaluated in terms of impacts reduction.

Once the detailed knowledge of each urban service has been acquired through available data, past experiences, and simulation results, then the interdependencies between them and the cascade effects due to failures or extreme climate events can be studied. Within RESCCUE, this is done with two different approaches characterized by a different level of detail (Figure 2):

1. Detailed approach: advanced models and tools to describe specific cascading effects produced by extreme climate events on several urban services are developed. Then, the analysis of certain impact events could be achieved via the use of loosely coupled models and tools (integrated models), using the outputs of one as inputs of the other, being able to simulate cascading effects in a detailed but simple way. In this case, adaptation strategies and measures will be proposed

and prioritized based on hazard and risk reduction but, also, through multi-criteria analysis, providing an overview of other kinds of co-benefits

2. Holistic approach: using a methodology for holistic resilience assessment, the relations and the cascading effects among the different urban services can be analyzed. In this case, adaptation measures and strategies will be focused on the recovery of the normal functioning of the city and, specifically, of its strategic urban services and infrastructures. This concept will be expressed by the concept of recovery time and the efficiency of the measures and strategies, in terms of decrease of recovery.

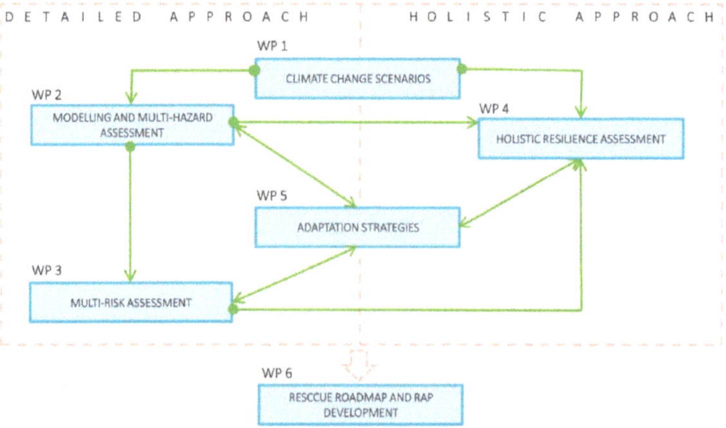

Figure 2. Summary of RESCCUE framework. Source: [7].

With the detailed approach, the analysis of hazard and risk produced by complex interactions and cascade effects involving different urban sectors is done. Then, as not all sectorial models are studied in detail and coupled with others, the whole spectrum of interdependencies and cascading effects is then covered by the holistic approach with a minor level of detail.

The two approaches presented before, coexist in the several different work packages studied in the RESCCUE project. Whereas some tasks are only part of the detailed approach, some others only focus on the holistic one, while there are a few that belong to both, linking the two and allowing to combine them.

The combination of both approaches allows to understand the functioning of the city as a whole, while focusing on some very detailed impacts that are crucial to understand how the several city services affect each other.

By having this detailed–holistic approach, the RESCCUE project has been able to deliver a very useful resilience roadmap for the cities in the form of a RAP, where the strategic lines in which the city must focus are also fed with concrete measures that will be applied to solve specific problems.

3. Special Issue

During the four years that the RESCCUE project has lasted, many outputs have been produced related to urban resilience. Obviously, during the first years of the project, the main results generated were related to climate change scenarios and modelling of climatic variables [8–10], hazard assessment for several specific urban services [11,12], impact assessment methodologies and implementations to specific sectors [13–15], and the preparation of the resilience assessment framework to be used in the project [16,17].

As the project advanced, some of the initial WPs finished, and thus, right now there are no new results related to climate change scenarios, for example. This is precisely why the current Special

Issue has been mainly dealing with the topics of impact assessments, urban resilience assessments, adaptation strategies, flood risk and urban services.

Due to the complexity of cities, urban resilience is a transversal and multi-sectorial issue, affecting different urban services, several hazards and all the steps of the risk management cycle. This is precisely why the topics contained in this Special Issue overlap (Figure 3), which is why all the papers presented deal with at least two of these topics (Table 1).

Table 1. List of the papers published in this Special Issue, classified by the several topics addressed in them.

Title of the Paper	Flood Risk	Adaptation Strategies	Urban Resilience Assessments	Urban Services	Impact Assessment
Assessment of Urban Flood Resilience in Barcelona for Current and Future Scenarios. The RESCCUE Project	X		X		X
Flood Risk Assessment in an Underground Railway System under the Impact of Climate Change—A Case Study of the Barcelona Metro	X			X	X
Methodology to Prioritize Climate Adaptation Measures in Urban Areas. Barcelona and Bristol Case Studies		X			X
Socio-Economic Assessment of Green Infrastructure for Climate Change Adaptation in the Context of Urban Drainage Planning	X	X			
Interlinking Bristol Based Models to Build Resilience to Climate Change	X			X	X
Flood Depth-Damage Curves for Spanish Urban Areas	X				X
The Contribution of NBS to Urban Resilience in Stormwater Management and Control: A Framework with Stakeholder Validation	X	X			
RAF Resilience Assessment Framework—A Tool to Support Cities' Action Planning			X		
Investigating the Effects of Pluvial Flooding and Climate Change on Traffic Flows in Barcelona and Bristol	X			X	X
Urban Resilience to Flooding: Triangulation of Methods for Hazard Identification in Urban Areas	X	X			
Climate Change Implications for Water Availability: A Case Study of Barcelona City				X	X
Electrical Grid Risk Assessment Against Flooding in Barcelona and Bristol Cities	X				X

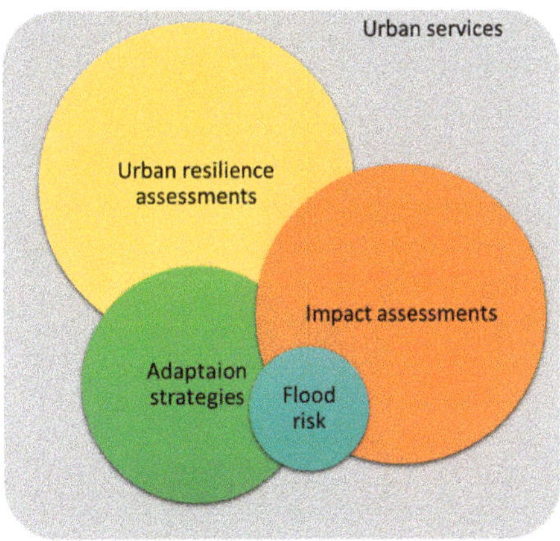

Figure 3. Main topics analyzed in the papers of this Special Issue.

4. RESCCUE Toolkit

Throughout the project's lifetime, RESCCUE has placed much effort on dissemination tasks, aiming at allowing general and specialized audiences to access information about the project progress and its outcomes, as well as promoting the widest application of the RESCCUE outcomes in other cities.

In this regard, in addition to the scientific production generated by the project partners, such as this Special Issue, RESCCUE has also developed the RESCCUE toolkit, an interactive space where the main project's results are gathered, along with a set of guidelines outlining the steps to be taken to make your city resilient. In there, all the tools, datasets and methodologies developed within RESCCUE can be found, sorted by topic and by the three case studies: Barcelona, Lisbon and Bristol.

5. Final Remarks from the RESCCUE Coordinator: Marc Velasco

During the last four years, the RESCCUE project has been my life. For quite a long time, I have been repeating over and over one of the mottos of the project: cities are complex systems of interconnected systems. The RESCCUE consortium, with a varied canvas of partners from different worlds, has been exactly like a city, a complex system of interconnected entities. Managing this has been a challenging but very satisfying task, as we jointly managed to overcome all the problems that appeared along the way.

As you will see while reading this Special Issue, RESCCUE has been a challenging and successful project, that has only been possible due to the hard work of all people involved. I would like to take this opportunity to acknowledge the work and support from all my colleagues who worked in the project. RESCCUE project has been quite a journey for all of us, and I hope that it can also become something special to you.

Along the 4 years of the project, we have learnt that the only way to make our cities stronger and more prepared is by working together in a holistic and transversal way. This is why I would like to tell decision makers and urban service operators that you are not alone in this. Therefore, take advantage of all the work that we have done, so you do not have to start from square one. If you have not done it yet, now is the time to check the RESCCUE Toolkit, where you will be able to find the key RESCCUE results to replicate the work in your city.

What we have started in RESCCUE is only the beginning. Now it is time to pass on the baton, so you can move forward to transform your city to be more and more prepared for the coming challenges.

Author Contributions: M.V., has been the Coordinator of the RESCCUE Project and the main contributor to this Editorial. B.R. and E.M.-G., Work Package leaders of RESCCUE and editors of this Special Issue, have mainly contributed to Section 3 of this editorial, presenting the work included in this Special Issue. All authors have read and agreed to the published version of the manuscript.

Funding: This research was funded by Horizon2020 Programme, Grant Agreement No. 700174.

Acknowledgments: This Special Issue presents some of the results achieved in the framework of the RESCCUE project (Resilience to Cope with Climate Change in Urban Areas—a multisectoral approach focusing on water) (www.resccue.eu). RESCCUE is a research project funded by the European Commission under the H2020 program, and its main goal is to provide methodologies and tools for the evaluation, planning and management of urban resilience in the context of climate change.

Conflicts of Interest: The authors declare no conflict of interest.

References

1. United Nations, Department of Economic and Social Affairs, Population Division. *World Urbanization Prospects: The 2014 Revision, Highlights (ST/ESA/SER.A/352)*; UN: New York, NY, USA, 2014.
2. ARUP, International Development and the Rockefeller Foundation. *City Resilience and the City Resilience Framework*; ARUP: London, UK, 2015.
3. UN-Habitat. Resilience. Available online: https://unhabitat.org/urban-themes/resilience/ (accessed on 30 June 2018).

4.   Rockefeller Foundation. 100 Resilient Cities. Available online: http://www.100resilientcities.org/ (accessed on 30 June 2018).

5.   Walloth, C.; Gurr, J.M.; Schmidt, J.A. *Understanding Complex Urban Systems: Multidisciplinary Approaches to Modeling*; Springer International Publishing: Basel, Switzerland, 2014.

6.   Watts, D.; Ren, H. Classification and discussion on methods for cascading failure analysis in transmission. In Proceedings of the 2008 IEEE International Conference on Sustainable Energy Technologies, Singapore, 24–27 November 2008; pp. 1200–1205.

7.   Velasco, M.; Russo, B.; Martínez, M.; Malgrat, P.; Monjo, R.; Djordjevic, S.; Fontanals, I.; Vela, S.; Cardoso, M.A.; Buskute, A. Resilience to cope with climate change in Urban Areas—A multisectorial approach focusing on water—The RESCCUE Project. *Water* **2018**, *10*, 1356. [CrossRef]

8.   Redolat, D.; Monjo, R.; Lopez-Bustins, J.A.; Martín-Vide, J. Upper-Level Mediterranean Oscillation index and seasonal variability of rainfall and temperature. *Theor. Appl. Clim.* **2019**, *135*, 1059–1077. [CrossRef]

9.   Redolat, D.; Monjo, R.; Paradinas, C.; Pórtoles, J.; Gaitán, E.; Prado-Lopez, C.; Ribalaygua, J. Local decadal prediction according to statistical/dynamical approaches. *Int. J. Climatol.* **2020**, 1–17. [CrossRef]

10.   Monjo, R.; Royé, D.; Martin-Vide, J. Meteorological drought lacunarity around the world and its classification. *Earth Syst. Sci. Data* **2020**, *12*, 741–752. [CrossRef]

11.   Forero-Ortiz, E.; Martínez-Gomariz, E. Hazards threatening underground transport systems. *Nat. Hazards* **2020**, *100*, 1243–1261. [CrossRef]

12.   Russo, B.; Velasco, M.; Monjo, R.; Martínez-Gomariz, E.; Sánchez, D.; Domínguez, J.L.; Gabàs, A.; Gonzalez, A. Evaluación de la resiliencia de los servicios urbanos frente a episodios de inundación en Barcelona. El Proyecto RESCCUE. *Water Eng.* **2020**, *24*, 101. [CrossRef]

13.   Evans, B.; Chen, A.S.; Prior, A.; Djordjevic, S.; Savic, D.A.; Butler, D.; Goodey, P.; Stevens, J.R.; Colclough, G. Mapping urban infrastructure interdependencies and fuzzy risks. *Procedia Eng.* **2018**, *212*, 816–823. [CrossRef]

14.   Martínez-Gomariz, E.; Russo, B.; Gómez, M.; Plumed, A. An approach to the modelling of stability of waste containers during urban flooding. *J. Flood Risk Manag.* **2019**, *13*. [CrossRef]

15.   Martínez-Gomariz, E.; Guerrero-Hidalga, M.; Russo, B.; Yubero, D.; Gómez, M.; Castán, S. Desarrollo y aplicación de curvas de daño y estanqueidad para la estimación del impacto económico de las inundaciones en zonas urbanas españolas. *Water Eng.* **2019**, *23*, 229. [CrossRef]

16.   Cardoso, M.A.; Brito, R.S.; Almeida, M.C. Approach to develop a climate change resilience assessment framework. *H2Open J.* **2020**, *3*, 77–88. [CrossRef]

17.   Cardoso, M.A.; Brito, R.S.; Pereira, C.; David, L.M. Avaliação da resiliência dos serviços urbanos de águas face às alterações climáticas. *A&R* **2020**, *6*, 63–77. [CrossRef]

 *sustainability*

Article

Assessment of Urban Flood Resilience in Barcelona for Current and Future Scenarios. The RESCCUE Project

Beniamino Russo [1,2,3,*], Marc Velasco [1], Luca Locatelli [1], David Sunyer [1], Daniel Yubero [1], Robert Monjo [4], Eduardo Martínez-Gomariz [3,5], Edwar Forero-Ortiz [5], Daniel Sánchez-Muñoz [6], Barry Evans [7,8] and Andoni Gonzalez Gómez [9]

[1] AQUATEC (SUEZ Advanced Solutions), Paseo de la Zona Franca, 46-48, 08038 Barcelona, Spain; marc.velasco@suez.com (M.V.); luca.locatelli@aquatec.es (L.L.); dsunyer@aquatec.es (D.S.); dyuberop@aquatec.es (D.Y.)

[2] Grupo de Ingeniería Hidráulica y Ambiental (GIHA) (Group of Hydraulic and Environmental Engineering), Escuela Politécnica de La Almunia (EUPLA, Universidad de Zaragoza) (Technical College of La Almunia, University of Zaragoza), Calle Mayor, 5, 50100 Zaragoza, Spain

[3] Flumen Research Institute, Universitat Politècnica de Catalunya, Jordi Girona 1-3, 08034 Barcelona, Spain; eduardo.martinez-gomariz@upc.edu or eduardo.martinez@cetaqua.com

[4] Fundación de Investigación del Clima (FIC) (Climate Research Fundation), Calle Gran Vía, 22, 28019 Madrid, Spain; rma@fic.es

[5] Cetaqua, Water Technology Centre, Carretera d'Esplugues, 75, 08940 Barcelona, Spain; eaforero@cetaqua.com

[6] IREC, Power Systems department, Jardins de les Dones de Negre, 1, 2ª pl., 08930 Barcelona, Spain; dsanchezm@irec.cat

[7] Centre for Water Systems, University of Exeter, Exeter EX4 4QF, UK; b.evans@exeter.ac.uk

[8] School of Built Environment, College of Sciences, Massey University, Auckland 0745, New Zealand

[9] Ajuntament de Barcelona (Barcelona Municipality), Carrer de Torrent de l'Olla 218, 08012 Barcelona, Spain; agonzalezgom@bcn.cat

* Correspondence: brusso@unizar.es; Tel.: +34-932-479-869

Received: 26 June 2020; Accepted: 6 July 2020; Published: 13 July 2020

Abstract: The results of recent climate projections for the city of Barcelona show a relevant increment of the maximum rainfall intensities for the period 2071–2100. Considering the city as a system of systems, urban resilience is strictly linked to the proper functioning of urban services and the knowledge of the cascading effects that may occur in the case of the failure of one or more critical infrastructures of a particular strategic sector. In this context, the aim of this paper is to assess urban resilience through the analysis of the behavior of the main urban services in case of pluvial floods for current and future rainfall conditions due to climate change. A comprehensive flood risk assessment including direct, indirect, tangible and intangible impacts has been performed using cutting edge sectorial and integrated models to analyze the resilience of different urban services (urban drainage, traffic, electric and waste sectors) and their cascade effects. In addition, the paper shows how the information generated by these models can be employed to feed a more holistic analysis to provide a general overview of the city's resilience in the case of extreme rainfall events. According to the obtained results, Barcelona could suffer a significant increase of socio-economic impacts due to climate change if adaptation measures are not adopted. In several cases, these impacts have been geographically distributed showing the specific situation of each district of the city for current and future scenarios. This information is essential for the justification and prioritization of the implementation of adaptation measures.

Keywords: urban resilience; cascading effects; climate change; pluvial floods; 1D/2D coupled models

---

1. Introduction

Urban resilience refers to the ability of an urban system—and all its constituent socio-ecological and socio-technical networks across temporal and spatial scales—to maintain or rapidly return to desired functions in the face of a disturbance, to adapt to change and to quickly transform systems that limit current or future adaptive capacity [1].

In this context, a city can be considered as a system of systems, and its urban resilience is strictly linked to the proper functioning of urban services and the knowledge of the cascading effects that may occur in the case of the failure of one or more critical infrastructures of a particular strategic sector [2]. Moreover, urban areas are complex systems that cannot be understood by sectorial and disciplinary approaches alone [3,4], and the focus of smart cities models on strengthening different sectors with technological advancement could contribute to building upon a city's resilience in terms of dealing with natural hazards [5].

This paper shows how pluvial flood urban resilience can be assessed by analyzing the behavior of critical urban services and the related cascading effects in the case of failures due to heavy storm events. With this aim, sectorial and integrated models have been developed and calibrated to analyze the resilience of several urban services in Barcelona for current (baseline scenario) and future (business as usual scenario) rainfall conditions up to the horizon of 2100 [6]. In addition, the information generated by these models, together with the historical information available for each urban service, has been used to feed a more holistic model which covers all the urban services of the city. This twofold approach, including risk treatment (implementation of adaptation strategies) in a comprehensive flood risk management process, is presented in Figure 1.

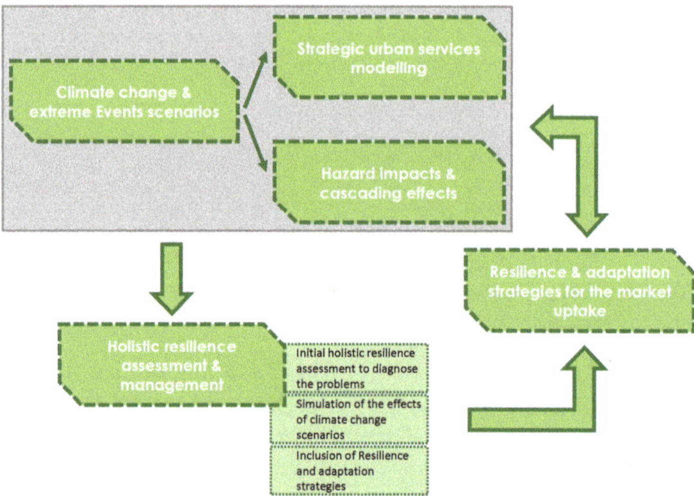

Figure 1. Twofold approach to achieve an urban resilience assessment for current and future scenarios.

A flood-resilient city can be defined as a city which is able to resist, absorb, accommodate and recover from the effects of a flood hazard in a timely and efficient manner, including through the preservation and restoration of its essential basic structures and functions [7]. In this context, flood resilience assessment has been performed in Barcelona through a 1D/2D urban drainage model linked to other urban services models to evaluate the cascade effects produced by urban floods on traffic, electric and waste collection systems. The employment of a coupled 1D/2D urban drainage model providing flow variables (flow depths, flow velocity and flood extension) on urban surfaces during pluvial flood events is essential to perform tangible and intangible risk assessments. Moreover, the results of these integrated models have been used to feed a holistic tool to assess the resilience of the city as a whole.

The paper proposes specific and holistic approaches to assess pluvial flood resilience in urban areas. The approaches are complementary and interconnected and can be used to understand the interrelations between urban services and infrastructures, as well as representing a valuable tool for decision making.

2. Materials and Methods

2.1. The Effects of Climate Change on Maximum Rainfall Intensity in Barcelona

Recently, the Climate Research Foundation (Fundación de Investigación del Clima; hereafter, FIC from the acronym in Spanish) provided climate projections and predictions for different climate variables in Barcelona, the results of which are summarized in Figure 2. These results confirm the same trends as other previous studies developed for the city and are in line with the data from the last Climate Plan published by Barcelona City Council [8]. According to the data provided by FIC, phenomena such as extreme rainfall, heatwaves and droughts could experience significant increases due to an acceleration of the hydrological cycle [6,9].

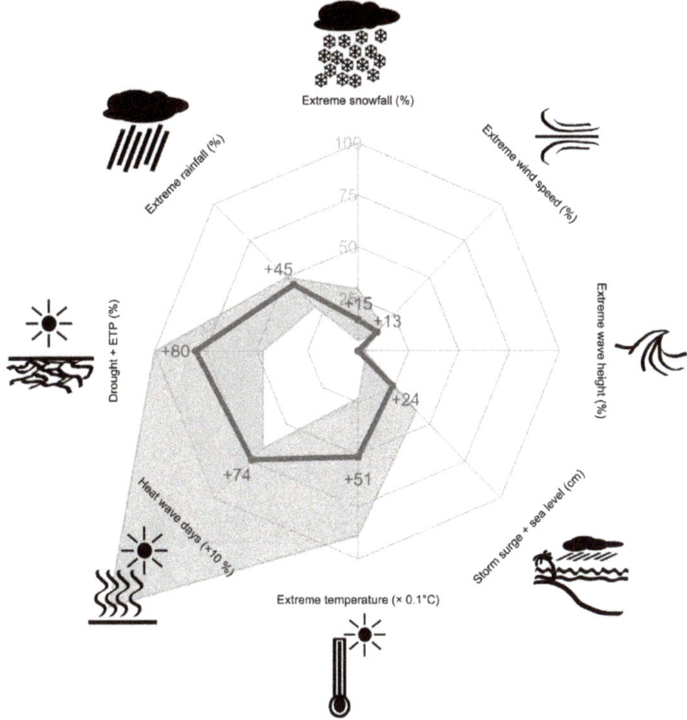

Figure 2. Extremes compass rose for Barcelona: maximum point change in extreme climate events over the century, taking into account return periods between 2 and 100 years. The center represents no changes, and the edge corresponds to an increase of 100% for every variable except for heat wave days (the border is +1000%) and extreme temperature (the border is +10 °C). Thick lines represent the median scenario, and the shaded area is the uncertainty region (5–95%).

Particularly, in the case of maximum rainfall intensities and the horizon of 2071–2100 for the city of Barcelona, the value of the coefficient of climate change (defined as the ratio between future and current maximum intensities, for certain return periods and time intervals) [10,11] was found to be in a range between 1.07 and 1.26 depending on the frequency and duration of each maximum rainfall intensity (Figure 3).

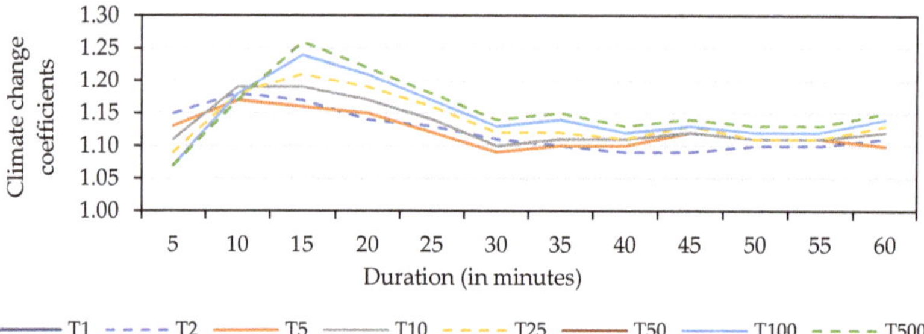

Figure 3. Fiftieth percentiles of the climate change coefficients obtained for different rainfall durations and return periods for the horizon 2071–2100 for the city of Barcelona [6].

These results were obtained using statistical spatial and temporal downscaling techniques on 20 future pluviometric series provided by 10 general atmospheric circulation models, forced by Representative Concentration Pathway (RCP) 4.5 and 8.5 scenarios and previously validated for a historical control period (1976–2005) [6]. The climate change coefficients in Figure 3 represent the 50th percentile of the results obtained.

Once these climate change coefficients were obtained, they were applied to synthetic storms with different return periods (T1, T10, T50, T100 and T500) used for the last drainage master plan of the city of Barcelona [11] obtained through the alternating blocks method. Figure 4 shows the urban drainage Barcelona project design storm for a return period of 10 years with a duration of approximately 2 h and 30 min after the application of climate change coefficients for each different rainfall duration.

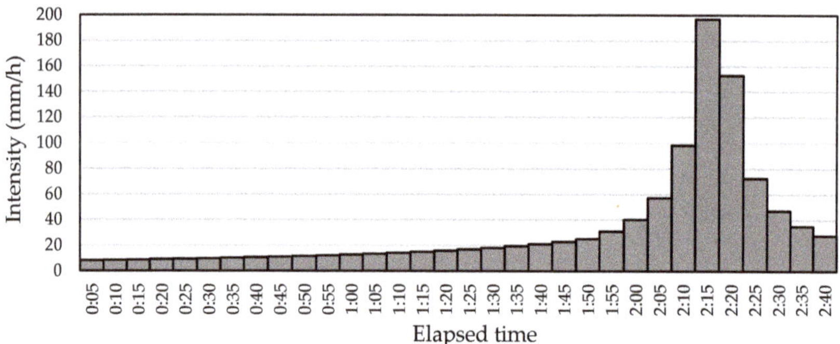

Figure 4. Urban drainage Barcelona project design storm with a return period of 10 years and a duration of 2 h and 35 min [6].

2.2. 1D/2D Coupled Approaches for Urban Pluvial Modelling

Urban areas have a complex topography and contain small-scale elements such as streets and buildings that are usually not taken into account in standard river floodplain studies [12]. Therefore, a higher resolution is required to represent features at the city scale, although this may lead to larger computational time, notwithstanding the fact that urban model areas are generally smaller than a river floodplain. For all of these reasons, urban, pluvial flooding requires a different modeling approach than the one used for fluvial flooding [12].

During the last two decades, several authors have published papers about the need to develop and use urban stormwater models (USMs) based on coupled approaches (the modeling of the surface and sewer flows at the same time by 1D/1D or 1D/2D models) to represent adequately urban flood caused by surcharged sewers [13–16] and carry out realistic flood risk assessments [17].

Although the choice between using a 1D or a 2D surface overland flow model (to be coupled to a 1D sewer model) determines the accuracy of results and the computational time required to obtain them, when the flow overtops street curbs and does not remain within the street profile, using a 2D model is crucial [12,18].

In 1D/2D USMs, the underground sewer network is represented by a 1D sewer model while the surface flow is computed using a 2D model. The 2D model reproduces the urban surface topography and is essential to achieve a more realistic simulation of the flow spreading across complex urban surfaces, with results such as flow depths and velocities anywhere in the urban model area [12].

USM can be semi-distributed (SD) or fully distributed (FD). SD models, commonly applied in urban stormwater modeling, are based on subcatchment units where rainfall is applied, while runoff is estimated and routed according to specific hydrological losses and rainfall–runoff transformation methods. FD models, which are generally more detailed and theoretically more realistic, are based on the two-dimensional (2D) discretization of the overland surface, where runoff volumes are estimated and directly routed by the 2D overland flow module [19]. Both kinds of approaches can be followed to create 1D/2D coupled models that are able to simulate, at the same time, the behavior of the sewer system and the urban surfaces and their mutual interaction in case of pluvial flooding events (Figure 5). Finally, hybrid models (H) can account for runoff produced by rainfall which is directly applied from subcatchment units formed by building areas (roofs, terraces and courtyards) and directly conveyed into the sewer systems; for the other impervious (streets, sidewalks, squares, etc.) and pervious (parks and natural areas) urban surfaces, the 2D overland flow model computes and routes the runoff produced by the rainfall directly applied to these surfaces [20]. These approaches are represented in Figure 5.

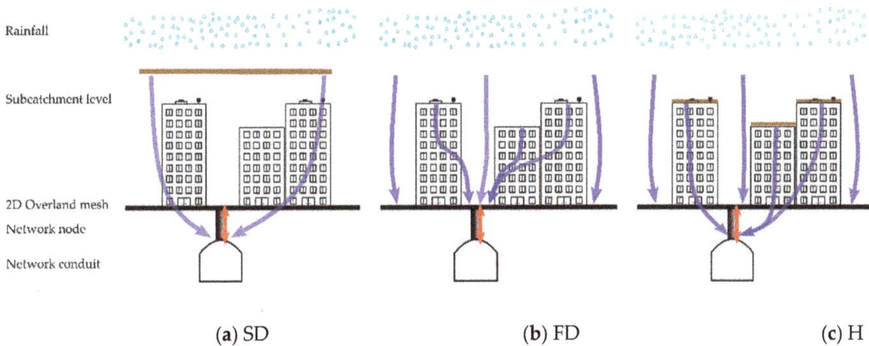

Figure 5. Scheme of semi-distributed (SD) (**a**), fully-distributed (FD) (**b**) and hybrid (H) (**c**) 1D/2D coupled urban stormwater model (USM) approaches (adapted from [19]). In brown, subcatchment units are represented, while blue lines and arrows indicate the pathway of the runoff from the source (subcatchment units or discretized 2D surface) to the sewer system.

The amount of runoff entering the underground sewer network is limited by the hydraulic efficiency of surface drainage structures (inlets, transversal grates, etc.) [21–23] and their state of maintenance and clogging [23,24], although these aspects are often neglected in urban drainage models [19]. Generally, SD models apply all the runoff estimated in a given subcatchment directly into the selected computational node of the sewer system, without accounting for the hydraulic capacity of surface drainage capacity. With this assumption, this kind of model only considers the flooding that occurs when the sewer system surcharges and neglects urban floods produced by the poor capacity of inlet systems [19]. On the contrary, FD modelling packages, such as Infoworks ICM (Integrated Catchment Modeling) software [25], can take into account the hydraulic performance of surface drainage systems connecting network nodes with the 2D overland surface mesh by weirs, orifices and other experimental equations [20,21].

2.3. Barcelona Semi-Distributed 1D/2D USM

After the two first investigations concerning the development and calibration of a detailed 1D/2D USM in Barcelona, covering approximately half of the administrative land of the city (more than 50 km^2) [20,26], a new SD model has been developed and calibrated in the framework of this work [2] and the new drainage master plan of the city (PDISBA, from the acronym in Spanish) [11,27].

The large amount of effort related to the analysis of the deficit of surface drainage systems through the last two drainage master plans of the city and the consequent implementation of thousands of inlets in all the urban catchments allow the assumption that stormwater could be quickly introduced into the sewer system by avoiding uncontrolled runoff circulation and aiming to develop an SD 1D/2D USM (referred to as 1D/2D USM in the following).

Moreover, the new 1D/2D USM presents two relevant improvements with respect to the previous ones: the model includes the main and secondary sewer network, reaching a total length of 1650 km of pipes, and covers the whole hydrological area of the city (administrative land and upstream surfaces), exceeding 120 km^2 of model domain.

The model, with more than 85,000 nodes and a discretized 2D domain in an unstructured mesh of more than 1,360,000 cells, was developed through the Infoworks ICM (Integrated Catchment Modeling) software (www.innovyze.com) [25] and was calibrated and validated using the historical data recorded by more than 100 flow depth gauges located in the city's sewage network and more than 20 rain gauges distributed in the analyzed domain using Thiessen polygons [11,27]. The average size of the 2D cells for overland flow modeling is in the range of 25–100 m^2.

The model required high-quality topographic information (physical data from the network, digital terrain model 2 × 2 m with a resolution in height of approximately 15 cm) and phenomenological information (rainfall data and flow level for the calibration phase), in addition to an adequate hardware configuration to reduce computation time during numerical simulations [20].

The new 1D/2D USM allowed the estimation of flow variables (flow depth, flow velocity and flood extension) on the surface prone areas by several numerical simulations of historic events and synthetic storm hyetographs for current and future scenarios. These values were used for the flood hazard and intangible risk assessment (concerning pedestrian and vehicular circulation) and the evaluation of tangible direct and indirect impacts. The same outputs were also used to feed other integrated models of critical urban services to assess the cascading effects of floods in these sectors.

2.4. Modeling of the Effects of Pluvial Floods on Several Urban Services

Projections of rainfall and sea level rise were used to feed the 1D/2D coupled USM to analyze the hydraulic behavior of the underground sewer system and the overland flood-prone areas in the case of pluvial floods for current and future rainfall conditions.

Concerning the assessment of multiple hazards and risks, the proposed methodology is based on the development of coupled models and tools ("loosely or integrated models"); thus, the outputs of certain models are used as inputs in others according to the scheme presented in Figure 6 [27].

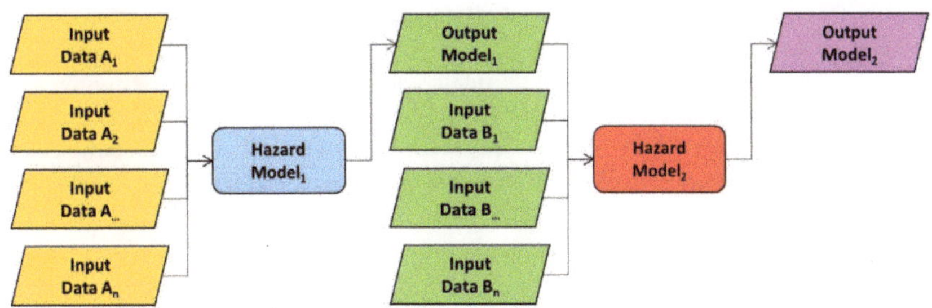

Figure 6. Scheme of loosely coupled model approach used for the multi-hazards assessment [27].

The aim of the developed loosely coupled models was the assessment of multi-hazards and multi-risks (including direct and indirect impacts) produced by urban pluvial floods and the cascading effects on other urban services (electrical system, waste collection system and surface traffic system). Table 1 summarizes the analyzed services affected by pluvial floods in Barcelona, the behavior of which was based on the developed loosely coupled models [27]. Figure 7 shows the analyzed interrelationships to assess the resilience of some main city services in the case of urban flooding episodes [28].

Table 1. Loosely coupled models developed and used for the flood resilience analysis in Barcelona.

Loosely Coupled Model	Involved Sectors	Main Purposes
1D/2D coupled model	Urban drainage	Flood hazard assessment and socio-economic flood risk assessment for people and properties
Flooding—traffic model	Urban drainage and surface traffic	Assessment of flood hazard and flood impacts on traffic system
Flooding—electric model	Urban drainage and electric system	Assessment of flood hazard and flood impacts on electric system
Flooding—waste collecting model	Urban drainage and waste collecting model	Assessment of flood hazard on waste collecting system

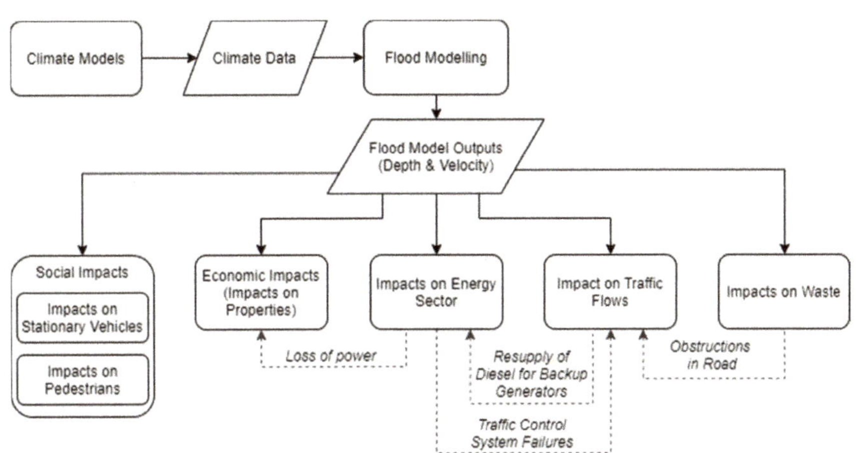

Figure 7. Diagram of the impact analyses carried out within this paper and the potential cascading effects.

2.5. Social Flood Impacts Model

Pluvial flood impacts can be classified into tangible and intangible impacts and direct and indirect impacts [29]. In this study, socio-economic impacts produced by pluvial flooding have been assessed according to comprehensive and detailed methodologies carried out and implemented in previous investigations in several urban areas [29,30].

In the social field, for the assessment of the intangible impacts, human risk focuses on the safety of pedestrians and vehicles exposed to pluvial flooding events. Risk is defined as the combination of hazard and vulnerability according to the approach proposed by Turner et al. [31] and implemented in previous studies [29,30]. According to this, hazard assessment is based on the severity and frequency of the hydrodynamic variables and is classified based on specific flood hazard experimental criteria regarding pedestrian and vehicular stability in urban flooded areas (Figure 8) [27,32–34].

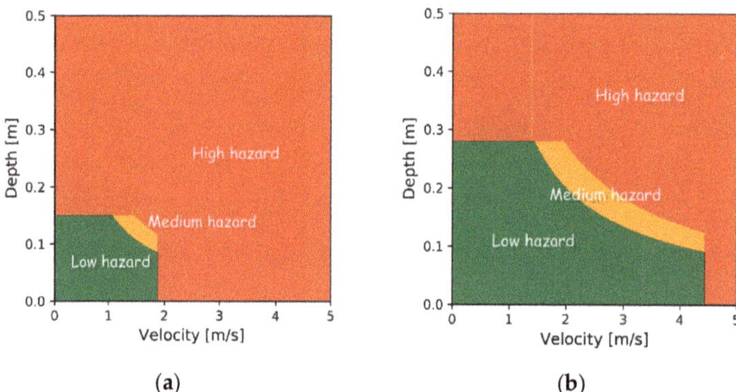

Figure 8. Experimental flood hazard criteria for (**a**) pedestrians and (**b**) vehicles.

Regarding flood vulnerability for pedestrians, it is considered to be a function of exposure and sensitivity, taking into account several indicators such as demographic density, the percentage of people of a critical age and of foreign inhabitants and the number of critical infrastructures. By setting thresholds for the proposed indicators, the vulnerability of each census district can be qualitatively scored and classified as low, medium and high. On the other hand, in order to assess the vulnerability for vehicular circulation, the exposure for each urban street, expressed in terms of vehicular daily intensity, is considered. Based on this value, flood vulnerability regarding vehicular circulation is qualitatively scored and classified as low, medium and high, in a similar manner to pedestrian vulnerability [28–30].

Methods for risk determination can be qualitative or quantitative, with both having limitations. If we define risk as the probability or threat of a hazard occurring in a vulnerable area, flood risk can be assessed through a flood risk map related to a determined scenario and return period by combining hazard and vulnerability maps [29,30]. Pedestrians and vehicles are expected to be the most potentially affected by floods in Barcelona. Their risk is related to their stability, and in Barcelona, this is assessed for the present (baseline) and also for the future (business as usual (BAU)) scenarios according to the rainfall variable projections for different return periods.

Qualitative risk assessment defines hazards, vulnerability and risk levels by significance levels such as "high", "medium" and "low" and evaluates the resultant level of risk against qualitative criteria. In this case, hazard and vulnerability maps are generally elaborated through specific criteria and indexes, and so risk maps will be created by multiplying the vulnerability index (1, 2 or 3, corresponding to low, medium and high vulnerability) by the hazard index (1, 2 or 3, corresponding to low, medium and high hazard). Finally, the total risk varies from 1 to 9, where higher levels indicate higher risk according to the following risk matrix (Figure 9) previously employed in other works [29,30].

<table>
<tr><td rowspan="3" colspan="2"></td><td colspan="3" align="center">Hazard</td></tr>
<tr><td align="center">Low</td><td align="center">Medium</td><td align="center">High</td></tr>
<tr></tr>
<tr><td rowspan="3">Vulnerability</td><td align="center">Low</td><td align="center">Low</td><td align="center">Low</td><td align="center">Medium</td></tr>
<tr><td align="center">Medium</td><td align="center">Low</td><td align="center">Medium</td><td align="center">High</td></tr>
<tr><td align="center">High</td><td align="center">Medium</td><td align="center">High</td><td align="center">High</td></tr>
</table>

Figure 9. Proposed flood risk matrix for pedestrians and vehicles [28,30].

2.6. Economic Flood Impacts Models

Regarding economic flood risk assessment in Barcelona, tangible direct and indirect damage were considered [28]. Specifically, tailored flood depth–damage curves were developed for the case of Barcelona [35,36] and used to feed a detailed damage model regarding properties and vehicles (the two most affected assets by pluvial floods in the city). The model was already successfully applied for the city of Badalona [30,37] and validated using insurance claims according to the data received from the Spanish public insurance company "Consorcio de Compensación de Seguros (CCS)" [28]. This public entity covers all the damage produced by extraordinary events, such as damage related to natural hazards (e.g., pluvial floods). On the other hand, indirect damages were assessed by an econometric model, achieving a constant relation between direct and indirect tangible damage. This model was also validated using field data [28].

2.7. Integrated Flooding–Surface Traffic Model

The simulations related to the integrated flooding–surface traffic model in Barcelona were carried out using a mesoscale model and the TransCAD Transportation Planning software (https://www.caliper.com/tcovu.htm) adopted by city council's mobility department (https://www.caliper.com/tcovu.htm) [38]. The mesoscale model simulated the vehicular flow in each link of the street network; each link contained detailed information regarding the volume of traffic, its typology (for example: number of cars, trucks, bicycles, etc.), travel time, the residual capacity of the section, etc. Flood maps produced through the city's 1D/2D USM were used as inputs for a dynamic traffic model to estimate the effects within the city's surface transportation network produced by pluvial floods. A recent study developed by Pyatkova et al. [39] analyzed how the flow depth information can be used as criteria to approximate the reduction of vehicular free-flow speeds to 20 kmh^{-1} along streets that have standing water, with a reduction to 0 kmh^{-1} where the water depths exceed a threshold value (Table 2). As previously mentioned, the mesoscale traffic model contained information on a wide number of parameters relating to traffic flows for each road section, including the maximum speeds allowed on each section. To simulate the effect of flooding within the traffic model, we needed to adjust the maximum allowable speed parameters based on food model outputs. This approach involved using geospatial analysis as a precursor to modify the input data of a traffic model, as outlined in Evans et al. [38], where the outputs from the 1D/2D USM were used to spatially define vehicular speed restrictions along the road network. The results from the traffic model run under flooded conditions were compared to the benchmark traffic model (the traffic model run under dry weather conditions) and the impacts in terms of relative disruption to traffic flows were analyzed.

Table 2. Flood hazard effects on traffic flow (from [38,39]).

Flood Depth Range (m)	Hazard Classification	Maximum Vehicle Speed (km/h)
Flow depth < 0.1	Low	Road speed limit
0.1 < Flow depth < 0.3	Medium	20
Flow depth > 0.3 m	High	0 (Road closed)

Here, flood hazard analysis was performed by a GIS (Geographic Information System) spatial analysis of the flooded road links; the rules applied in relation to traffic speed reductions are outlined in Table 2, and the flow depths were provided by the 1D/2D coupled USM. The results of this analysis were used to select and modify the maximum allowable speed limits for flooded roads within the traffic model based on the flood hazard. Using these new input parameters, the traffic model was run and the results compared to the normal (dry weather) traffic modelling conditions. Detailed information about the approach can be found in the work published by Evans et al. [38].

2.8. Integrated Flooding–Electric System Model

This model analyzed the potential effects of pluvial floods on the electrical system of the city of Barcelona, with special emphasis on critical infrastructures such as high and medium-voltage substations, as well as distribution centers, taking into account the possible effects of climate change.

The model was designed for the hazard, risk and cost assessment of the electrical assets, as shown in Figure 10.

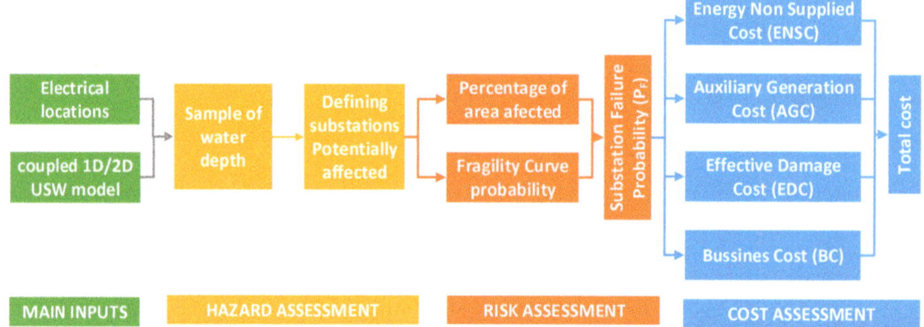

Figure 10. Integration of GIS spatial analysis for flood assessment on the electrical model.

In particular, the flooding hazard level of each electrical infrastructure was assessed on the basis on flood influence areas of 5 m, 25 m and 30 m in radius with respect to their location depending on the asset type (distribution center (DC), medium-voltage (MV) and high-voltage (HV) substation respectively), as well as considering the flow depths values every 2 m in order to avoid local errors and potential uncertainties of the electrical asset location and of the source data provided by the 1D/2D USW model. In addition, a 10 cm threshold was used to consider significant local flooding. Using these parameters, the flood affections were classified as complete, partial or null, quantifying the percentage of flooded surface in each area of influence of each electric infrastructure according to the methodology proposed by Sánchez et al. [28,40].

One of the most important uncertainties of this model was the lack of knowledge about the specific location of critical electrical infrastructures (sometimes located on surfaces and at other times underground or with self-protection elements which were not always known).

For the impacts analysis, a vulnerability curve (known as a fragility curve in the energy sector) of the electrical infrastructure proposed by the Federal Emergency Management Agency [41] was used. The curve relates the probability of failure of an electrical infrastructure to the flood depth. Furthermore, this curve was partially modified to carry out a sensitivity analysis of the final results regarding this input [40]. The results obtained from the analysis of the percentage of flooding surface in each area and from the fragility curve were computed to obtain a probability of failure, later categorized into four different risk categories as shown in Table 3.

Table 3. Failure probabilities for electrical assets.

Probability Range	Categorical Description
$P_F < 0.01$	Low Failure Probability (LFP)
$0.01 < P_F \leq 0.1$	Moderate Failure Probability (MFP)
$0.1 < P_F \leq 0.5$	High Failure Probability (HFP)
$P_F > 0.5$	Non-Acceptable Failure Probability (NAFP)

The cost assessment was based on estimations based on GIS computing; furthermore, we established the supply area of each electrical location using Thiessen polygons and obtained the power supplied through an estimation of the consumers per area based on the census of the city.

Based on these estimations, it was possible to extract the number of consumers affected and the time needed to repair the substation as well as the cost of the energy not supplied, the cost incurred by businesses, auxiliary generation and the damage received by the location [40].

2.9. Integrated Flooding–Waste Collection System Model

In the case of pluvial floods, waste containers can lose their stability due to buoyancy, dragging or overtopping, thereby allowing debris and leachate to escape from the containers and contaminate the floodwater and the environment [42]. On the other hand, the containers displaced by the flow can obstruct superficial drainage pathways or obstruct narrow streets, exacerbating the effects of the flood. Consequently, waste containers' stability in the case of pluvial flooding is definitely an environmental, safety and health concern which needs to be addressed in a context of urban flood resilience assessment.

In order to analyze the significance of this problem in Barcelona, an integrated flooding–waste collection system model was developed and validated. In the city, there are more than 27,000 containers, which can be classified according to the type of waste they contain (waste, organic, paper and cardboard, plastic and packaging and glass), their volume in liters (3200, 3000, 2400, 2200 and 1800 L) or the manner in which they can be loaded (lateral, bilateral, rear or underground) (Figure 11) [27,42]. In order to study the stability of these containers, stability curves depending on the type of container, their filling degree and the overland flow parameters (flow depth and velocity) were created [42].

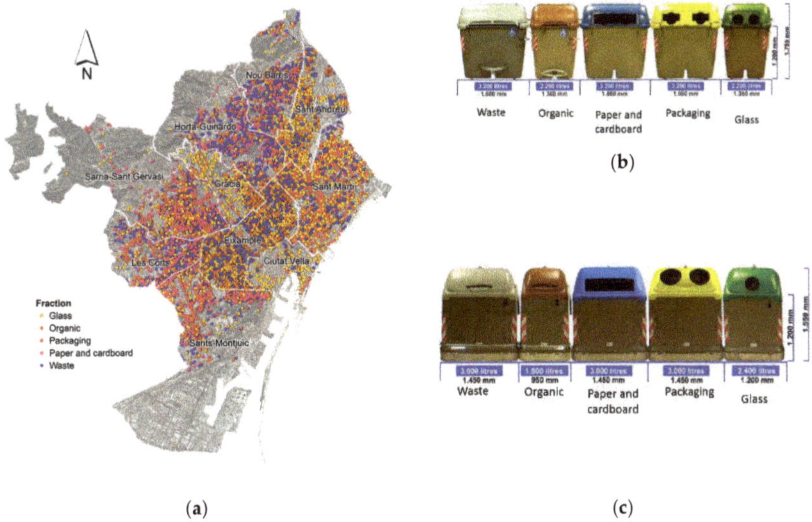

Figure 11. (**a**) Container distribution in Barcelona classified according to fraction type and types of containers as they are loaded by the bin lorry, (**b**) lateral load and (**c**) bilateral load. Adapted from Martínez-Gomariz et al. [42].

Finally, on the basis on the location of the containers (Figure 11) and the flow parameters provided by the 1/D/2D USM model, flood hazard maps showing the unstable waste containers were created for an historical storm event to validate the model and several synthetic project storms of 1, 10 and 50 years [27,42].

2.10. Holistic Model of Urban Resilience

The information and the results provided by the 1D/2D USM were used as inputs to the HAZUR® holistic tool for the evaluation of the potential cascading effects produced by pluvial floods on several main urban services (as well as others not contemplated by the integrated flood models previously described) and to estimate their recovery time for current and future scenarios. This analysis involved 34 urban services grouped into nine sectors with 563 critical infrastructures. The main urban sectors and services analyzed included the water cycle, energy, telecommunications, transport, emergencies, public health, environment and green infrastructures, waste and citizens. The HAZUR® tool is capable of analyzing cascading effects generated from certain impacts (Figure 12). In the case of pluvial floods, impacts and cascade effects on electric and transport sectors were assessed for several synthetic project storms with different return periods (T1, T10, T150, T100 and T500) for current (baseline) and future (BAU) scenarios.

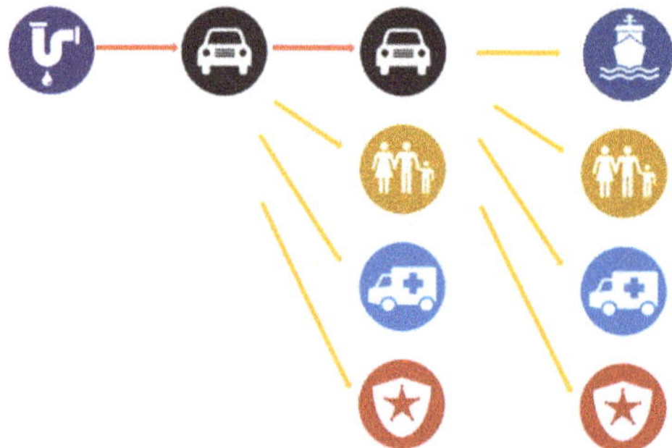

Figure 12. Cascading effects simulated by HAZUR® for a pluvial flood with a 10 year return period in Barcelona.

Taking into account the down-times included in the HAZUR® tool (obtained from the sectorial models or by expert assessment) and considering the interdependencies which exist between several services and infrastructures, the cascading effects can be simulated [43]. Figure 12 presents an example of cascading effects generated by a 10 year return period flood event in Barcelona. In this case, as can be seen in Figure 12, the tool allows us to see that the lack of capacity of the drainage system in the upper part of the city generates a flood on the high-speed ring "Ronda de Dalt", stopping the circulation of vehicles. This affects many other services (such as medical emergency services, the local police or the citizens), and it also causes the failure of the other high-speed ring "Ronda Litoral", which in turn would affect the same services as before (in another area of the city) as well as the port of the city due to the connection of this highway with that infrastructure.

3. Results

The 1D/2D USM and the derived loosely coupled (integrated) models described in the previous section were developed and validated using field data provided by sensors and historic collected information to estimate the potential effects of climate change on the urban drainage sectors and the cascading effects on other main urban services [27]. The use of this modeling approach allowed us to achieve valuable results in terms of flood hazards, as well as in terms of socio-economic risk and impacts on other sectors of the city. In this section, the specific results directly related to the urban drainage sector and the other analyzed urban services (surface transport, electric system and waste collection) are presented.

3.1. Assessment of Social Impacts Produced by Pluvial Floods

3.1.1. Flood Risk for Pedestrians

The hydraulic behavior of the urban drainage system of the city (considering both the hydraulic response of the sewer network and the overland flow on the urban surfaces) was simulated using the 1D/2D USM for a large set of synthetic projects storms with return periods T of 1, 10, 50, 100 and 500 years for current and future scenarios. As our first results, following the specific flood hazard criteria for pedestrians presented in Section 2.5, detailed flood hazard maps were created.

The results also allowed the estimation of flood hazards for pedestrians for each district of the city (with a total area of approximately 102 km^2) and their evolution in the case of a climate change scenario. The results concerning the current scenario show that areas classified as having high flood hazard conditions have reduced risk for low return periods (null for T = 1 and less than 5% for T = 10, with this last one being the designed return period for the sewer system of the city), which progressively increases for higher return periods. This notwithstanding, the results show that climate change scenarios could produce an increase in high flood hazard areas of between 20% and 50%. Additionally, for the simulation corresponding to return period T1 and BAU scenario, the high flood hazard area is null for each district [28].

To assess the flood risk for pedestrians, flood hazard was combined with the human vulnerability, which was achieved according to the indicators mentioned in Section 2.5 (details can be found in [28]). Vulnerability was qualitatively assessed for each census district in low, medium and high levels. These classification ranges resulted in the vulnerability map presented in Figure 13, which was also considered for BAU due to the mainly consolidated urbanistic characteristics of the city [28].

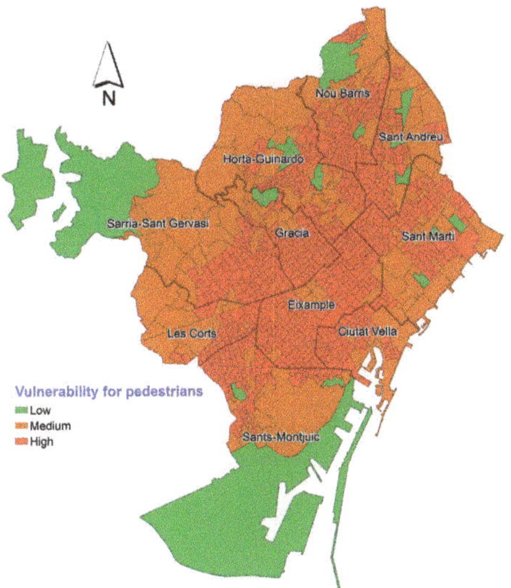

Figure 13. Vulnerability maps for pedestrians.

As stated above, the flood risk results were presented in terms of flood risk maps for all the considered return periods and scenarios (baseline and BAU). Figure 14 shows the flood risk maps related to a rainfall storm event with a return period of 10 years for both scenarios.

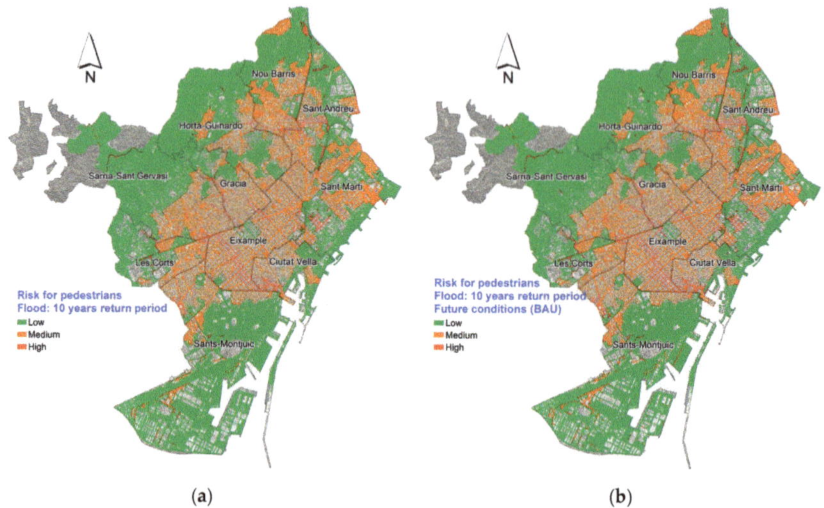

(a)   (b)

Figure 14. Example of flood risk maps for pedestrians for a synthetic 10 year return period projected storms related to (**a**) baseline and (**b**) business as usual (BAU) scenarios.

Furthermore, the high-risk area (in percentage) for pedestrians was broken down into districts in order to observe the riskiest districts in terms of pedestrians' stability. Moreover, in order to highlight the effect of climate change in terms of the increase of high-risk areas in Barcelona, we also present the variation of high flood risk areas for pedestrians according to the 10 districts into which Barcelona is administratively divided (Figure 15). It is possible to observe the major increases of high flood risk areas (around 30% for the whole district of Barcelona) with respect to the climate change coefficients (from 12% to 16%) for the same return periods [28].

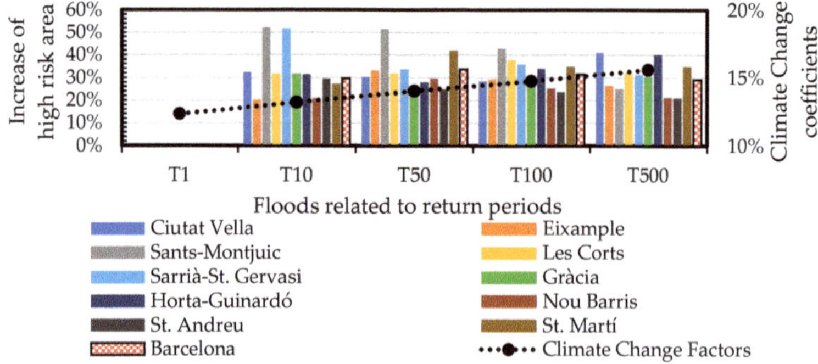

Figure 15. Expected increase of high-risk areas according to the future conditions.

3.1.2. Flood Risk for Vehicles

The flow variables (flow depths and flow velocity) provided by the 1D/2D USM were also used to generate flood hazard maps for vehicles for each district of the city and their evolution in the case of a climate change scenario. The results concerning the current scenario show that areas classified with high flood hazard conditions have reduced risk compared to the case of pedestrians; in particular, high hazard is null for T = 1 and is less than 5% for T = 10, with this last one being the designed return period for the sewer system of the city, which progressively increases for higher return periods. This notwithstanding, the results show that climate change scenarios could produce an average increase of 30% for the whole city with a peak of 50% for specific districts. Additionally, for the simulation corresponding to return period T1 and the BAU scenario, the high flood hazard area is null for each district [28].

In order to assess the vehicles' vulnerability, three levels were also proposed based on a unique indicator: The vehicular flow intensity (VFI) expressed in veh/day. Depending on this value and defined thresholds, the vulnerability of each urban road was classified into three levels (low, medium and high) [28]. The final vulnerability map is shown in Figure 16.

Furthermore, for vehicles, flood risk was assessed through the elaboration of flood risk maps for all the considered return periods and scenarios (baseline and BAU). Figure 17 shows the flood risk maps related to a rainfall storm event with a return period of 10 years for both scenarios.

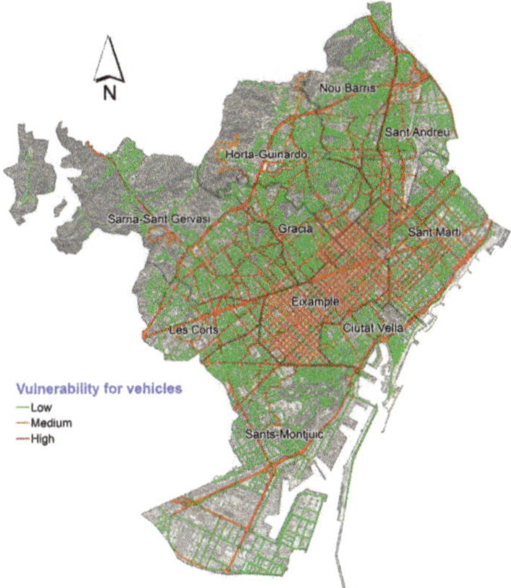

Figure 16. Vulnerability map for vehicles. Green, orange and red colors indicate low vulnerability (vehicular flow intensity (VFI) < 100), medium (100 < VFI < 1000) and high (VFI > 1000), respectively.

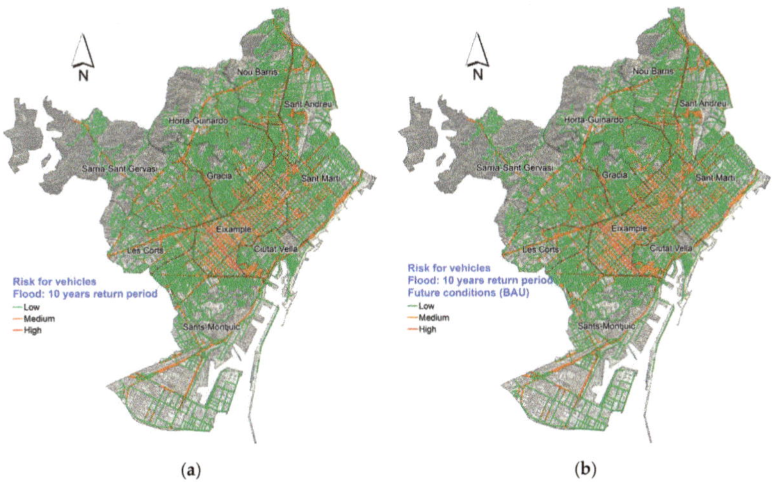

(**a**)   (**b**)

Figure 17. Example of flood risk maps for vehicles for synthetic 10 year return period projected storms related to (**a**) baseline and (**b**) BAU scenarios.

In this case, the assessment has also been broken down into districts in order to observe the riskiest districts in terms of vehicles' stability. Moreover, in order to highlight the effect of climate change in terms of the increase of high-risk areas in Barcelona, we present the variation of high-flood risk areas for vehicles in all of the districts (Figure 18). In this case, it is also possible to observe a major increase of high flood risk areas (from 20% to 40% for the whole city area) with respect to the climate change coefficients (from 12% to 16%) for the same return periods.

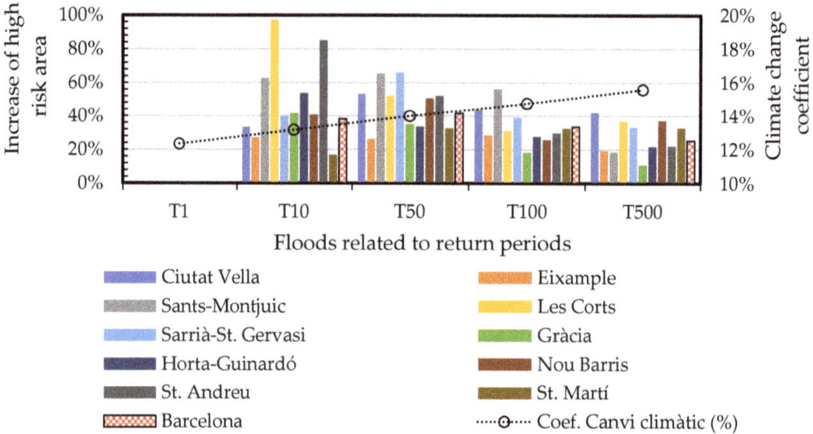

Figure 18. Expected increase of high-risk areas according to the future conditions.

3.2. Assessment of Economic Impacts Produced by Pluvial Floods

For the estimation of tangible direct damages caused by pluvial floods generated by urban floods, both properties and vehicles were considered in the economic assessment. According to the claims data provided by the Spanish re-insurance company (CSS), these two risk categories are the most significant.

According to the developed methodology to estimate property damage, flow depths on the streets provided by the 1D/2D USM were properly reduced to achieve flood depths for properties using specific sealing coefficients, which were collected for 14 land uses in Barcelona [35,36]. As a second step, flood damages suffered by the properties were evaluated on the basis of tailored flood depth damage curves for all the 14 land uses; a detailed flood damage model was developed and validated in previous studies [28,30]. Models considered different typologies of properties: without basements, with a basement and with up to two basements. On the other hand, configurations with or without parking access were considered [28,30].

Regarding the evaluation of vehicle damage, a novel methodology—also based on the concept of damage curves—was implemented. The methodology tried to reduce the uncertainty due to the mobility of vehicles, proposing heterogeneous vehicular occupation for several areas of the city based on the information provided by aerial photographs [28,37]. For this assessment, flood damage curves developed by the Army Corps of the United States of America [44] for five types of vehicles were adapted for the case study of Barcelona [28,37]. These curves were converted into a single damage curve weighted according to the percentage of vehicle types in Barcelona, also taking into account their depreciation according to statistical information concerning vehicle types and their age [28,30,37].

For both properties' and vehicles' flood damage assessment, damage maps were achieved for the return periods T1, T10, T50, 100 and T500 and current (baseline) and future (BAU) scenarios and aggregated for each district of the city (Figures 19 and 20). These figures show how future rainfall conditions for a projected storm of 10 years significantly worsen the situation in several districts of the city. Specifically, it can be observed that all the districts of the downtown would suffer high losses, and the better situation of several districts upstream would be exacerbated due to climate change.

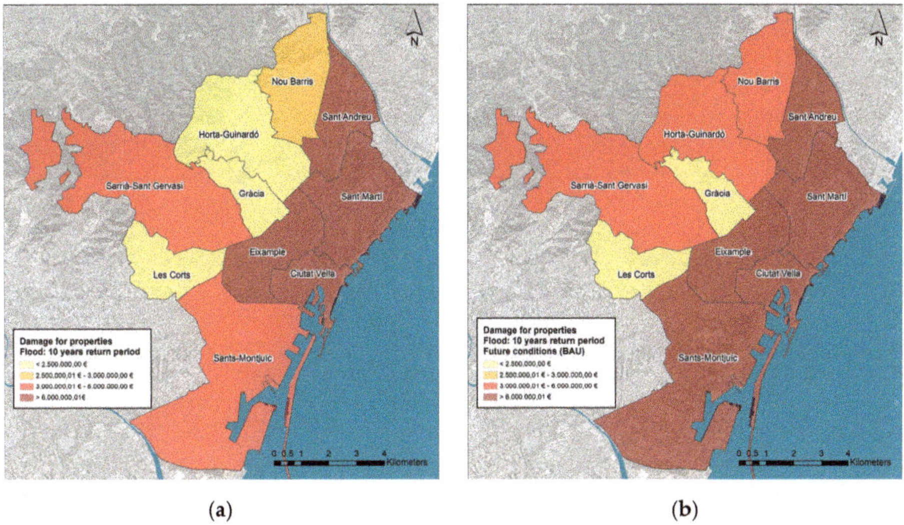

(**a**)   (**b**)

Figure 19. Example of economic flood damage maps for properties for synthetic projected storms of 10 years related to baseline (**a**) and BAU scenarios (**b**) indicating aggregated damages for districts.

Moreover, for both scenarios, the expected annual damage (EAD) [29] for the whole city including flood damages related to properties and vehicles [30] was calculated. The results indicate that, due to climate change, the EAD would grow from € 39.8 M to € 54.7 M [28].

Finally, the methodology for the estimation of indirect damages produced by pluvial floods based on an econometric method of input–output (IO) tables indicated a linear relationship between direct and tangible losses. Specifically, according to the obtained results, indirect tangible damages produced by pluvial floods in Barcelona could represent around 29% of direct damages. This increase could be taken into account in the previously reported EAD [28].

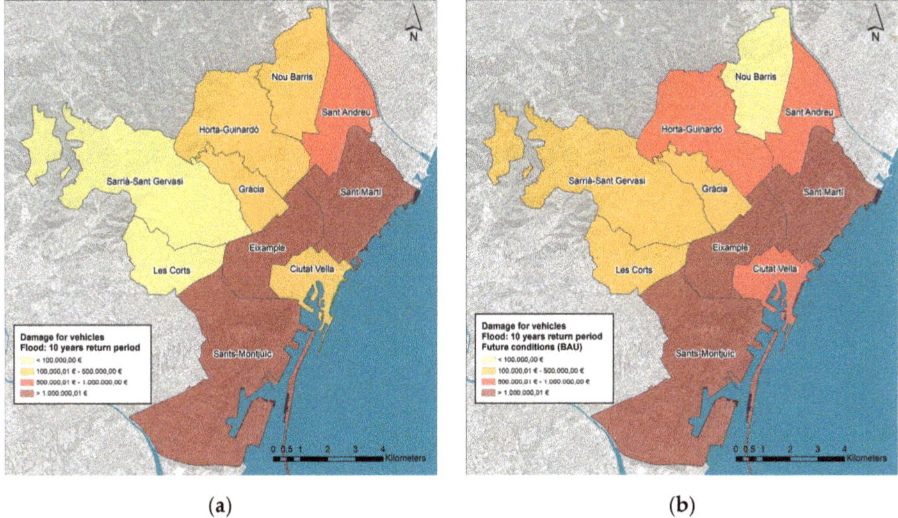

(**a**)   (**b**)

Figure 20. Example of economic flood damage maps for vehicles for synthetic projected storms of 10 years related to baseline (**a**) and BAU scenarios (**b**) indicating aggregated damages for districts.

3.3. Assessment of the Effects of Pluvial Floods on the Surface Traffic Service

The climate-related resilience of a city depends on its capacity to maintain the correct functioning of the main urban services during extreme weather events such as pluvial floods. The results of the impacts produced by this kind of floods on the surface traffic system were analyzed according to the methodology presented in Section 2.7. In this case, flood hazard was assessed through flood hazard maps elaborated on the basis of flood depths provided by the 1D/2D USM and the specific hazard criteria previously presented. Hazard maps were elaborated for the return periods T1, T10, T50, T100 and T500 and current (baseline) and future (BAU) scenarios. Examples of flood hazard maps are shown in Figure 21. Comparing the results for both scenarios, it can be observed that, for the total amount of 1492 km, the increase of the road links that could be affected by speed reduction ranged between 3% and 30% depending on the return period, while the increase in terms of closed road links could be around 20% for all the considered return periods (Figure 22).

Finally, through the TransCAD mesoscalar traffic model, the increase in transit time for all the synthetic storm events was assessed and monetized following the methodology proposed by the Multi-Color Handbook [45]. The monetization of the increase of traveling time for the whole city allowed the estimation of a specific EAD for baseline (1.82 M€) and BAU (2.0 M€) [28,38].

(a) (b)

Figure 21. Example of flood hazard maps for surface traffic for synthetic projected storms of 10 years related to baseline (**a**) and BAU scenarios (**b**).

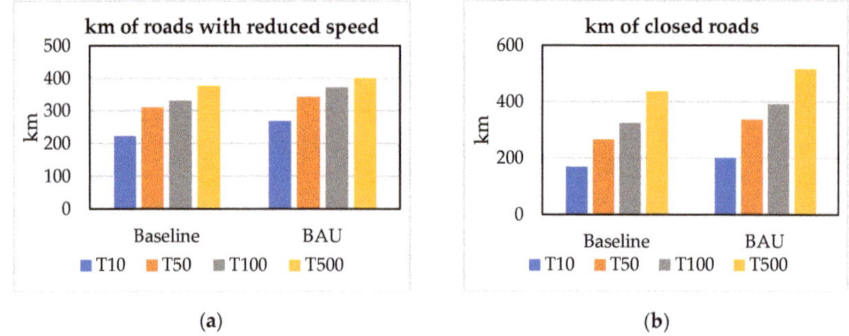

(a) (b)

Figure 22. Representation of the effects produced by pluvial flood on the surface transport system in Barcelona for current (baseline) and future (BAU) scenarios in terms of km of roads with reduced speed (**a**) and km of closed roads (**b**).

3.4. Assessment of the Effects of Pluvial Floods on the Electric System

Through the maximum flow depths provided by the 1D/2D USM and the geolocation of electrical infrastructures, an impact analysis was carried out according to the methodology presented in Section 2.8.

Figure 23 shows an example of the risk assessment carried out for all the electrical assets for return periods of T10 and T100. Here, it is possible to see the most affected areas and electrical assets, which are classified within the categories specified in Table 3 and sized with respect to their failure risk. It is possible to observe that the area near the Besòs riverside is clearly the most affected by pluvial flooding, showing the densest cloud of affected locations. Additionally, the figure shows the increase of the failure probability from the blue-colored baseline scenario (BAS) to the BAU scenario, colored in green.

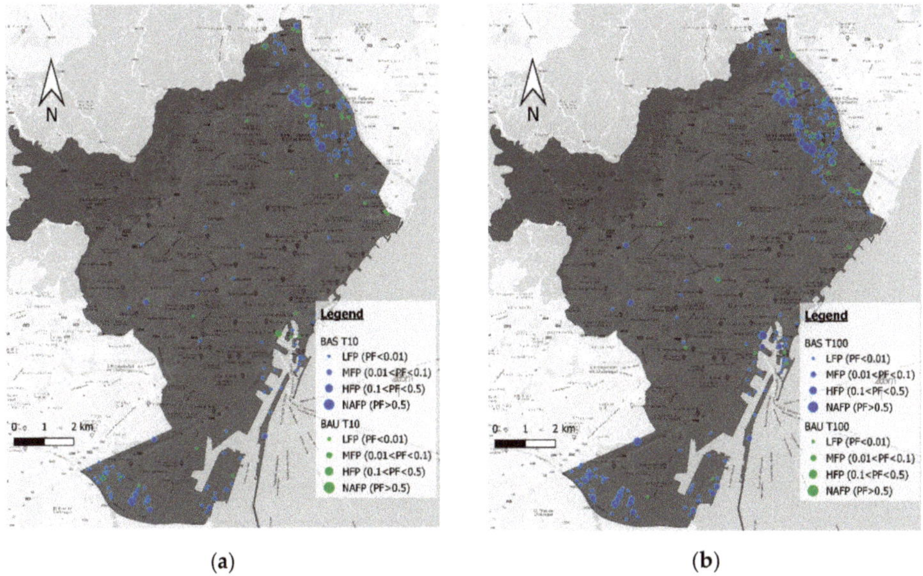

Figure 23. Example of risk maps of all the electrical assets studied for T10 (**a**) and T 100 (**b**).

Table 4 shows the number of electrical infrastructures potentially affected by pluvial flood in Barcelona for baseline and BAU scenarios and their potential level of impact. The table also shows the social impact that each type of flooding provokes in society by counting the number of people affected in each case (reaching, in the worst case, 725,119 out 1,620,343 total people in Barcelona) and the losses provoked for each case, which in the worst scenario amounts to 771,129.01 €. It should be noted that the high (HV) and medium-voltage (MV) substations with a potential flood risk have been studied throughout the city, while only the distribution centers (DCs) in the vicinity of Besós and Llobregat rivers and coastal areas were considered [28,40].

Table 4. Electrical infrastructure potentially affected by pluvial flood in Barcelona for baseline (BAS) and BAU scenarios. DC: distribution center; HV: high-voltage substation; MV: medium-voltage substation.

Return Period	Scenario	Type of Location	Number of Locations Affected	Customers Affected	Costs Provoked
T10	BAS	DC	165	14,984	90,403.68 €
		HV	6	116,872	2377.23 €
		MV	11	94,231	5585.61 €
	BAU	DC	187	290,613	192,823.10 €
		HV	6	116,872	3709.35 €
		MV	13	150,723	2231.57 €
T50	BAS	DC	227	295,490	304,720.21 €
		HV	6	116,872	11,267.27 €
		MV	13	372,311	6627.44 €
	BAU	DC	254	314,932	476,756.76 €
		HV	7	116,872	20,367.44 €
		MV	15	372,311	18,549.21 €
T100	BAS	DC	249	314,044	451,294.19 €
		HV	7	116,872	19,438.12 €
		MV	13	372,311	12,771.98 €
	BAU	DC	272	315,991	556,183.29 €
		HV	8	116,872	28,873.49 €
		MV	15	581,566	41,375.86 €
T500	BAS	DC	296	318,232	633,795.69 €
		HV	9	215,368	56,870.91 €
		MV	17	582,487	28,035.45 €
	BAU	DC	324	320,679	771,129.01 €
		HV	11	215,368	66,869.66 €
		MV	18	725,119	53,948.15 €

3.5. Assessment of the Effects of Pluvial Floods on Waste Collection System

The integrated flood–waste collection model allowed the estimation of the potential number of unstable containers and their location on specific hazard maps based on their typology and degree of filling for the return periods T1, T10 and T50 [27,28,42]. This analysis showed that, for the most extreme episode (T50), some districts of the city such as Ciutat Vella and l'Eixample could have between 20% and 25% of their containers dragged due to the flow and that, in some cases, this amount could increase to values above 30% for the BAU scenario. Figure 24 shows, as an example, the computed number of containers which are potentially unstable for each district under the assumptions of current and future rainfall conditions for a designed 10 year return period storm.

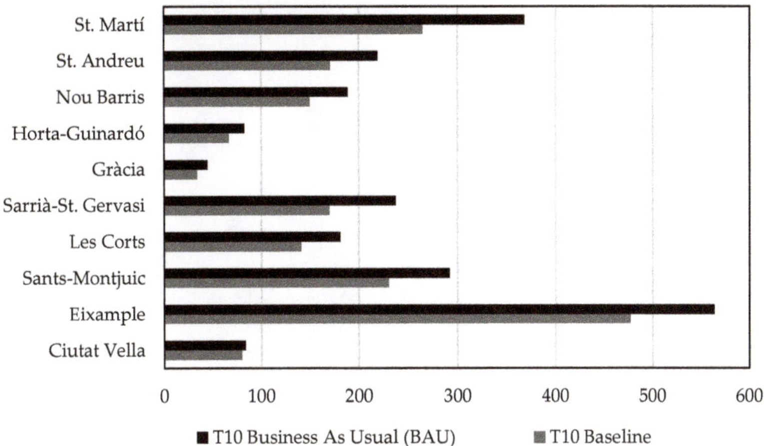

Figure 24. Distribution of computed number of containers which are potentially unstable for each district under current (baseline) and future (BAU) rainfall conditions due to a flooding corresponding to a designed 10 year storm.

3.6. Assessment of Flood Resilience through a Holistic Approach

The holistic model was used to determine the recovery time of the city in the case of extreme episodes of pluvial flooding produced by extreme rain events. The analysis of the holistic simulations allowed the estimation of a recovery time of approximately 1.5 h (calculated as an average value for all the events with return periods T1, T10, T50, 100 and T500), while for the BAU scenario, this value increased up to 2 h.

4. Discussion

The potential increase of maximum rainfall intensities in Barcelona due to climate change could produce a significant increase of tangible and intangibles impacts due to pluvial floods. This paper aimed to perform a comprehensive multi-risk assessment using a detailed 1D/2D USM and several loosely coupled models in order to estimate direct impacts not only due to the poor efficiency of the drainage systems of the city but also due to several cascading effects on other critical urban services. This kind of analysis represents a key tool for decision makers to achieve a reliable estimation of the cost of not acting and to propose and justify correct adaptation measures which are able to reduce a large set of tangible and intangible impacts. For the case of Barcelona, the development and calibration of a 1D/2D USM and its integration in several loosely coupled (or integrated) models allowed us to perform a multi-risk analysis whose main important outputs are shown in Table 5. Moreover, the geographic detailed analysis of the potential flood impacts could help in the prioritization of the implementation of adaptation measures [46]. For example, the results provided by some impact models concerning intangible (safety for pedestrians and vehicles, stability of containers) and tangible (economic losses for properties and vehicles) damage indicate that the highest economic and social risks are concentrated in the districts located in the downtown of the city (near the sea).

Table 5. Potential pluvial flood impacts due to climate change assessed by loosely coupled models. EAD: expected annual damage.

Model	Type of Impact	Indicator (BAU vs. Baseline)	Values for T/EAD
1D/2D USM	Intangible	Increase (%) of high flood risk area for pedestrian and vehicles	Pedestrians: +30 (T10), +34 (T50), +32 (T100), +30 (T500) Vehicles: +38 (T10), +42 (T50), +34 (T100), +25 (T500)
1D/2D USM + Damage model	Tangible	Increase (%) of EAD (including properties, vehicles and indirect damages)	42%
1D/2D USM + Traffic model	Tangible & Intangible	Increase (%) of km of closed roads; EAD due to travelling time rise	+31 (T10), +60 (T50), +66 (T100), +116 (T500); + 0.18 M€
1D/2D USM + Electric model	Tangible & Intangible	Increase (%) of the number of flooded electric infrastructures; related EAD	+31 (T10), +60 (T50), +66 (T100), +116 (T500); + 0.18 M€
1D/2D USM + Waste model	Intangible	Increase (%) of the number of unstable waste containers	+13 (T10), +12 (T50), +11 (T100), +10 (T500); 0.012M€

5. Conclusions

This paper demonstrates how the integration of a detailed and calibrated 1D/2D USM with other models and tools which are able to describe the behavior of other urban services can be useful to simulate the response of these services during pluvial floods produced by heavy storm events.

Furthermore, through the development of these loosely coupled models, socio-economic impacts related to these events can be estimated and the cascading effects can be fully analyzed, as well as the interrelationships between services and critical infrastructures.

In this study, the effects of floods in the potential context of climate change for the city of Barcelona have been analyzed through a multi-risk approach, and the results of this assessment, in terms of tangible and intangible impacts, have been presented for the whole city and with a geographic discretization (i.e., in terms of city districts).

The results demonstrate that Barcelona could suffer a significant increase in these impacts due to climate change if adaptation measures are not adopted. It was demonstrated that increments of maximum rainfall intensity of 12–16% could cause increments of more than 25–30% in terms of social impacts (e.g., intangible damages such as the increase of areas classified with high hazard conditions in case of pluvial flood events) and of 42% of economic losses (including tangible direct and indirect damages) expressed in monetary terms through the concept of EAD that has been calculated for each analyzed urban district. Economic losses related to traffic disruption due to pluvial floods could also increase by 9%, while for the electric system, the increase of economic damage could be 70%, although the final EAD result was shown to be quite low.

Moreover, the average recovery time of the city (defined as the time in which urban services do not recover their normal functioning) could increase from 1.5 to 2 h due to climate change effects.

Finally, the paper shows the geographical distribution of the socio-economic impacts. This information could be very useful for the prioritization of implementation of adaptation measures.

Author Contributions: Conceptualization, B.R.; methodology, B.R., M.V., L.L., E.M.-G., R.M., B.E. and D.S. and D.S.-M; validation, B.R. and M.V.; formal analysis, B.R. and E.M.-G.; investigation, B.R., M.V., L.L., R.M., D.S.-M., E.M.-G., and B.E.; resources, D.S., E.M.-G., B.E. and D.S.-M.; data curation, D.Y., D.S., E.M.-G., E.F.-O., B.E., D.S.-M., A.G.G., writing—original draft preparation, B.R.; writing—review and editing, B.R., M.V., L.L., D.S., E.M.-G., B.E. and D.S.-M; visualization, E.M.-G.; supervision, M.V.; project administration, B.R.; funding acquisition, B.R. and E.M.-G. All authors have read and agreed to the published version of the manuscript.

Funding: This research was funded by Horizon2020 Programme, Grant Agreement No. 700174.

Acknowledgments: This paper presents some of the results achieved in the framework of the RESCCUE project (Resilience to Cope with Climate Change in Urban Areas—a multisectoral approach focusing on water) (www.resccue.eu). RESCCUE is a research project funded by the European Commission under the H2020 program, and its main goal is to provide methodologies and tools for the evaluation, planning and management of urban resilience in the context of climate change.

Conflicts of Interest: The authors declare no conflict of interest.

References

1. Meerow, S.; Newell, J.P.; Stults, M. Defining urban resilience: A review. *Landsc. Urban Plan.* **2016**, *147*, 38–49. [CrossRef]

2. Velasco, M.; Russo, B.; Martínez, M.; Malgrat, P.; Monjo, R.; Djordjevic, S.; Fontanals, I.; Vela, S.; Cardoso, M.A.; Buskute, A. Resilience to cope with climate change in urban areas—A multisectorial approach focusing on water—The RESCCUE project. *Water* **2018**, *10*, 1356. [CrossRef]

3. Walloth, C.; Gurr, J.M.; Schmidt, J.A. *Understanding Complex Urban Systems: Multidisciplinary Approaches to Modeling*; Springer International Publishing: Cham, Switzerland, 2014; ISBN 978-3-319-02996-2.

4. Visvizi, A.; Lytras, M. *Smart Cities: Issues and Challenges. Mapping Political, Social and Economic Risks and Threats,* 1st ed.; Lytras, A., Visvizi, M., Eds.; Elsevier: Amsterdam, The Netherlands, 2019; ISBN 9780128166390.

5. Calzada, I.; Almirall, E. Data ecosystems for protecting European citizens' digital rights. *Transform. Gov. People Process Policy* **2020**, *14*, 133–147. [CrossRef]

6. Monjo, R.; Paradinas, C.; Gaitán, E.; Redolat, D.; Prado, C.; Pórtoles, J.; Torres, L.; Russo, B.; Velasco, M.; Pouget, L.; et al. Report on Extreme Events Prediction. Deliverable 1.3, RESCCUE EU H2020 Project. Available online: http://www.resccue.eu/ (accessed on 24 April 2020).

7. Hammond, M.J.; Chen, A.S.; Djordjević, S.; Butler, D.; Mark, O. Urban flood impact assessment: A state-of-the-art review. *Urban Water J.* **2015**, *12*, 14–29. [CrossRef]

8. Barcelona City Council. *Climate Plan 2018–2030*; Urban Ecology: Barcelona, Spain, 2018.

9. Monjo, R.; Gaitán, E.; Pórtoles, J.; Ribalaygua, J.; Torres, T. Changes in extreme precipitation over Spain using statistical downscaling of CMIP5 projections. *Int. J. Climatol.* **2016**, *36*, 757–769. [CrossRef]

10. Arnbjerg-Nielsen, K. Quantification of climate change effects on extreme precipitation used for high resolution hydrologic design. *Urban Water J.* **2012**, *9*, 57–65. [CrossRef]

11. Russo, B. *Multi-Hazards Assessment Related to Water Cycle Extreme Events for Future Scenarios-Business As Usual. Deliverable 2.3*; RESCCUE EU H2020 Project. Internal report.

12. Hénonin, J.; Russo, B.; Mark, O.; Goubersville, P. Real-time urban flood forecasting and modelling—A state of the art. *J. Hydroinform.* **2013**, *15*, 717–736. [CrossRef]

13. Phillips, B.C.; Yu, S.; Thompson, G.R.; Silva, N. De 1D and 2D Modelling of Urban Drainage Systems using XP-SWMM and TUFLOW. In Proceedings of the 10th International Conference on Urban Drainage, Copenhagen, Denmark, 21–26 August 2005; pp. 21–26.

14. Lipeme Kouyi, G.; Fraisse, D.; Rivière, N.; Guinot, V.; Chocat, B. 1D modelling of the interactions between heavy rainfall-runoff in urban area and flooding flows from sewer network and river. In Proceedings of the 11th International Conference on Urban Drainage, Edinburgh, UK, 31 August–5 September 2008.

15. Obermayer, A.; Guenthert, F.W.; Angermair, G.; Tandler, R.; Braunschmidt, S.; Milojevic, N. Different approaches for modelling of sewer caused urban flooding. *Water Sci. Technol.* **2010**, *62*, 2175–2182. [CrossRef]

16. Leandro, J.; Chen, A.S.; Djordjević, S.; Savić, D.A. Comparison of 1D/1D and 1D/2D coupled (sewer/surface) hydraulic models for urban flood simulation. *J. Hydraul. Eng.* **2009**, *135*, 495–504. [CrossRef]

17. Kandori, C.; Willems, P. Impact of the two-directional interaction of sewer and river systems on the flood risk. In Proceedings of the 11th International Conference of Urban Drainage, Edinburgh, UK, 31 August–5 September 2008; pp. 1–10.

18. Mark, O.; Weesakul, S.; Apirumanekul, C.; Aroonnet, S.B.; Djordjevic, S. Potential and limitations of 1D modelling of urban flooding. *J. Hydrol.* **2004**, *299*, 284–299. [CrossRef]

19. Pina, R.D.; Ochoa-Rodriguez, S.; Simões, N.E.; Mijic, A.; Marques, A.S.; Maksimović, Č. Semi- vs. Fully-distributed urban stormwater models: Model set up and comparison with two real case studies. *Water* **2016**, *8*, 58. [CrossRef]

20. Russo, B.; Sunyer, D.; Velasco, M.; Djordjevic, S. Analysis of extreme flooding events through a calibrated 1D/2D coupled model: The case of Barcelona (Spain). *J. Hydroinform.* **2015**, *17*, 473–491. [CrossRef]

21. Gómez, M.; Russo, B. Methodology to estimate hydraulic efficiency of drain inlets. *Proc. Inst. Civ. Eng. Water Manag.* **2011**, *164*, 81–90. [CrossRef]

22. Gómez, M.; Russo, B. Hydraulic efficiency of continuous transverse grates for paved areas. *J. Irrig. Drain. Eng.* **2009**, *135*. [CrossRef]

23. Gómez, M.; Parés, J.; Russo, B.; Martínez-Gomariz, E. Methodology to quantify clogging coefficients for grated inlets. Application to SANT MARTI catchment (Barcelona). *J. Flood Risk Manag.* **2019**, *12*. [CrossRef]

24. Gómez, M.; Rabasseda, G.H.; Russo, B. Experimental campaign to determine grated inlet clogging factors in an urban catchment of Barcelona. *Urban Water J.* **2013**, *10*. [CrossRef]

25. Innovyze. *InfoWorks Integrated Catchment Modelling (ICM)*, version 10.0; Innovyze: Monrovia, CA, USA, 2020.

26. Russo, B.; Pouget, L.; Malgrat, P. Evaluacion del impacto del cambio climatico en un caso de estudio de Barcelona a través de una modelización 2D-1D del drenaje dual. In Proceedings of the Jornadas de Ingeniería del Agua (JIA), Barcelona, Spain, 5–6 October 2011.

27. Russo, B. *Multi-Hazards Assessment Related to Water Cycle Extreme Events for Current Scenario. (Public Summary) Deliverable 2.4.* RESCCUE EU H2020 Project. Available online: http://www.resccue.eu/sites/default/files/d2.4._multi-hazards_assessment_related_to_water_cycle_extreme.pdf (accessed on 24 April 2020).

28. Evans, B. *Impact Assessments of Multiple Hazards in Case Study Areas. Deliverable 3.4*; RESCCUE EU H2020 Project. Internal report.

29. Velasco, M.; Russo, B.; Cabello, À.; Termes, M.; Sunyer, D.; Malgrat, P. Assessment of the effectiveness of structural and nonstructural measures to cope with global change impacts in Barcelona. *J. Flood Risk Manag.* **2018**, *11*. [CrossRef]

30. Martínez-Gomariz, E.; Locatelli, L.; Guerrero, M.; Russo, B.; Martínez, M. Socio-Economic Potential Impacts Due to Urban Pluvial Floods in Badalona (Spain) in a Context of Climate Change. *Water* **2019**, *11*, 2658. [CrossRef]

31. Turner, B.L.; Kaspersonb, R.C.; Matsone, P.A.; McCarthy, J.; Corell, R.; Christensene, L.; Eckley, N.; Kasperson, J.X.; Luers, A.; Martello, M.L.; et al. A framework for vulnerability analysis in sustainability science. *Proc. Natl. Acad. Sci. USA* **2003**, *100*, 8074–8079. [CrossRef]

32. Russo, B.; Gómez, M.; Macchione, F. Pedestrian hazard criteria for flooded urban areas. *Nat. Hazards* **2013**, *69*, 251–265. [CrossRef]

33. Martínez-Gomariz, E.; Gómez, M.; Russo, B. Experimental study of the stability of pedestrians exposed to urban pluvial flooding. *Nat. Hazards* **2016**, *82*, 1259–1278. [CrossRef]

34. Martínez-Gomariz, E.; Gómez, M.; Russo, B.; Djordjević, S. A new experiments-based methodology to define the stability threshold for any vehicle exposed to flooding. *Urban Water J.* **2017**, *14*, 930–939. [CrossRef]

35. Martínez-Gomariz, E.; Guerrero-Hidalga, M.; Russo, B.; Yubero, D.; Gómez, M.; Castán, S. Desarrollo y aplicación de curvas de daño y estanqueidad para la estimación del impacto económico de las inundaciones en zonas urbanas españolas. *Ingenieria del Agua* **2019**, *23*, 229. [CrossRef]

36. Martínez-Gomariz, E.; Forero-Ortiz, E.; Guerrero-Hidalga, M.; Castán, S.; Gómez, M. Flood Depth—Damage Curves for Spanish Urban Areas. *Sustainability* **2020**, *12*, 2666. [CrossRef]

37. Martínez-Gomariz, E.; Gómez, M.; Russo, B.; Sánchez, P.; Montes, J.A. Methodology for the damage assessment of vehicles exposed to flooding in urban areas. *J. Flood Risk Manag.* **2019**, *12*, 1–15. [CrossRef]

38. Evans, B.; Chen, A.S.; Djordjevi, S.; Webber, J.; Gonzalez, A.; Stevens, J. Investigating the Effects of Pluvial Flooding and Climate Change on Traffic Flows in Barcelona and Bristol. *Sustainability* **2020**, *12*, 2330. [CrossRef]

39. Pyatkova, K.H. *Flood Impacts on Road Transportation*; University of Exeter: Exeter, UK, 2019.

40. Sánchez-Muñoz, D.; Domínguez-García, J.L.; Martínez-Gomariz, E.; Russo, B.; Stevens, J.; Pardo, M. Electrical grid risk assessment against flooding in Barcelona and Bristol cities. *Sustainability* **2020**, *12*, 1527. [CrossRef]

41. FEMA. *Multi-Hazard Loss Estimation Methodology, Flood Model: Hazus-MH MR4 Technical Manual*; FEMA: Washington, DC, USA, 2009.
42. Martínez-Gomariz, E.; Russo, B.; Gómez, M.; Plumed, A. An approach to the modelling of stability of waste containers during urban flooding. *J. Flood Risk Manag.* **2019**. [CrossRef]
43. Russo, B.; Velasco, M.; Monjo, R.; Martínez-Gomariz, E.; Sánchez, D.; Domínguez, J.L.; Gabàs, A.; Gonzalez, A. Evaluación de la resiliencia de los servicios urbanos frente a episodios de inundación en Barcelona. El Proyecto RESCCUE. *Ingenieria del Agua* **2020**, *24*, 101. [CrossRef]
44. U.S. Army Corps of Engineers (USACE). *Economic Guidance Memorandum, 09-04, Generic Depth-Damage Relationships for Vehicles*; USACE: Washington, DC, USA, 2009.
45. Penning-Rowsell, E.; Viavattene, C.; Pardoe, J.; Chatterton, J.; Parker, D.; Morris, J. *The Benefits of Flood and Coastal Risk Management: A Handbook of Assessment Techniques*; Flood Hazard Research Centre, Middlesex University: London, UK, 2010.
46. Guerrero-Hidalga, M.; Martínez-Gomariz, E.; Evans, B.; Webber, J.; Termes-Rifé, M.; Russo, B.; Locatelli, L. Methodology to Prioritize Climate Adaptation Measures in Urban Areas. Barcelona and Bristol Case Studies. *Sustainability* **2020**, *12*, 4807. [CrossRef]

Correction

Correction: Russo, B., et al. Assessment of Urban Flood Resilience in Barcelona for Current and Future Scenarios. The RESCCUE Project. *Sustainability* 2020, *12*, 5638

Beniamino Russo [1,2,3,*], Marc Velasco [1], Luca Locatelli [1], David Sunyer [1], Daniel Yubero [1], Robert Monjo [4], Eduardo Martínez-Gomariz [3,5], Edwar Forero-Ortiz [5], Daniel Sánchez-Muñoz [6], Barry Evans [7,8] and Andoni Gonzalez Gómez [9]

[1] AQUATEC (SUEZ Advanced Solutions), Paseo de la Zona Franca, 46-48, 08038 Barcelona, Spain; marc.velasco@suez.com (M.V.); luca.locatelli@aquatec.es (L.L.); dsunyer@aquatec.es (D.S.); dyuberop@aquatec.es (D.Y.)

[2] Grupo de Ingeniería Hidráulica y Ambiental (GIHA) (Group of Hydraulic and Environmental Engineering), Escuela Politécnica de La Almunia (EUPLA, Universidad de Zaragoza) (Technical College of La Almunia, University of Zaragoza), Calle Mayor, 5, 50100 Zaragoza, Spain

[3] Flumen Research Institute, Universitat Politècnica de Catalunya, Jordi Girona 1-3, 08034 Barcelona, Spain; eduardo.martinez-gomariz@upc.edu or eduardo.martinez@cetaqua.com

[4] Fundación de Investigación del Clima (FIC) (Climate Research Fundation), Calle Gran Vía, 22, 28019 Madrid, Spain; rma@fic.es

[5] Cetaqua, Water Technology Centre, Carretera d'Esplugues, 75, 08940 Barcelona, Spain; eaforero@cetaqua.com

[6] IREC, Power Systems Department, Jardins de les Dones de Negre, 1, 2ª pl., 08930 Barcelona, Spain; dsanchezm@irec.cat

[7] Centre for Water Systems, University of Exeter, Exeter EX4 4QF, UK; b.evans@exeter.ac.uk

[8] School of Built Environment, College of Sciences, Massey University, Auckland 0745, New Zealand

[9] Ajuntament de Barcelona (Barcelona Municipality), Carrer de Torrent de l'Olla 218, 08012 Barcelona, Spain; agonzalezgom@bcn.cat

* Correspondence: brusso@unizar.es; Tel.: +34-932-479-869

Received: 27 October 2020; Accepted: 28 October 2020; Published: 25 November 2020

The authors would like to make the following corrections about the published paper [1]. The changes are as follows:

Replacing Table 5.

Table 5. Potential pluvial flood impacts due to climate change assessed by loosely coupled models. EAD: expected annual damage.

Model	Type of Impact	Indicator (BAU vs. Baseline)	Values for T/EAD
1D/2D USM	Intangible	Increase (%) of high flood risk area for pedestrian and vehicles	Pedestrians: +30 (T10), +34 (T50), +32 (T100), +30 (T500) Vehicles: +38 (T10), +42 (T50), +34 (T100), +25 (T500)
1D/2D USM + Damage model	Tangible	Increase (%) of EAD (including properties, vehicles and indirect damages)	42%

Table 5. *Cont.*

Model	Type of Impact	Indicator (BAU vs. Baseline)	Values for T/EAD
1D/2D USM + Traffic model	Tangible & Intangible	Increase (%) of km of closed roads; EAD due to travelling time rise	+31 (T10), +60 (T50), +66 (T100), +116 (T500); + 0.18 M€
1D/2D USM + Electric model	Tangible & Intangible	Increase (%) of the number of flooded electric infrastructures; related EAD	+31 (T10), +60 (T50), +66 (T100), +116 (T500); + 0.18 M€
1D/2D USM + Waste model	Intangible	Increase (%) of the number of unstable waste containers	+13 (T10), +12 (T50), +11 (T100), +10 (T500); 0.012M€

With:

Table 5. Potential pluvial flood impacts due to climate change assessed by loosely coupled models.

Model	Type of Impact	Indicator (BAU vs. Baseline)	Values for T/EAD
1D/2D USM	Intangible	Increase (%) of high flood risk area for pedestrian and vehicles	Pedestrians: +30 (T10), +34 (T50), +32 (T100), +30 (T500) Vehicles: +38 (T10), +42 (T50), +34 (T100), +25 (T500)
1D/2D USM + Damage model	Tangible	Increase (%) of EAD (including properties, vehicles and indirect damages)	+42%
1D/2D USM + Traffic model	Tangible and Intangible	Increase (%) of km of closed roads; EAD due to travelling time rise	+31 (T10), +60 (T50), +66 (T100), +116 (T500); +0.18 M€
1D/2D USM + Electric model	Tangible and Intangible	Increase (%) of the number of flooded electric infrastructures; related EAD	+13 (T10), +12 (T50), +11 (T100), +10 (T500); +0.12M€
1D/2D USM + Waste model	Intangible	Increase (%) of the number of unstable waste containers	Empty: +27 (T10), +28 (T50) 50% full: +28 (T10), +32 (T50) 100% full: +28 (T10), +36 (T50)

Reference

1. Russo, B.; Velasco, M.; Locatelli, L.; Sunyer, D.; Yubero, D.; Monjo, R.; Martínez-Gomariz, E.; Forero-Ortiz, E.; Sánchez-Muñoz, D.; Evans, B.; et al. Assessment of Urban Flood Resilience in Barcelona for Current and Future Scenarios. The RESCCUE Project. *Sustainability* **2020**, *12*, 5638. [CrossRef]

Publisher's Note: MDPI stays neutral with regard to jurisdictional claims in published maps and institutional affiliations.

sustainability

Article

Electrical Grid Risk Assessment Against Flooding in Barcelona and Bristol Cities

Daniel Sánchez-Muñoz [1,*], José L. Domínguez-García [1], Eduardo Martínez-Gomariz [2,3], Beniamino Russo [4,5], John Stevens [6] and Miguel Pardo [7]

1   IREC, Power Systems department, Jardins de les Dones de Negre, 1, 2ª pl., 08930 Sant Adrià de Besòs, Barcelona, Spain; jldominguez@irec.cat
2   Cetaqua, Water Technology Centre, Carretera d'Esplugues, 75, 08940 Cornellà de Llobregat, Spain; eduardo.martinez@cetaqua.com
3   FLUMEN Research Institute, Universitat Politècnica de Catalunya, Jordi Girona 1-3, 08034 Barcelona, Spain; eduardo.martinez-gomariz@upc.edu
4   AQUATEC (SUEZ Advanced Solutions), Paseo de la Zona Franca, 46-48, 08038 Barcelona, Spain; brusso@aquatec.es
5   Grupo de Ingeniería Hidráulica y Ambiental (GIHA), Escuela Politécnica de La Almunia (EUPLA), Universidad de Zaragoza, Calle Mayor 5, 50100 La Almunia de Doña Godina, Zaragoza, Spain; brusso@unizar.es
6   Bristol City Council. 100 Temple Street, P.O. Box 3176, Bristol BS1 6AG, UK; john.stevens@bristol.gov.uk
7   E-distribución. Av. de Vilanova, 12, 08018 Barcelona, Spain; miguel.pardo@enel.com
*   Correspondence: dsanchezm@irec.cat; Tel.: +34-933-56-26-15

Received: 28 January 2020; Accepted: 15 February 2020; Published: 18 February 2020

Abstract: Climate change is increasing the frequency and intensity of extreme events and, consequently, flooding in urban and peri-urban areas. The electrical grid is exposed to an increase in fault probability because its infrastructure was designed considering historical frequencies of extreme events occurred in the past. In this respect, to ensure future energy plans and securing services is of great relevance to determine and evaluate the new zones that may be under risk and its relation to critical infrastructures for such extreme events. In this regard, the electrical distribution system is one of the key critical infrastructures since it feeds the others and with the future plans of zero-emissions (leading to the electrification of transport, buildings, renewable energies, etc.) will become even more important in the short term. In this paper, a novel methodology has been developed, able to analyze flood hazard maps quantifying the probability of failure risk of the electrical assets and their potential impacts using a probabilistic approach. Furthermore, a process to monetize the consequences of the yielded risk was established. The whole method developed was applied to the Barcelona and Bristol case study cities. In this way, two different examples of application have been undertaken by using slightly different inputs. Two main inputs were required: (1) the development of accurate GIS hazard flooding models; and (2) the location of the electrical assets (i.e., Distribution Centers (DCs)). To assess and monetize the flood risk to DCs, a variety of variables and tools were required such as water depths (i.e., flood maps), DCs' areas of influence, fragility curves, and damage curves. The analysis was performed for different return periods under different scenarios, current (Baseline) and future (Business As Usual (BAU)) rainfall conditions. The number of DCs affected was quantified and classified into different categories of risk, where up to 363 were affected in Barcelona and 623 in Bristol. Their risk monetization resulted in maximums of 815,700 € in Barcelona and 643,500 € in Bristol. Finally, the percentage of risk increases when considering future rainfall conditions (i.e., BAU) when calculated, resulting in a 2.38% increase in Barcelona and 3.37% increase in Bristol, which in monetary terms would be an average of a 22% increase.

Sustainability **2020**, *12*, 1527

Keywords: RESCCUE project; Electrical distribution network; Flooding; Risk Assessment; city resiliency; GIS model

---

1. Introduction

The future projections for climate change augur severe scenarios for extreme climate events, especially flooding. The predictions indicate increases of frequency in high flows by 10%–30%, while also increasing in magnitude as well [1,2]. The Climate Research Foundation (FIC) as part of the RESCCUE project has studied the changes in terms of rainfall intensity for two European cities, Barcelona and Bristol, with an expected increase of up to 40% [3]. An increase in the rainfall intensity will provoke consequently higher flood depths in the surface of the cities because of the exceedance of the drainage and sewer system capacity. Consequently, current flood prone areas will be covered by higher depths and new flood prone areas will arise [4], which will increase the likelihood of affecting critical city infrastructures.

As critical infrastructures, the electric power systems are considered the backbone of the city due to the increase of power-dependent utilities and devices. The water supply through water pumps, telecommunication centers, transport (e.g., tramway, underground, electric buses, traffic lights, etc.), and a large list of city services depend on the electrical infrastructure [5], therefore a general system failure may end in the collapse of a city until emergency equipment is installed [6]. Due to this, the resilience of cities is of extreme relevance.

Like any other kind of infrastructure, the electrical was designed considering certain return periods of events that could affect the system at any point, and they were protected and isolated accordingly. However, the problem arises when the intensity of the considered return period increases due to climate change, generating unexpected extreme occurrences that increase the likelihood of damaging the infrastructures that are not prepared for it [2].

Taking into account the aforementioned points and mixing all ideas together, a plausible problem is presented; the increased probability of electric blackout provoked by flooding due to more frequent extreme rainfall events caused by climate change, and thus generating the effect of cascading failures in other urban services.

Although the problem presented above has not been extensively studied, there exist some other investigations studying similar problems but following different perspectives or focusing either on the impact assessment or on economical assessment. The most complete study found during the literature review was a GIS-based method assessing electrical grid and gas network through fragility curves focused on seismic events [7]. When focusing on flooding events a methodology to assess the flooding impact probability of the electrical assets was proposed in [8] where through spatial network models identified and compared the risk of critical infrastructures on flooded lands. Also, [9] proposed a method to investigate quantitatively the robustness of the grid against flooding events based on the Hazus methodology [10] providing a detailed risk analysis. The last relevant method found during the literature review was [11], presenting an integrated modelling framework combining geospatial information on infrastructure and flood hazard and geospatial modelling of businesses and economic activities. Additionally, in [12,13] was proposed a methodology to assess the economic losses caused by flooding events to electrical assets that in fact, has been used partially on this study.

In this context, this study aims to identify first the hazards and to assess later the potential impacts caused in the electrical sector in Barcelona and Bristol cities that inevitably affect the population of the cities. This impact assessment is carried out to evaluate the probability of power system failures after a flooding event occurs, which allows identification of the most critical locations in order to implement, if necessary, adaptation measures effectively. The impact assessment will also allow an estimate of the potential cost of the energy lost during blackout periods, and damages caused to the electrical assets. Therefore, an analysis of the consequences caused by flooding to population and the electrical Distribution Systems Operators (DSO) considering failures within the electrical sector is presented here, together with the description of a novel method to assess risk and estimate losses in the distribution centers (DCs) of the power network (example given in Figure 1). The application of these tools has been carried out for the city of Barcelona and Bristol.

(a)

(b)

Figure 1. Distribution Center example. (a) Outside view, (b) Inside view.

The paper starts with an overview of the study areas and the data used to later explain the probabilistic GIS-based method developed and how the data was used to conduct the DC risk analysis and its corresponding risk monetization. The results are presented in different sub-sections discussing later all the details to finally conclude with the main findings and evidence drawn.

2. Study Areas and Data Used

The considered case studies have been two different cities, Bristol (UK) (Figure 2a) and Barcelona (Spain) (Figure 2b). Both cities are very different in almost every possible feature to consider, in the city design as house grouping, drainage systems, terrestrial topography, electrical grid, etc. and even more different in the weather conditions. Although they have a similar areal extent (Bristol 111 km^2 and Barcelona 102 km^2) the population density in Barcelona is three times bigger than in Bristol, having around 1,621,000 inhabitants while Bristol has only 460,000. However, the studied area for this energy analysis in Barcelona has been reduced up to 33% due to key data availability, therefore considering only the inhabitants living in the main city area (i.e., 326,000 inhabitants).

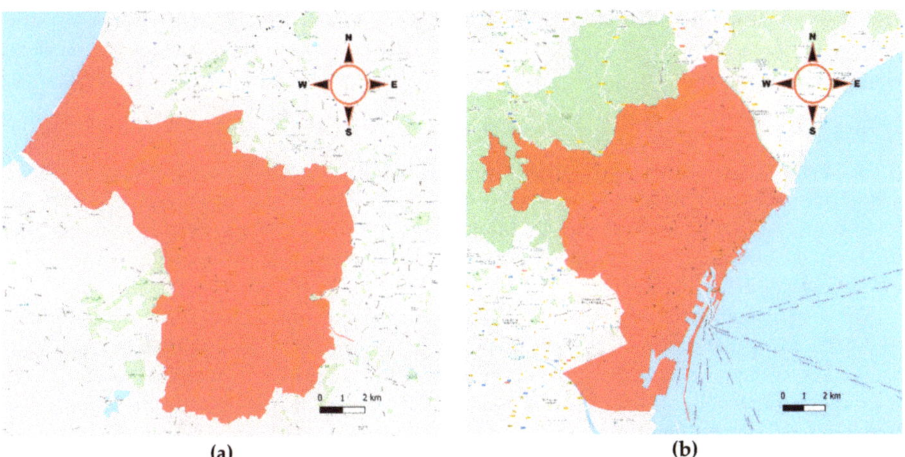

(a)

(b)

Figure 2. Areas of study considered. (a) Bristol city area, (b) Barcelona city area.

2.1. The Bristol Context

Bristol is recognized as one of the most susceptible cities within the top 10 Flood risk areas in the UK [14]. This is mainly caused by the influence of the tidal river "Avon", directly connected to the sea by mean of the "Severn Estuary", where tides have a great effect on the river water depth. This tidal effect is transmitted to a lesser extent to the other rivers that flow into the river Avon. Bristol has undergone severe floods caused by the combination of storm surges and spring tides, but also others provoked by heavy rainfall.

Some great historical floods have occurred in Bristol throughout time. The first example is an historical flood that took place in 1607 when the death of around 2000 people was estimated. A second more recent example took place in 1896 when a tidal flood event caused a 1m water depth in the city center, and in 1968 when a combination of a great rainfall with a river flood event killed 7 people and flooded around 800 properties in the Bristol area. Also, an extreme rainfall event in 2012 caused the flooding of 25 properties due to surface water flooding [14]. All these examples give an idea of the flooding experienced in Bristol and present a real problem that this city has suffered over time.

2.1.1. Bristol Flooding Data

From the flooding problems occurred in the past, some city plans have been born and the flooding records act as a great tool to be used in studies to try to assess the risk to the city in any sector, which in this case will be the electrical sector.

The study carried out in this paper has used two of the most recent city studies conducted by Bristol City Council; the Central Area Flood Risk Assessment (CAFRA) focused on the river and tidal flooding and the Surface Water Management Plan (SWMP), focused on the flooding caused by rainfall extreme events. In Table 1, the return periods given for each study and scenario are introduced.

Table 1. Different return periods and scenarios run under the flood risk studies.

	Return Period	
Flood risk study	SWMP	CAFRA
Current Scenario	T20, T100	T20, T100
BAU Scenario	T20, T100	T20, T100

2.1.2. Bristol Electrical Data

Regarding the electrical grid, Western Power Distribution is the Distribution System Operator (DSO) responsible in Bristol city who has provided part of the suitable data necessary for this study, which comprises the locations of all 11 kV DCs involved in the assessment, substantiating the risk that these assets can be exposed to.

2.2. The Barcelona Context

In Barcelona, although two main rivers (Llobregat and Besòs) cross the city and flow into the sea, the tidal influence of the Mediterranean Sea is not relevant, in that the main flooding events have been caused by heavy rainfall resulting in overtopping banks and river overflow. Every year, one or more serious rainfall events take place in Barcelona during summer and autumn, normally due to the cold drop phenomenon.

For what the Catalonia water agency (ACA) has recorded for Barcelona, in February of 1920 there was great flooding that caused the loss of around 600 properties [15]. After that, the worst flooding of Barcelona occurred on the 25th September 1962 when 200 l/m^2 were registered in less than 3h. This caused the overflow of Besòs and Llobregat rivers making entire wards disappear through inundation, destroying whole factories and causing more than 12,000 victims and 617 people dead in several places of Barcelona province [16]. This disaster boosted the channeling of both Llobregat and Besòs Rivers, preventing further flooding caused by river overflow. Thereafter, another example of severe flooding took place in October of 1987 when Barcelona city services such as the underground, roads, train, tram, airport, gas, communications and the electric services were blocked until the progressive restoration in the following days [15].

Taking into account this historical record, the tidal assessment seems irrelevant, as there were no occurrences for tidal flooding. On the other hand, the extreme rainfall is the phenomenon responsible for causing the most important flooding events that critically affected the city, hence the importance of the surface cover flood risk assessment of the different city services in Barcelona.

2.2.1. Barcelona Flooding Data

The entire drainage system of Barcelona has been modelled, providing GIS layers with detailed information about flooding areas of Barcelona for different return periods and scenarios. Layers with different return periods (T1, T10, T50, T100 and T500) were generated for a current climate scenario [17] and for a climate change scenario by considering BAU conditions [18]. The current scenario is based on historical rainfall data. However, the BAU scenario was created by simulating extreme rainfall events considering RCP 4.5 and 8.5 by the year 2100.

2.2.2. Barcelona Electrical Data

In the Barcelona case, the electrical grid is managed by Endesa DSO, who provided the location of the DCs of three important areas: Besòs riverside, Llobregat riverside, and seashore. Due to the non-availability of DCs location within the entire city, it is not possible to estimate the losses in the entire city, which makes the focus of the study on the three areas mentioned. These areas were selected by the DSO as the most relevant for their infrastructure. In addition, to keep the study based on the real data obtained, no estimations on the other zones have been made.

3. Methodology

To calculate the failure probability of the DC to assess and calculate the potential cost of flooding, a new methodology has been designed. Such methodology provides an estimation of the assets impacted as well as the economic evaluation of such faulty conditions. The methodology is structured in different steps as inputs, probability evaluation, number of assets counting and finally the impact outputs. A graphical explanation of the method explained in detail in the following sections is given in Figure 3.

3.1. Risk Assessment

For the electrical asset risk assessment, a novel procedure has been established. The first step was to create a sampling shape layer by using an average diameter of 20 meters for the electrical DCs wanted for evaluation, with the diameter based on recommendation in the Energy Networks Association article [19]. The diameter was used to define the influence area for each location and after that a uniformly distributed cloud of 106 sampling points was created all across the extent (Figure 4). It is worth noting that such areas allow the tool to cope with location uncertainty in GIS data.

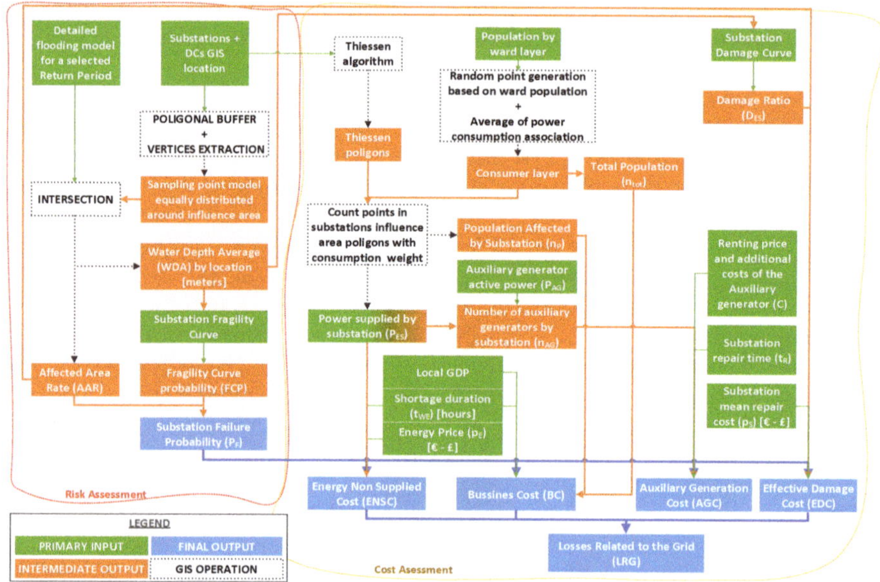

Figure 3. Flowchart of the method followed in the analysis.

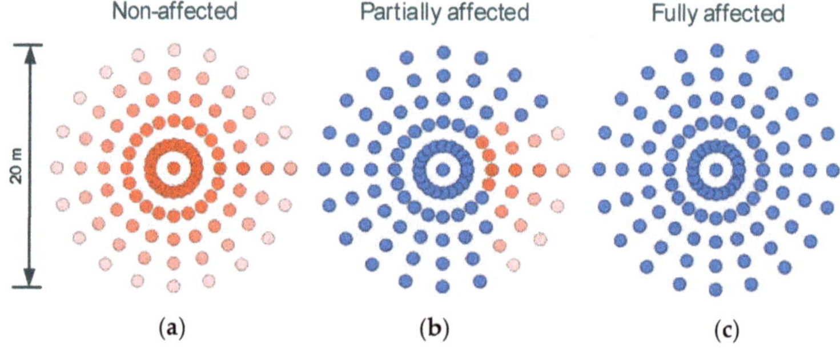

Figure 4. Non-affected area (**a**), partially affected area (**b**), fully affected area (**c**).

The second step consisted of the intersection of the sampling layer with the detailed flooding map that contained the water depth of each flooded area. After crossing both shapes, the following parameters were extracted:

- Water depth (in meters).
- Flooding occurrence (Y/N). For this, the condition to get a positive answer was to have a Water Depth ≥ 10 cm (Flooding was assumed to occur once this this threshold was reached).

The third step was to process the information obtained for every sampling point to calculate a representative figure for each location. In this way, the rate of the affected area and an average for the water depth for each location was calculated.

To calculate the Affected Area Rate (AAR), the number of sampling points affected were counted (n_Y) and later divided by the total number of sampling points (n_{total}) (Equation (1)).

$$AAR = \frac{n_Y}{n_{total}} \times 100 \tag{1}$$

Afterwards, the Water Depth Average (WDA) was calculated for the sampling points flood depth, obtaining a general representative number of each location (Equation (2)):

$$WDA = \frac{1}{n} \sum_{i=1}^{n} Y_i \tag{2}$$

Once the water depth for each location was calculated, the fourth step was to introduce this parameter in the X-axis of each fragility curve represented in Figure 5. The failure probability was obtained and represented from 0 to 1 in the Y-axis.

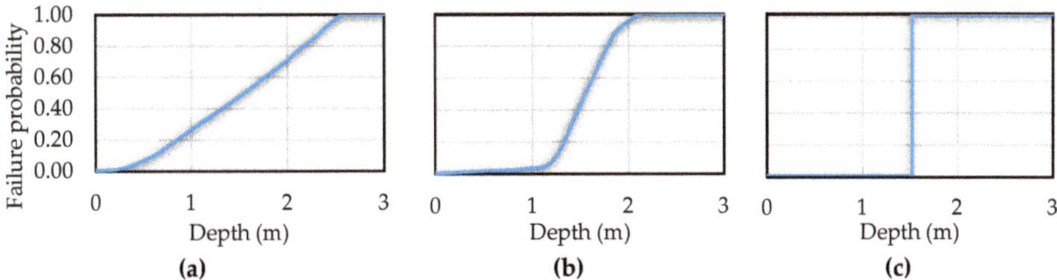

Figure 5. Original flooding fragility curves for HV, MV and LV electrical substations and distribution centers (**b**) (adapted from FEMA, 2009). Softened fragility curve (**a**) and Hardened fragility curve for sensitivity analysis (**c**).

The original fragility curve (Figure 5b) used in this study was adapted from that of the Federal Emergency Management Agency of the United States [10], previously formed through data gathered from important disasters occurred in the US electrical grid. It must be remarked that the US grid can have different standards and protective measures compared to Europe and a different substation and DC topology. However, in further studies with more data available, the curves can be rebuilt to fit with the real conditions and features. Taking into account this dissimilarity with the grid established in Bristol or Barcelona, a sensitivity analysis was performed to assess the possible error caused by this. In this manner, it is possible to offer a better resolution by contemplating a wider spectrum (Figure 5a,c).

After obtaining the result from the fragility curves, the fifth step was to multiply the Fragility Curve Probability (FCP) by the AAR, obtaining a final probability of failure for each analysis performed for each return period given (Equation (3)).

$$P_F = AAR \times FCP \tag{3}$$

The last step was to classify each location studied according to the P_F calculated by following the categories established in Table 2.

Table 2. Different categories set for ranges of failure probabilities.

$P_F > 0.01$	Low Failure Probability (LFP)
$0.01 < P_F \leq 0.10$	Moderate Failure Probability (MFP)
$0.10 < P_F \leq 0.50$	High Failure Probability (HFP)
$P_F > 0.50$	Non-Acceptable Failure Probability (NAFP)

3.2. Economical Losses Caused by Electrical Asset Failure

Flooding can cause extensive potential economic losses due to the impact caused to the electrical assets described in the previous sections. The losses considered in this study are those caused by non-supplied electricity, damages provoked to the electrical assets, the expenditures associated with the renting of emergency electrical supply appliances, and the businesses earning losses provoked by the shortage. The methodology followed for the calculation of all mentioned losses is explained below.

3.2.1. Effective Damage Cost (EDC)

To monetize the potential damages caused to the electrical assets, a damage curve adapted from (FEMA-HAZUS) has been applied (Figure 6). The curve initially was given for a 3m depth, but it has been interpolated from the original one up to a water depth of 9m according to the maximum water depths obtained in the flooding maps. In Figure 6, the original curve is shown in blue, and the one that was used for the analysis in green.

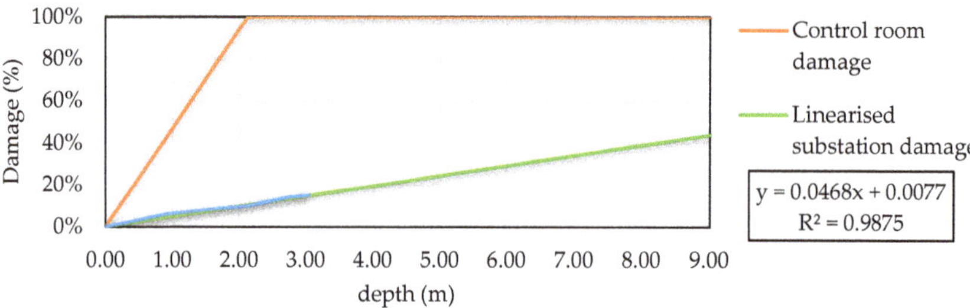

Figure 6. Damage curve used for Effective Damage Cost calculation (Adapted and interpolated from FEMA-HAZUS [10]).

After introducing the WDA (obtained by Equation (2)) in the damage curve, a percentage of damage (D_n) in each DC analyzed was obtained. After that, the effective damage cost (EDC) was calculated according to Equation (4) by multiplying the damage ratio (D_{ES}), by the failure probability (P_F) and by the price of the corresponding DC (p_{SC}) that has been estimated based on the substation voltage given in Table 3. The price of a DC assumed for the cost calculation corresponds to an 11 kV substation.

$$EDC = P_F \times D_{ES} \times p_S \tag{4}$$

Table 3. Cost of the different substations analyzed based on the voltage (p_S) (Adapted from "Climate change and critical infrastructure-floods" [12]).

Voltage (kV)	225	132	63	33	25	11
Substation cost (€)	9,000,000	5,560,000	3,000,000	1,890,000	1,590,000	1,070,000

3.2.2. Cost Associated with Businesses Losses (BC)

First, a GIS consumer layer based on the ward's population of each city was created, by using databases of 2018 obtained from the open data portal of both cities studied [20,21].

A cloud of random points based on the ward population was then generated, therefore representing the potential consumers distributed along all the study extent. After that, a Thiessen polygons layer was generated for each electrical asset, representing the supply coverage area of each DC (Figure 7). The next step was the association of the number of consumers of each random point to the overlapping Thiessen polygon, getting a total of associated consumers for each DC. It is important to indicate that such polygons provide an averaged estimation of the population that for global risk assessment can be accepted as a reasonable result, allowing its application to larger areas.

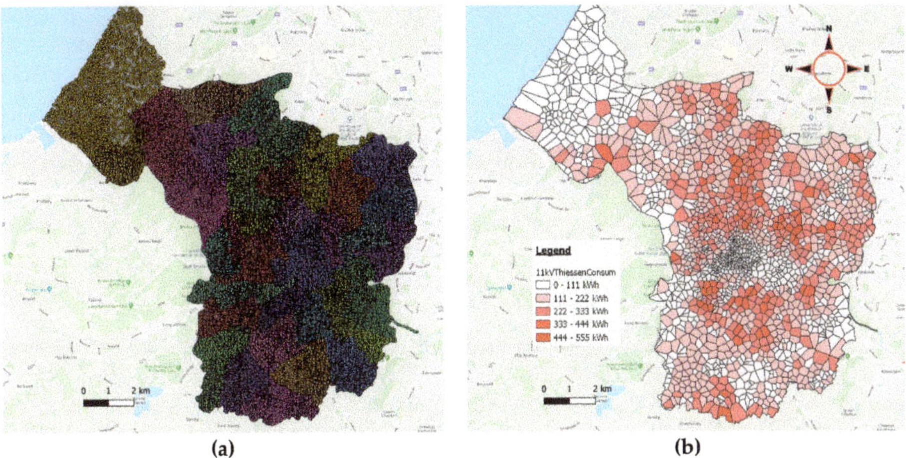

(a)　　　　　　　　　　　**(b)**

Figure 7. (a) Consumer layer based on the different wards of the area studied; (b) Thiessen polygon layer representing the area of influence of each DC and colored by the total power distributed.

Equations (5) and (6) are taken from a European Commission Joint Research Center (JRC) technical report [13] analyzing climate change and critical infrastructure. These equations evaluate the losses accounted by city businesses provoked by electrical shortages. Equation (5) uses the Gross Domestic Product (GDP) of the city and multiplies it by the probability of failure of the DC (P_F), by a ratio of the part of population affected (number of people affected by the shortage (n_P) divided by the total population (n_{tot})), by the fraction of the year that the shortage takes place (shortage duration in days (t_{WE}) divided by 365). In this way, the previous DC failure analysis allows the calculation of this formula that will estimate the cost of the shortage to the businesses.

$$Bussiness\ Cost\ (BC) = GDP \times P_F \times \frac{n_P}{P_{tot}} \times \frac{t_{WE}}{365} \tag{5}$$

Equation (6) will give the total cost of the losses associated with local businesses by adding the values previously calculated in Equation (5).

$$Total\ Bussiness\ Cost\ (TBC) = \sum_k BC_i \tag{6}$$

3.2.3. Energy Non-Supplied Cost (ENSC)

In this section, the electrical supply losses caused by the shortage duration have been estimated by following some different steps for the two cities studied due to the data availability.

The first step was to estimate the power supplied at each DC. In Barcelona, this step was not necessary because the real parameter was known, making the analysis performed in the selected areas quite accurate.

However, in Bristol, the average electrical demand of each DC studied was estimated based on the GIS layer generated in "3.2.2. Cost Associated with Businesses Losses (BC)" where the consumers were associated with the different DCs. The power estimation consisted of the multiplication of the total number of consumers by an average consumption of 531 kWh gathered from the world data portal [22]. In this case, the power losses could be underestimated, due to the business locations, the industries and other possible sources of consumption were neglected.

Another important parameter required for the calculation of ENSC is the repair time (t_R), which was calculated by associating the different damage categories to the repair time obtained from "power grid recovery after natural hazard impact" [6], creating in such a way a damage-time curve based on these categories (Figure 8). Because in this study the damage will never overpass the 50% mark (see Figure 8), the categories that could exceeded that mark were discarded for a better equation curve-fitting, and the rest of categories were represented in a scatter chart, looking for a trend line that really fits the curve. In this case, it was found that a polynomial curve fitted almost to the perfection with the damage-time curve (R^2 = 0.9987). Hence, it was possible to adapt from a categorical scale to continuous by using the trend line equation.

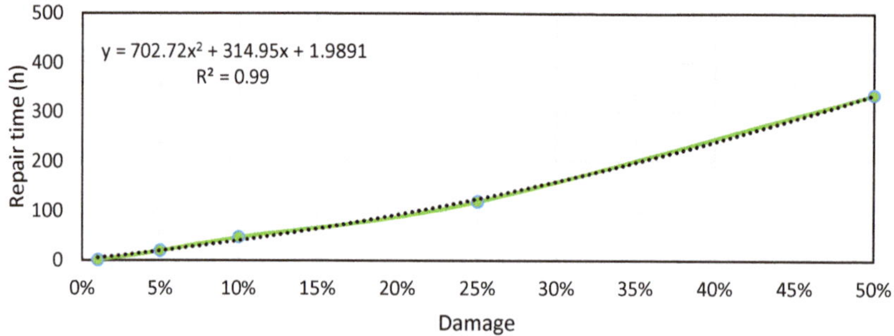

Figure 8. Repair time-damage curve obtained from deriving data from "power grid recovery after natural hazard impact" [6].

For those future cases in which 100% damage can be reached, it is suggested to consider a total replacement.

Thus, the equation obtained from Figure 8 was applied to the damage percentage calculated in the previous step for all locations analyzed to obtain the corresponding repair time of each one.

Thus, D_{ES} was applied to the trend line equation, obtaining the corresponding repair time of each location damaged.

Once the DCs power and the repair time were calculated, the energy losses were derived according to Equation (7).

$$ENSC \begin{cases} P_F \cdot P_{ES} \cdot t_R \cdot p_E, & t_R \leq t_{WE} \\ P_F \cdot P_{ES} \cdot t_{WE} \cdot p_E, & t_R > t_{WE} \end{cases} \tag{7}$$

where P_F is the failure probability, P_{ES} is the DC power, p_E is the energy price and t_{WE} is the period without energy.

In such a way, the energy blackout will be extended up to the Auxiliary Generation systems that are put into operation. From this point onwards, the cost will be directly attached to the Auxiliary Generation Cost.

3.2.4. Auxiliary Generation Cost (AGC)

In this section, the cost attached to the energy generation to avoid the prolongation of the shortage was estimated. The use of diesel emergency power generator support was assumed for when the repair time exceeded 9h of duration. This period was chosen as a hypothesis assuming the rainfall duration given for the flooding maps calculation (i.e., 3 hours) and 6h of response time to rent, transport and install the equipment. In this way, the calculation of the cost of penalties to the DSO company can be neglected in the analysis, although this, depending on the severity of the event experienced could be an underestimation.

It has been considered the transport of the equipment (C_{AGT}) to the affected DC as 20€, and three different tranches of renting price depending on the number of days required. The first renting tranche (C_{Rt1}) has been set as 100€/day when the period is less than 1 week. The second tranche (C_{Rt2}) takes part when the problem is extended from one week to three weeks with a renting price of 50€/day and the third tranche (C_{Rt3}) when the repair tasks need more than three weeks, the price is reduced up to 40€/day. The prices used for this calculation have been taken from a company of machinery renting in Barcelona [23]. In addition, the fuel consumption cost (C_{FC}) was added to the calculation as well as the number of auxiliary generators (n_{AG}) (Equation (8)).

$$AGC \begin{cases} 0, t_R \leq t_{WE} \\ P_F{\cdot}C_{AGT}{\cdot}n_{AG} + C_{FC} + C_{Rt1}{\cdot}n_{AG}, \ t_{WE} < t_R < 1 \ week \\ P_F{\cdot}C_{AGT}{\cdot}n_{AG} + C_{FC} + C_{Rt2}{\cdot}n_{AG}, \ 1 \ week < t_R < 3 \ weeks \\ P_F{\cdot}C_{AGT}{\cdot}n_{AG} + C_{FC} + C_{Rt3}{\cdot}n_{AG}, \ t_R > 3 \ weeks \end{cases} \tag{8}$$

The number of auxiliary generators was calculated by dividing the DC power consumption (P_{ES}) by the maximum active power given by the generator (P_{AG}) and rounding up the result to the whole number (Equation (9)).

$$n_{AG} = \lceil \frac{P_{ES}}{P_{AG}} \rceil \tag{9}$$

3.2.5. Estimation of the Total Losses

The total losses were calculated by adding all the cost values calculated in the previous sub-sections (Equation (10)).

$$Losses \ Related \ to \ the \ Grid \ (LRG) = EDC + BC + ENSC + AGC \tag{10}$$

4. Results

4.1. Risk Assessment Results

After the application of the methodology presented in Section 3, based on hydraulic modelling, probabilistic functions, and GIS processes, the results are presented for both case studies in Barcelona and Bristol, and both scenarios presented, current (Baseline) and future (BAU) rainfall conditions.

4.1.1. Barcelona

In Barcelona city, the results obtained for the risk analysis are given in Figures 9 and 10. These results are the output of the different analyses made with the different fragility curves FC1 (Softened fragility curve), FC2 (Original fragility curve) and FC3 (Hardened fragility curve) presented in Section 3.1.

The number of DCs affected in FC1 and FC2 was the same in the corresponding return periods, but with the difference that these DCs were allocated in different risk categories. While FC2 allocated more DCs in LFP and NAFP, FC1 did in MFP and HFP. However, FC3 diminishes the number of DCs affected due to those under LFP and MFP being dismissed (Figure 9).

Also, between scenarios, a percentage of increase over the total of elements analyzed was extracted (i.e., 1342 DCs).

Depending on the return period analyzed and the fragility curve observed, there is a different increase from current scenario to BAU. With increases of up to 32 DCs affected in the BAU scenario means 2.38% of the total number of DCs, although this happens in the lowest importance category (i.e., LFP). In MFP, an increase of 22 DCs was found in T500 return period, meaning 1.64% over the total and in HFP 7 DCs, that is a percentage of 0.52%. All the increases found for the maximum return period analyzed T500, are depicted in the map of Figure 10, colored in green.

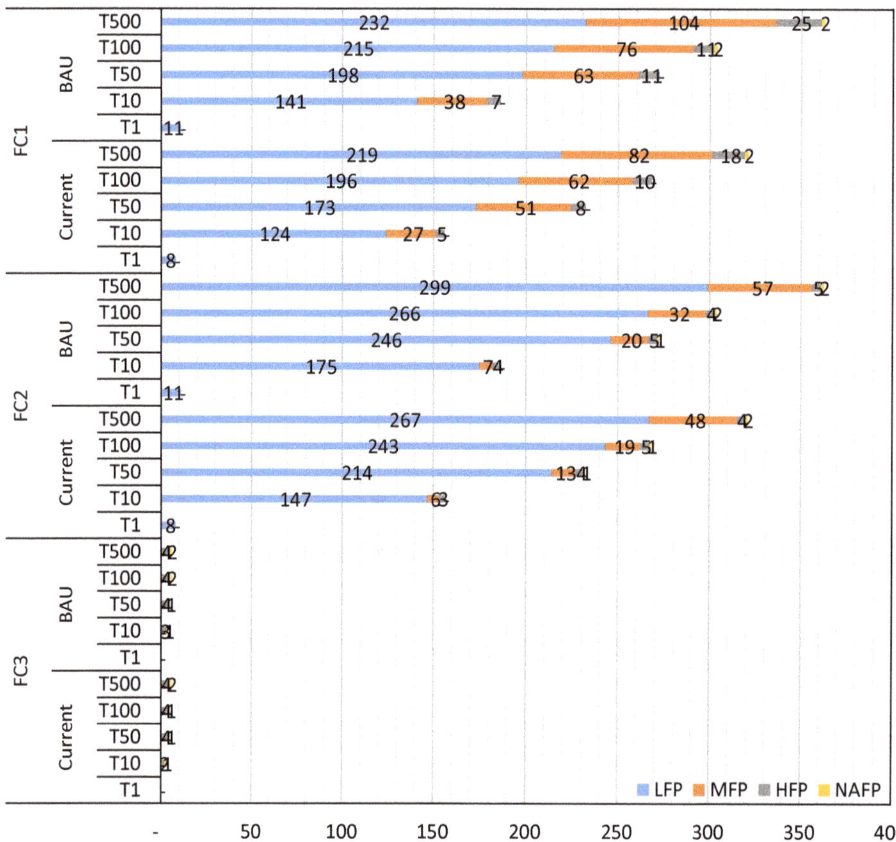

Figure 9. Summary chart of the DCs at potential risk found by the implementation of the method developed in Barcelona.

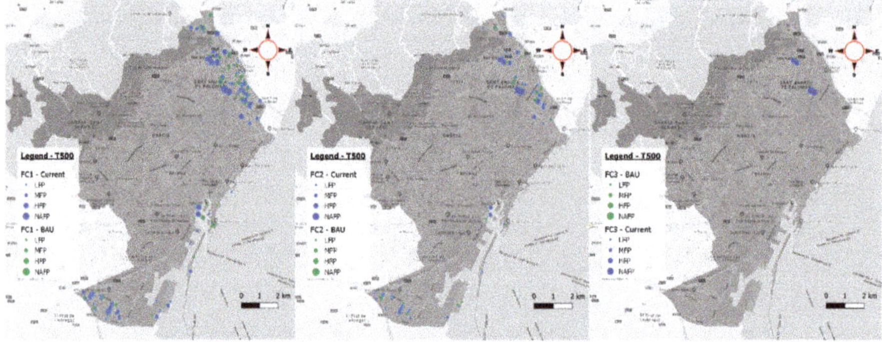

Figure 10. Multiple Barcelona map representation of the DCs analyzed, regarding the different fragility curves (FCs) studied and the different scenarios analyzed for return period T500.

4.1.2. Bristol

In Bristol city, the effect of FC1 and FC2 is the same as in Barcelona. The number of DCs detected at risk is the same, but this is through allocating them in different risk categories and showing the same allocation pattern. Otherwise, the results obtained by means of FC3 are almost null, presenting only one DC at risk in NAFP (Figure 11).

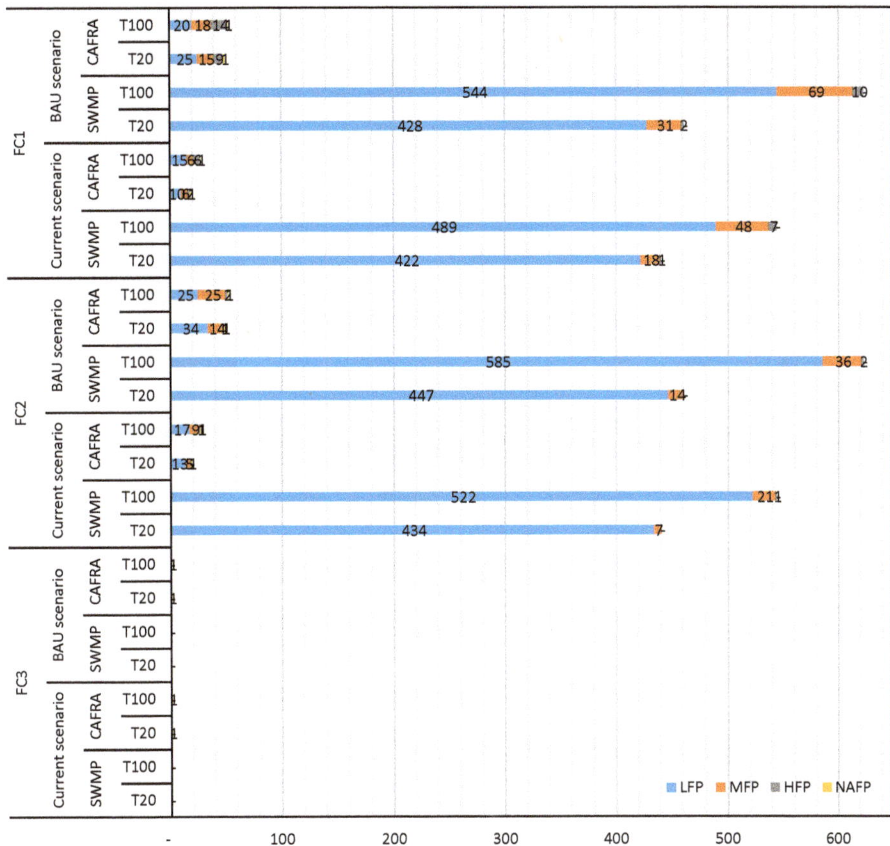

Figure 11. Summary chart of the DCs at potential risk in Bristol found by the implementation of the method developed.

The analysis between scenarios was also made in Bristol, where a total of 1869 DCs were analyzed. From current to BAU, it was found in LFP that there was a maximum increase of 63 DCs, representing 3.37% over the original total. Also, in MFP there was a maximum increase of 1.12% (21 DCs) and 0.43% in HFP with eight more DCs affected. In Figure 12, the increases are represented for the maximum return period analyzed (i.e., T100).

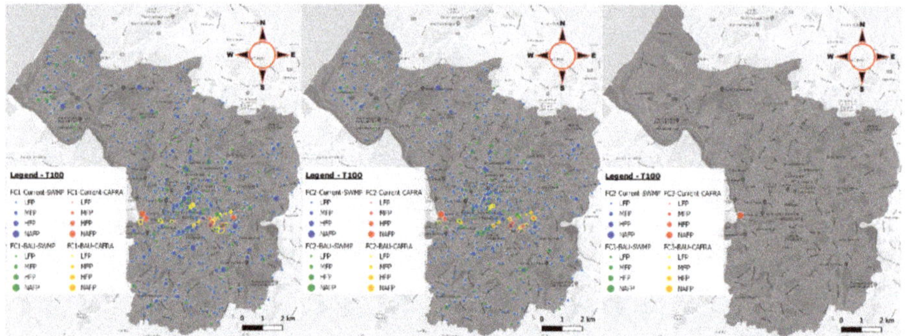

Figure 12. Multiple map representation of the DCs analyzed in Bristol, covering the different FCs studied and the different scenarios analyzed for return period T100.

4.2. Monetization Results

In general, the effective damage cost (EDC) is the most significant part of the cost with around 90% of the total gross, followed by the business cost (BC) with an average percentage of 8%.

In contrast, the AGC signified 1% of the total while the ENSC was practically negligible.

The detailed results for both study cases analyzed are explained with a bit more detail in the following paragraphs.

4.2.1. Barcelona

In Barcelona, after calculating the flooding cost with the methodology described in Section 3.2, the different results obtained for the three different fragility curves (FC) cases show that in general, there is a rising increase in the cost incurred from the lowest to highest return periods calculated. In return periods of T1, there are no significant costs associated but from T10 to T500 it is possible to find costs from around 150,000 € up to 860,000€ at its highest (Figure 13).

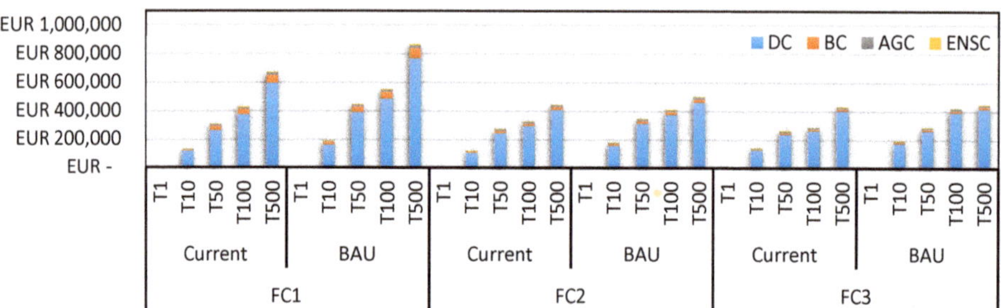

Figure 13. Summary chart of the monetization results for Barcelona study case. It is differentiated between the return period studied (T), different fragility curves applied (FC) and scenarios applied.

The fragility curves applied show a significant difference between the application of FC1 and FC2, decreasing the costs associated with DCs failure. However, between FC2 and FC3, the differences are not quite so evident, and even in T10 of FC3, there produced an increase in costs in comparison to FC2, but in general, the costs remained quite similar (Figure 13).

In Figure 14, the DCs susceptible to provoke the maximum losses can be identified. In general, the major losses come from the North-East side of Barcelona, which coincides with the Besòs riverside where several locations could potentially provoke costs up to 50,000€ in the case of the BAU scenario.

Also, in an area of the central part of the seashore, (i.e., the Raval ward) there is one DC prone to provoke losses up to 50,000€ as well and other several locations that could provoke costs up to 20,000€.

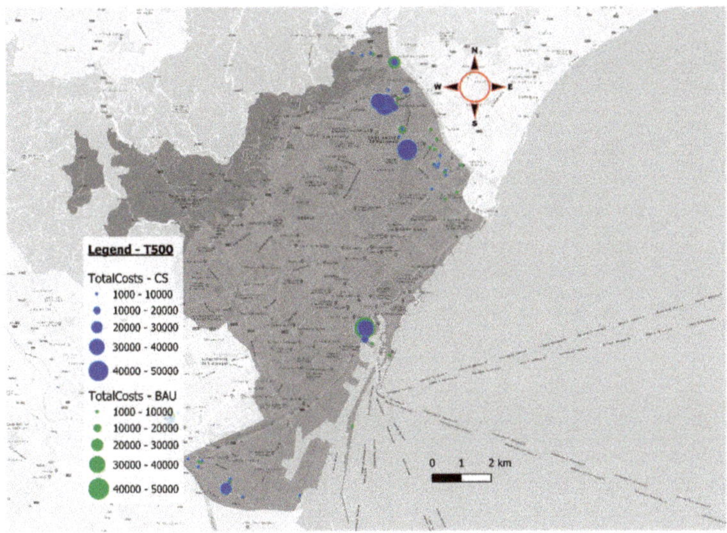

Figure 14. Barcelona map representing the locations that are prone to produce the major losses provoked by electrical shortages in the return period T500.

4.2.2. Bristol

In the Bristol case, the total losses range from 400,000€ to around 1,000,000 € maximum.

The patterns followed are virtually identical to the previous study case. The results produced for Bristol city with the application of FC2 and FC3 are quite similar while the FC1 generates slightly greater losses (Figure 15).

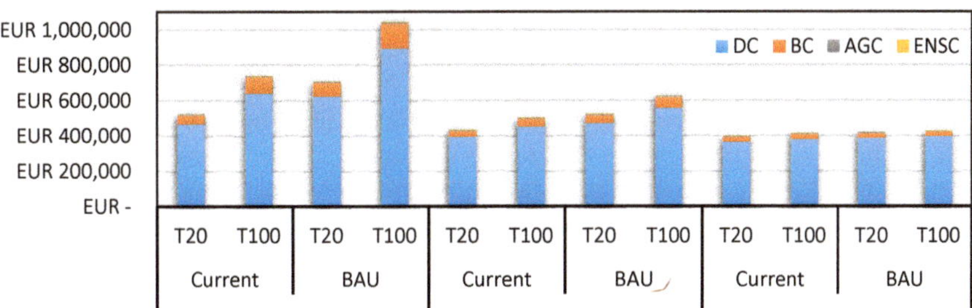

Figure 15. Summary chart of the monetization results for the Barcelona study case. It is differentiated between the return period studied (T), different fragility curves applied (FC) and scenarios applied.

In the map represented in Figure 16, it is possible to see where the costs from 1,000€ to 50,000€ are located and which type of flooding (i.e., tidal/fluvial (CAFRA) or pluvial (SWMP)) is causing the loss and the scenario in which is caused. The most severe cases are located around the city center and caused by tidal fluvial problems, with some cases that can provoke losses up to 40,000€ in the current scenario, or up to 50,000€ in the BAU scenario.

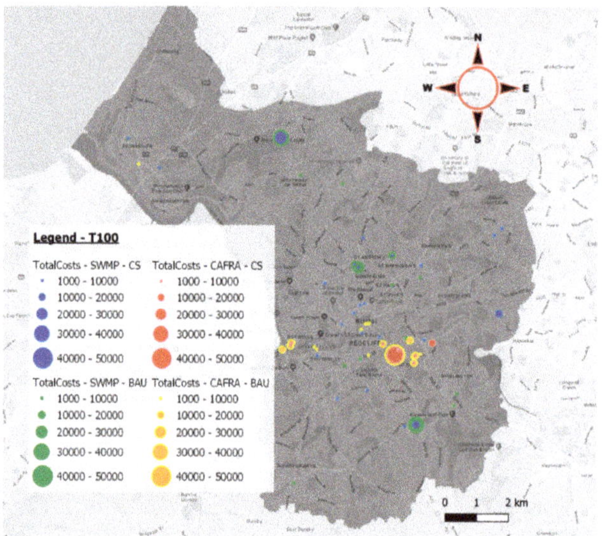

Figure 16. Bristol map representing the locations that are prone to produce the major losses provoked by electrical shortages in the return periods T100.

The losses caused by pluvial events are in other points of the Bristol area, more specifically in the Henbury ward with losses up to 40,000€ and in the Hengrove ward with losses up to 30,000€.

5. Discussion

The risk of assets has been quantified in other studies and in several ways, but not in the same way it was conducted here. For example, a GIS-based method was proposed by JRC [7] to assess the electrical grid and the gas network in case of seismic events using seismic fragility curves proposed by FEMA-HAZUS [10], which is similar to what was done on the failure assessment of this study. However, it was not taken into account how to assess the economic losses, later releasing other studies, with methodologies to evaluate the economic losses caused by storms and floods to the electrical infrastructure in [12,13], but never fusing them in a complete methodology and using it in case of flooding. The case of [9] builds on FEMA-HAZUS methodology [10] and follows a much more detailed analysis by using interdependencies but with the single problem of the huge requirement of data and the added difficulty of creating the network topology, and in most of the cases that information is confidential. Considering other methodologies as the proposed in [8] only the probability of flooding is assessed and not the probability of failure neither the economic losses. The most similar methodology can be found at [11] where a GIS-based approach is taken for the assessment of the electrical sector in flooding events determining the system exposure and vulnerability of the grid to flooding, with the particularity of not using fragility curves. A good point here is that this study takes into account other ways to assess the economic losses considering the economic sectors involved in each area of study.

Regarding the points highlighted above and the new procedures developed, this is a novel process that allows an analytical interpretation not only of the risk that electrical assets are exposed to but also about the potential costs that these assets could produce in many different ways to the population and to the DSO company.

This process enables the user to find the distribution center most exposed to flooding within a set region, allowing them to then take preventive measures if necessary on behalf of the responsible organization or authority.

This study evaluated several scenarios by using different parameters. This allowed a check on the effectiveness of the process and to establish comparisons between the different scenarios proposed.

5.1. Comparison between Fragility Curves

The results produced in the different analyses, carried out by applying the different fragility curves modelled, offer a comprehensive view of the effect of changing the curve shape, as demonstrated in the results produced. The analysis made with FC1 (softened fragility curve) may overestimate the results because high failure rates at low water depths are considered. On the contrary, when using FC3 (hardened fragility curve) the results can be underestimated due to the curve considering zero failure probability up to water depths of 1.4m. Taking into account the results presented, FC3 is neither representative nor realistic because it omits a high number of DCs with a failure probability in the other FCs. As the arrangement of the electrical elements can vary heavily between DCs, a zero probability would be a great underestimation. In fact, taking the experiences of electrical companies' employees, a DC can fail from very low water depths and the threshold to identify the flood risk was set to 10 cm in the risk analysis.

In general, taking FC2 as a reference point, FC1 results in a 19% cost average increase caused by the sum of all the possible small losses produced by low failure probabilities, while FC3 results in a 9% cost average reduction, as it neglects all of the small losses and keeps the gross.

In the end, the variation in the fragility curves can be taken as a pessimistic, neutral or optimistic view if choosing from FC1 to FC3, although in general, FC3 is not representative.

5.2. Comparison Between Current and BAU Scenario

The first comparison made was between a current scenario, where different return periods based on historical data were modelled in flooding shape layers, and a BAU scenario considering climate change with an RCP 8.5, estimating the flooding that could occur in the year 2100. As is normal, in the results obtained, the BAU scenario introduced higher risk and consequently higher costs.

Depending on the fragility curve applied, the differences between both scenarios change, but counting the maximum change experienced in both cities, the maximum increase in the number of DCs affected goes in the LFP category with a 2.38% increment in Barcelona and 3.37% increment in Bristol. Also, big increases are presented for MFP with 1.64% rise in Barcelona and 1.12% in Bristol. These increments seem to be very low but when translated into cost, the average increase is a 22% respective difference to the current scenario, but depending on the case analyzed, the percentage can vary (Figure 17).

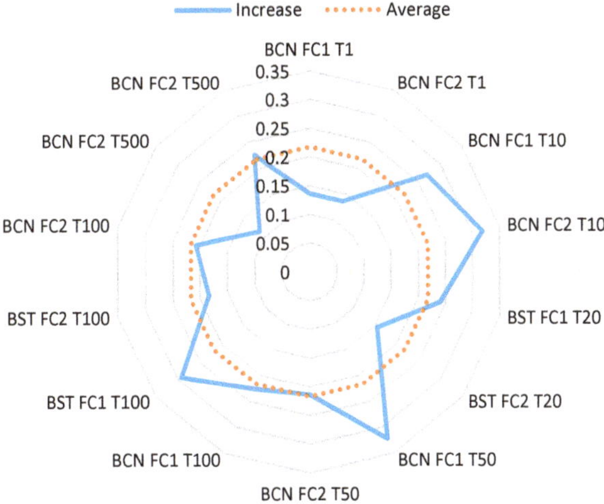

Figure 17. Radial chart that represents the cost increase provoked by climate change in a future scenario by taking into account all return periods for different fragility curves, scenarios and cities.

5.3. Comparison between Cities

It is difficult to make a complete comparison between both cities due to different return periods being analyzed (in Bristol these were lower than in Barcelona), and because of the context of each city. With regards to context, it has to be taken into account that the extent of both cities studied is more or less within the same scale (Bristol with 111 km^2 and Barcelona with 102 km^2) but the effective territory studied is much less in Barcelona (only 32% which means 33km^2) and also 326,000 inhabitants against 460,000 in Bristol. In this aspect, the number of inhabitants of both cities is also similar.

In addition, the hydraulic models that simulate the flooding are different. In Barcelona, the whole drainage system was included within the model, while in the underground Bristol sewer network only larger diameter pipes, although in Bristol this considered both pluvial and fluvial flood events, while in Barcelona, only the pluvial flooding was considered.

Bearing in mind the above statements and taking the unique return period (T100) run for both cities an analysis comparing the cities was performed, resulting in quite similar losses for each fragility curve applied but always resulting in Bristol being the city most affected by flooding. In Figure 18a the total costs are represented for both cities, for the two scenarios analyzed and the different fragility curves developed, showing a clear difference between cities in almost all cases analyzed. As this difference could be linked to the number of DCs analyzed in each city, the total cost has been unified taking into account the total number of DCs analyzed in each city (Figure 18b). In this case, the difference between cities is even more noticeable, which is something reasonable taking into account the great problem that Bristol has with flooding [14].

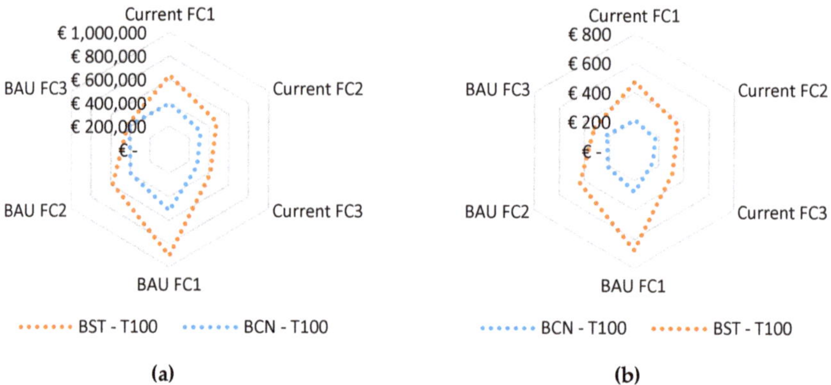

Figure 18. Radial chart that takes into account the return period T100 for different fragility curves and scenarios, comparing both cities taking into account (**a**) the total costs associated with risk and (**b**) the unified cost by DC.

6. Conclusions

This paper has gone through a methodology that aims to estimate and classify the DCs at risk of flooding in different classes set by failure probabilities, as well as the energy losses and their expenditures provoked by shortages caused by potential flooding. This methodology takes a probabilistic GIS-based approach to quantify the risk of electrical shortage in different areas caused by DCs flooded. In this research it has been demonstrated that it has become possible to implement this method to any city where the locations of the DCs and a flooding model are available.

The method used to go through different steps for each city, depending on the data availability for each one, offers several ways to estimate risk and electrical losses with, inevitably, different accuracy.

As a result, it can be drawn the potential losses incurred to put them into balance against the cost of taking protective measures if the assets analyzed are not already under protection.

Author Contributions: Conceptualization, D.S.-M. and J.L.D.-G.; Data curation, D.S.-M.; Formal analysis, D.S.-M.; Investigation, D.S.-M. and J.L.D.-G.; Methodology, D.S.-M. and J.L.D.-G.; Project administration, J.L.D.-G.; Resources, E.M.-G., B.R., J.S. and M.P.; Software, D.S.-M.; Supervision, J.L.D.-G.; Visualization, D.S.-M.; Writing—original draft, D.S.-M.; Writing—review and editing, J.L.D.-G., E.M.-G., B.R. and J.S. All authors have read and agreed to the published version of the manuscript.

Funding: This research was funded by the European Union's Horizon 2020 Research and Innovation Program (RESCCUE project), grant number 700174.

Acknowledgments: The authors want to acknowledge to the RESCCUE project where this research is framed. Also, to all the organizations that transferred the necessary data to carry out this study, in special, Western Power Distribution in UK and Endesa in Spain for the electrical data, Aquatec in Spain and Bristol City Council in UK for the flooding models and to the Bristol and Barcelona city councils for developing the Open Data Portals from where basic data was taken.

Conflicts of Interest: The authors declare no conflict of interest.

References

1. Energy Research Partnership. Future Resilience of the UK Electricity System. 2018. Available online: http://erpuk.org/wp-content/uploads/2018/11/4285_resilience_report_final.pdf (accessed on 15 October 2019).

2. Bisselink, B.; Bernhard, J.; Gelati, E.; Jacobs, C.; Mentaschi, L.; De Roo, A. *Impact of a Changing Climate, Land Use, and Water Usage on Water Resources in the Danube River Basin*; Publications Office of the European Union: Brussels, Belgium, 2018; ISBN 9789279858895.

3. Monjo, R.; Paradinas, C.; Gaitán, E.; Redolat, D.; Prado, C.; Pórtoles, J.; Torres, L.; Russo, B.; Velasco, M.; Pouget, L.; et al. Report on extreme events prediction. *Deliverable 1.3.* **2018**, 2–108.

4. National Academies of Sciences Engineering and Medicine. *Enhancing the Resilience of the Nation's Electricity System*; National Academies of Sciences Engineering and Medicine: Washington, DC, USA, 2017; ISBN 978-0-309-46307-2.

5. County of Santa Clara Office of Sustainability and Climate Action. *Silicon Valley 2.0 Climate Adaptation Guidebook*; The County of Santa Clara: Santa Clara, CA, USA, 2015.

6. Karagiannis, G.M.; Chondrogiannis, S.; Zehra, E.K.; Turksezer, I. *Power Grid Recovery after Natural Hazard Impact—A Science for Policy Report*; European Union: Brussels, Belgium, 2017; ISBN 978-92-79-74666-6.

7. Poljanšek, K.; Bono, F.; Gutiérrez, E. *GIS-Based Method to Assess Seismic Vulnerability of Interconnected Infrastructure A Case of EU Gas and Electricity Networks*; Office for Official Publications of the European Communities: Brussels, Belgium, 2010; ISBN 9789279152092.

8. Pant, R.; Thacker, S.; Hall, J.W.; Alderson, D.; Barr, S. Critical infrastructure impact assessment due to flood exposure. *J. Flood Risk Manag.* **2018**, *11*, 22–33. [CrossRef]

9. Vasenev, A.; Montoya, L.; Ceccarelli, A. A hazus-based method for assessing robustness of electricity supply to critical smart grid consumers during flood events. In Proceedings of the 2016 11th International Conference on Availability, Reliability and Security (ARES), Salzburg, Austria, 31 August–2 September 2016; pp. 223–228.

10. FEMA. *Multi-Hazard Loss Estimation Methodology, Flood Model: Hazus-MH MR4 Technical Manual*, 2009th ed.; FEMA: Washington, DC, USA, 2009.

11. Koks, E.; Pant, R.; Thacker, S.; Hall, J.W. Understanding Business Disruption and Economic Losses Due to Electricity Failures and Flooding. *Int. J. Disaster Risk Sci.* **2019**, *10*, 421–438. [CrossRef]

12. Karagiannis, G.M.; Turksezer, Z.I.; Alfieri, L.; Feyen, L.; Krausmann, E. *Climate Change and Critical Infrastructure—Floods*; European Union: Luxembourg, 2017; ISBN 9789279754456.

13. Karagiannis, G.M.; Cardarilli, M.; Turksezer, Z.I.; Spinoni, J.; Mentaschi, L.; Feyen, L.; Krausmann, E. *Climate Change and Critical Infrastructure—Storms*; European Union: Luxembourg, 2019; ISBN 9789279964039.

14. Bristol City Council. *Bristol—Local Flood Risk Management Strategy*; Bristol City Council: Bristol, UK, 2018.

15. Agència Catalana de l'Aigua. *Avaluació Preliminar del Risc D'inundació al Districte de Conca Fluvial de Catalunya*; Generalitat de Catalunya: Barcelona, Spain, 2011; Volume Annex.

16. Parreño, E. *El diluvio del Vallès. El Periódico*; El periódico: Barcelona, Spain, 2012; Available online: https://www.elperiodico.com/es/sociedad/20120923/50-aniversario-de-las-riadas-del-besos-2210462 (accessed on 17 February 2019).

17. Russo, B. *D2.2 Multi-Hazards Assessment Related to Water Cycle Extreme Events for Current Scenario*; RESCCUE project: Barcelona, Spain, 2018.

18. Russo, B. *D2.3 Multi-Hazards Assessment Related to Water Cycle Extreme Events for Future Scenarios—Business as Usual*; RESCCUE project: Barcelona, Spain, 2019.
19. Energy Networks Association. *Engineering Technical Report ETR 138 Resilience to Flooding of Grid and Primary Substations*; Energy Networks Association: London, UK, 2009.
20. Bcn.cat Superficie y Densidad de los Distritos y Barrios. Available online: https://www.bcn.cat/estadistica/castella/dades/anuari/cap01/C0101050.htm (accessed on 29 October 2019).
21. Mills, J. Population Estimates 2007–2018 (by LSOA11). Available online: https://opendata.bristol.gov.uk/explore/dataset/population-estimates-2005-2016-lsoa11/information/?disjunctive.ward (accessed on 29 October 2019).
22. Worlddata.info Energy Consumption in the United Kingdom. Available online: https://www.worlddata.info/europe/united-kingdom/energy-consumption.php (accessed on 30 October 2019).
23. Maquinas y Maquinas—Alquiler de Maquinaria, A. de Herramientas Alquiler de Grupo Electrógeno Insonorizado 250 KVA—400v. Available online: https://www.maquinas-maquinas.com/maquinaria/grupos-electrogenos/grupo-electrogeno-xxl-250-kva/ (accessed on 30 October 2019).

 *sustainability*

Article

Climate Change Implications for Water Availability: A Case Study of Barcelona City

Edwar Forero-Ortiz [1,2,*], **Eduardo Martínez-Gomariz** [1,2] **and Robert Monjo** [3]

1   Cetaqua, Water Technology Centre. Carretera d'Esplugues, 75, Cornellà de Llobregat, 08940 Barcelona, Spain

2   FLUMEN Research Institute, Universitat Politècnica de Catalunya, Calle del Gran Capità, 6, 08034 Barcelona, Spain; eduardo.martinez-gomariz@upc.edu

3   FIC- Climate Research Foundation, 28013 Madrid, Spain; rma@ficlima.org

*   Correspondence: eaforero@cetaqua.com; Tel.: +33-783-53-72-22

Received: 30 January 2020; Accepted: 26 February 2020; Published: 27 February 2020

Abstract: Barcelona city has a strong dependence on the Ter and Llobregat reservoir system to provide drinking water. One main concern for the next century is a potential water scarcity triggered by a severe and persistent rainfall shortage. This is one of the climate-driven impacts studied within the EU funded project RESCCUE. To evaluate potential drought scenarios, the Hydrologiska Byråns Vattenbalansavdelning (HBV) hydrological model reproduces the water contributions by month that have reached the reservoirs, regarding the accumulated rainfall over each sub-basin, representing the available historical-observed water levels. For future scenarios, we adjusted the input data set using climate projections of rainfall time series data of the project RESCCUE. Local outputs from 9 different climate models were applied to simulate river basins' responses to reservoirs' incoming water volume. Analyzing these results, we obtained average trends of the models for each scenario, hypothetical extreme values, and quantification for changes in water availability. Future water availability scenarios for Barcelona central water sources showed a mean decrease close to 11% in comparison with the period 1971–2015, considering the representative concentration pathway 8.5 (RCP8.5) climate change scenario in the year 2100. This research forecasts a slight downward trend in water availability from rainfall contributions from the mid-21st century. This planned future behavior does not mean that the annual water contributions are getting lower than the current ones, but rather, identifies an escalation in the frequency of drought cycles.

Keywords: drought; water scarcity; water availability; climate change; hydrological modeling; resilience

1. Introduction

Climate change estimates to affect all spheres of human activity in the natural environment, including water resources. Defined as a shortage in rainfall over an extended period, a season or more, drought affects both human activities and the environmental balance [1]. A significant proportion of the human population is currently experiencing restrictions on access to drinking water due to drought events, a vulnerable component of the natural and social action chains [2].

Increases in drought events' frequency and severity are forecast under the impact of climate change [3], examples include events in China (1991–1996), East Africa (2010–2011), United States of America (2011–2012), Australia (2013–2016), and Catalonia, Spain (2006–2008) [4]. Water resource availability shortage threatens urban areas due to factors such as rapid urbanization, increased water use, lack of climate change adaptation policies, and repeated drought events [5]; drought has historically affected 35% of the population hit by natural hazards [1].

Globally, the drinking water requirement for cities has increased due to rapid population growth in cities, pollution of water sources, stress on groundwater sources, and the impact of extreme weather conditions [5]. Researchers concur that drought events will be intense due to lower rainfall and higher evapotranspiration in some areas of Europe, above all impacting Spain [6,7]. Therefore, understanding the complexity of drought events is essential for the city of Barcelona in facing the next century, associated with rainfall shortage. However, drought is a slow process of shortage accumulation, and the sharpness of drought episodes is not only related to current rainfall, but much more related to the cumulative impact of previous hydrological balances.

One of the criteria for proper reservoir management as part of a water supply system is the rigorous design and implementation of the guidelines of reservoir operation along with environmental planning, allowing management to have the tools to cope with growing climate change influence on water scarcity [8]. Reservoir operation is "a large-scale multi-objective optimization problem" [9]. Therefore, this paper contributes to the understanding of the hydrological process in the Barcelona water supply reservoir system under climate change influence, as one of the factors involved in this process.

Barcelona and its metropolitan area dependent on the Ter and Llobregat reservoir system to provide the water demand throughout the year. According to the Catalonian river basin management plan [10], a document from Catalan Water Agency (ACA, Catalonian water resources administrator), Barcelona uses mainly Llobregat river water (38%), and Ter river water (55%), while for the remaining 7%–8% it employs groundwater. Both basins have their higher part controlled by reservoirs which modulate the required water resources. Barcelona is far from these reservoirs, but the drought situation depends on their stored water volumes.

When these volumes are lower than threshold levels (less than 30% of water stocks in the reservoirs) as set by the Drought Plan from the ACA [11], a drought contingency triggers concerning water use restrictions for activities such as irrigation, leisure, industrial purposes, etc., as happened in 2007 and 2008. Beyond environmental and social impacts generated by Barcelona 2006 and 2008 extreme drought events, a study estimated drought impacts valued at 1605 million euros, half a point of Catalonia's GDP [12].

Catalonian droughts' knowledge is most of all based on drought events' variability studies, either historically avoiding any future SPI (standardized precipitation index) and SPEI (standardized precipitation–evaporation index) indicators projection [13], or assessing climate change effects without considering the representative concentration pathways (RCPs) presented in the Intergovernmental Panel on Climate Change (IPCC) fifth assessment report [14].

Gallart et al. [15] estimated trends for the rivers' discharge in the Ter–Llobregat system analyzing their historical records. However, this approach did not attempt to consider the contextual factors that influence the availability of water resources in the future for the Ter Llobregat system. The research would have been relevant if a forecast had been considered, introducing a future water resources scenario.

The Drought Plan (Alert and eventual drought exceptional action plan) developed by the Catalonian government (Generalitat de Catalunya) and the ACA in 2016, proposes a Ter–Llobregat water resources evolution. This research applied the SIMGES hydrological model with the multivariate periodic autoregressive (MPAR) stochastic model, based on 68 years of historical monthly flow contribution contributions series and comparing them with the generated synthetic series for a 500-year return period.

Likewise, a synthetic series was designed to analyze critical episodes of drought and estimating probabilities of occurrence by extrapolating historical climatic conditions [11]. Therefore, this study focused on understanding how climate change plays a role in Barcelona's drought events as one of the significant nature-based concerns for the next century [10,16,17].

Our research aim is broadening future drought events' knowledge, considering climate change impacts. We defined the design and implementation of a model for water amounts reservoir balance at a month scale, analyzing basins rainfall. This paper, as a first of its two-fold aim, represents observed reservoir water levels implementing the HBV (Hydrologiska Byråns Vattenbalansavdelning) hydrological model and studied the application and validation of the SIMGES model and the

HBV model as appropriate tools to forecast drought frequency for Barcelona's case. After the historical model calibration and validation process, we obtained rainfall projections using nine Earth system models (ESM) and two representative concentration pathways' (RCP) scenarios—RCP4.5 and RCP8.5—belonging to the fifth Coupled Model Intercomparison Project (CMIP$_5$), provided by the Spanish Climate Research Foundation (FIC, accordingly to the Spanish acronym).

Second, we integrated these rainfall outputs within the hydrological model to simulate reservoir volumes as watershed responses, developing 30 different storage patterns. Outcomes of the models were analyzed to get average trends and extreme values for each scenario to estimate a single water availability trend for both reservoirs, to understand and analyze the water resource availability in Barcelona in the near future under different climate change situations. Our study outcomes provide additional support to plan water utility improvements, to evaluate extreme case scenarios, and to assess hazards related to water scarcity in further research.

2. Materials and Methods

2.1. Hydrological Model Background: Description and Setup

Drought, under a hydrological viewpoint, as usual, indicates below-normal levels of flow from lakes, streams, and reservoirs or groundwater with generally accepted indicators, such as the standardized runoff index (SRI), the surface water supply index (SWSI), the groundwater resources index (GRI), among others [4]. Precipitation patterns and changes in the precipitation–runoff relationship as essential variables to define drought which allows the assessment and simplification of climate change impact on urban water availability [3].

The study used precipitation pattern analysis to gain insight into rainfall influence over reservoirs' water availability. Simulations of the water volume at each reservoir applied HBV, an integrated hydrological modeling system, developed at the Swedish Meteorological Hydrological Institute. The model relies on three different reservoir modules: one that simulates the behavior of the soil; the second, the upper reservoir and, finally, the lower reservoir that accounts for the groundwater base flow [18].

Some researchers have highlighted [19,20] how the HBV model is accurate, reproducing present and future water processes in the Llobregat basin. Thus, the HBV model is suitable to simulate the reservoirs' contributions over the Llobregat and Ter's basin. The model requires the physical properties of the basin as well as the climatic inputs, including precipitation, temperature, and potential evapotranspiration. The time scale for the input data was day-to-day. Detailed discussion on the hydrological model's internal functioning falls outside the scope of this paper.

Using the Thornthwaite formula (ETP$_{raw}$) gave us an estimation of potential evapotranspiration. Then, we applied a correction to get a better adjustment to the results obtained according to two parameters, using the Penman evapotranspiration as a reference. The evapotranspiration calibration process with two meteorological stations data in Llobregat and Ter basins computed Penman evapotranspiration (ETP) equation according to data availability.

2.2. Model Calibration, Validation, and Sensitivity Analysis

Table 1 describes the seven calibrated parameters applied to characterize each rainfall event introduced into the model sub-basin. These are all conceptual parameters, not easy to estimate from basin physical properties. The choice of the seven calibration parameters followed a preliminary data analysis, checking their values' availability and validity for a monthly time-step; in addition, according to local conditions, snow-related parameters were discarded.

Table 1. The calibrated Hydrologiska Byråns Vattenbalansavdelning (HBV) model parameters for the Llobregat and Ter river basins.

HBV Model Parameter	Description {Unit}
β	A shape coefficient that determines the precipitation contribution to the runoff
FC	Field capacity {mm}
LP	Limit above Actual Evapotranspiration (AET) reaches ETP
K_1	Recession coefficient
K_2	Recession coefficient
UZL	Threshold parameter {mm}
Perc	Percolation ratio

A Montecarlo simulation was conducted examining 10,000 combinations where there was an available gauge station to calibrate the parameters at each basin, setting the established objective function as accurately as possible (Nash-coefficient, Equation (1)), at a daily timescale if conceivable, in addition, relating with monthly volumes used as a reference, based on the obtained hydrographs when the information was available. Validation of water volume contribution data used ACA's water contribution estimations from the Aquatool SIMGES module developed by IIAMA [21].

$$Nash\ coeffcient = 1 - \frac{\sum_{t=1}^{T}\left(Q_m^t - Q_o^t\right)^2}{\sum_{t=1}^{T}\left(Q_o^t - \overline{Q_0}\right)^2},\tag{1}$$

where Q_m^t = *simulated discharge*, Q_o^t = *observed discharge*, and $\overline{Q_0}$ = *mean observed discharges*. Nash-Sutcliffe efficiency ranges from $-\infty$ to 1. The closer to 1 the coefficient is, the more accurate the model is. An efficiency equal to 0 means that the approximation is as good as the mean of the observed data. Results are acceptable when positive values are higher than 0.2.

2.3. Framework for Assessing Future Water Resources Allocation

To define some potential situations, the representation of reservoir water contribution depends on the rainfall of the sub-basins over each dam. Outcomes include average trends of the models for each scenario, hypothetical extreme values, and the quantification of a possible number of times that reservoir systems could encounter warning events. Reservoir volumes relate to the ACA's 1999–2018 historical data and since 2006 climate model forecasting. Therefore, the historical data range was 1999–2005 and projections cover 2006 to 2100. Rainfall time-series projections from 9 distinct climate models (see Section 3.2) were employed to simulate reservoir input volumes' behavior. The outcomes of these nine models were averaged to find a single trend for each system obtaining four trends, two for each reservoir (RCP4.5 and RCP8.5).

3. A Case Study for Barcelona City

3.1. The Study Area

The Llobregat and Ter rivers and other small basins (defined as the Ter–Llobregat system) supply the Barcelona metropolitan area with drinking water. Llobregat river provides about 38%, and the Ter river supplies 55% of raw water for water drinking treatment plants for Barcelona [22,23]. The coupled basins' total drainage area is 4957 km^2, with a surface elevation variation from almost 2500 m (pre-Pyrenean mountain range) to the sea level, as Figure 1 shows.

A seasonal rainfall variability phenomenon in the two river basins led to water demand-supply fluctuations. Despite this infrastructure, water resource management is complex and involves groundwater extraction from aquifers and seawater desalination in extraordinary drought events [24].

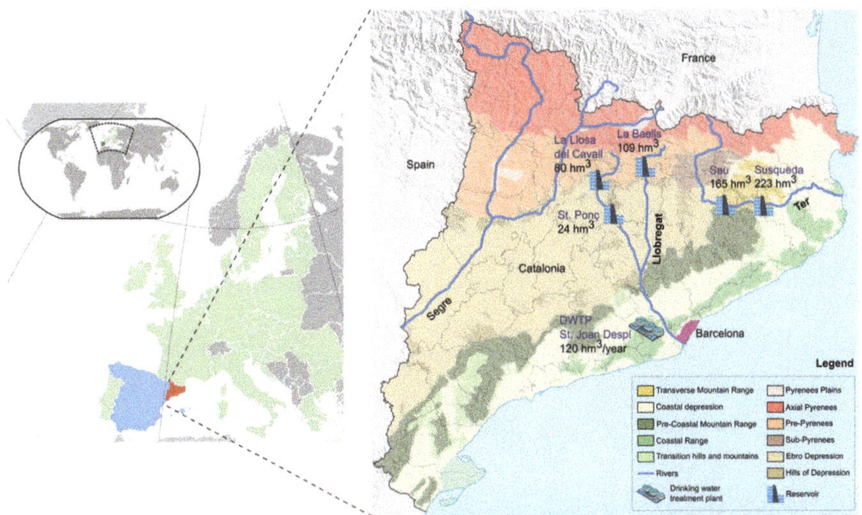

Figure 1. Location of the involved reservoirs, rivers, catchments, and morphology of the study area.

Some key aspects of each sub-basin over the Llobregat and Ter river basins (see Figure 2) can be listed as follows:

- La Baells: Llobregat river basin divides into four sub-basins. Those contained in Guardiola de Berguedà zone include aquifers 112 and 115, with an area of low permeability. This sub-basin catches the flow from the upstream sub-basins and adds its contribution.
- La Llosa del Cavall: Four different sub-basins constitute this Llobregat river basin. La Coma i La Pedra y La Llosa del Cavall represent a part of aquifer 116, with two additional sub-basins of low permeability downstream.
- Sant Ponç: Two sub-basins define the basin part of the Llobregat river. In this case, there is no aquifer over the area.
- Sau: This Ter river basin is divided into six different areas. Those located within the region up to the Ripoll gauge station correspond to the upper aquifers 110 and 115, which represent 75% of the total contribution reaching the Sau reservoir. Sub-basins' simulation was with the same properties but with different precipitation and temperature data.
- Susqueda: It considers one sub-basin of Ter river catchment. Besides, it employs a setting of aquifer 203 parameters for month-by-month water contributions.

3.2. Data

Research data was from three main sources. Rainfall and temperature records were from the Spanish Meteorological State Agency (AEMET), considering all available stations. Forty-four weather stations provide Ter basin 1980–2015 rainfall records, and eleven weather stations contribute Llobregat basin records. Each zone drawn with the same color represents an area with equal rainfall estimation by the Thiessen polygon method (Figure 2).

The calibration of the HBV model discharge results uses data from stream gauging stations of the ACA managed upper basins. As it is not possible to use gauging stations data in basin lower zones, we calibrated the ACA hydrological simulated data with the historical calibrated upper zone discharge data and compared the computed values against the HBV model dataset.

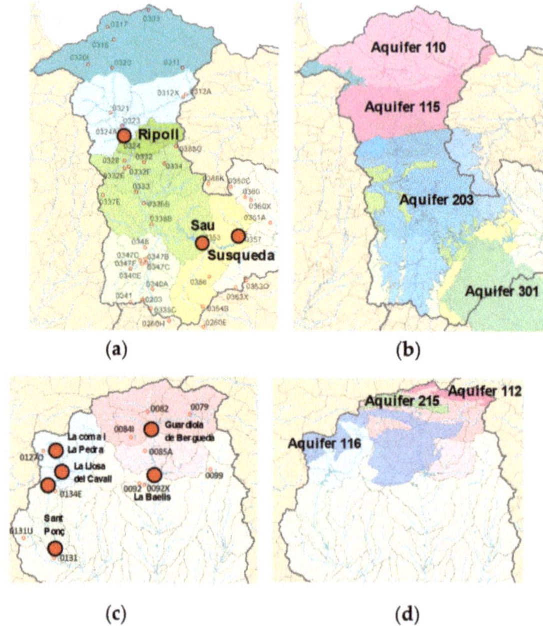

(a)   (b)

(c)   (d)

Figure 2. (**a**) Division of sub-basins for the reservoir system on the Ter river; (**b**) Aquifers in the Ter river basin; (**c**) Division of sub-basins for the reservoir system on the Llobregat river; (**d**) Aquifers in the Llobregat river basin. The red dots indicate the placement of the meteorological stations in the Ter and Llobregat basins. The large red circles with black edge indicate gauging stations that are named as the sub-basin where they are located.

In addition to the observed data, downscaled climate-model outputs were collected from the Climate Research Foundation database [25], as shown in Table 2. These local time series are outputs from a statistical downscaling method of Ribalaygua et al. [26] based on analog stratification and transfer functions. Data include local simulations of ERA-Interim reanalysis and nine CMIP$_5$ models under both historical experiments (1951–2005) and future projections (2006–2100) under RCP4.5 and RCP8.5.

Table 2. Available Coupled Model Intercomparison Project 5 (CMIP$_5$) climate models with outputs at a daily timescale. The table shows the responsible institution, climate model version, references, and spatial resolution for the atmospheric general circulation model (GCM).

Institution	CMIP$_5$ Model	Source	Resolution (Lon × Lat)
Commonwealth Scientific and Industrial Research Organisation (CSIRO), Bureau of Meteorology (BOM)	ACCESS1-0	[27]	1.87°×1.25°
Beijing Climate Center (BCC)	BCC-CSM1-1	[28]	2.8°×2.8°
Canadian Centre for Climate Modelling and Analysis (CC-CMA)	CanESM2	[29]	2.8°×2.8°
National Center for Meteorological Research, Météo-France and CNRS laboratory (CNRM-CERFACS)	CNRM-CM5	[30]	1.4°×1.4°
Geophysical Fluid Dynamics Laboratory (GFDL)	GFDL-ESM2M	[31]	2°×2.5°
Japan Agency for Marine-Earth Science and Technology (JAMSTEC), Atmosphere and Ocean Research Institute, the University of Tokyo (AORI), Japan National Institute for Environmental Studies (NIES)	MIROC-ESM-CHEM	[32]	1.4°×1.4°
Max Planck Institute for Meteorology (MPI-M)	MPI-ESM-MR	[33]	1.8°×1.8°
Meteorological Research Institute, Japan Meteorological Agency (MRI)	MRI-CGCM3	[34]	1.2°×1.2°
Norwegian Climate Centre (NCC)	NorESM1-M	[35,36]	2.5°×1.9°

4. Results

4.1. Evapotranspiration Calibration

The calibration process worked with two temperature stations' data (0085A for Llobregat basin and 0353 for Ter basin, see Figure 2), where the required data for computing Penman ETP was available. Over the Llobregat's temperature stations, we applied a correction with the 0085A station dataset. Likewise, for Ter's upper-temperature stations. Correction obtained with 0353 station dataset was applied at the downstream Ter's sub-basins due to its geographic location. Figure 3 shows the results at both stations. The left-hand side graph displays the considered monthly evapotranspiration, and the right-hand side one, its cumulative distribution.

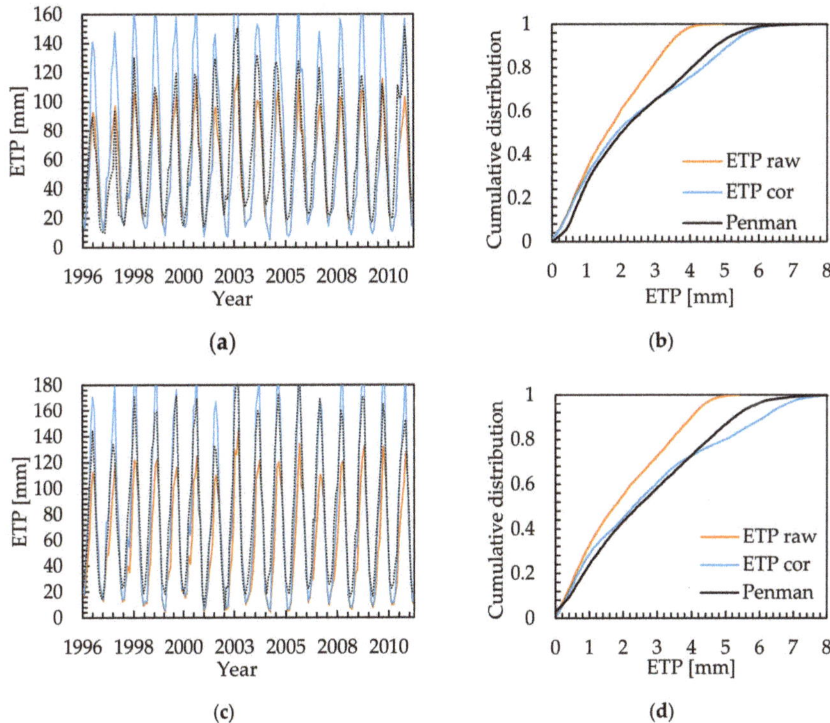

Figure 3. (**a**) Potential evapotranspiration calibration: Monthly ETP and (**b**) distribution of daily ETP at 0085A for the Llobregat basin; (**c**) Potential evapotranspiration calibration: Monthly ETP and (**d**) distribution of daily ETP at 0353 for the Ter basin.

To measure the efficiency of the correction, we computed the mean square error (MSE) for the Thornthwaite ETP, raw and corrected. Water volume contributions were checked for each season, and so, the ETP correction may change somewhat from one season to another.

4.2. Calibration and Validation of Hydrological Parameters

Figure 4 presents the response hydrographs at the locations where stream gauging stations provide records. These stations include EA078 in the la Baells sub-basin, EA087 and EA021 in the La Llosa del Cavall sub-basin and EA033 in the Sau sub-basin.

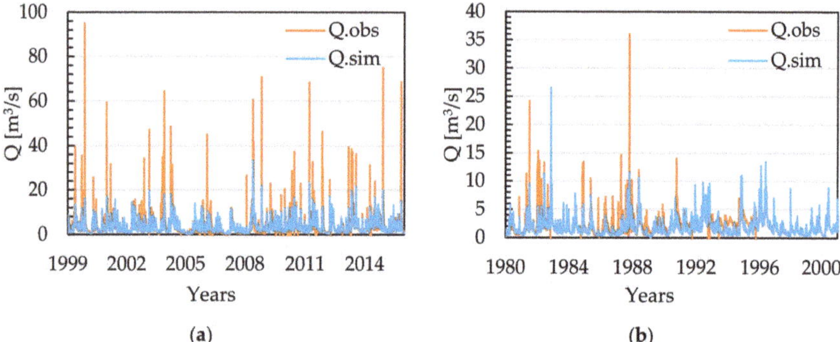

Figure 4. Observed (obs) and computed (sim) response hydrograph: (**a**) La Baells station; (**b**) La Llosa del Cavall station.

Figure 4a shows how response hydrograph performs a good base flows simulation but decreases accuracy in peak flows simulation. Nevertheless, the Nash-coefficient equal to 0.42 was enough to reproduce the behavior of the A78 sub-basin. Peak flows were undershot, but it was preferable to underestimate these contributions as a safety factor, considering weather station scarcity in this area, which directly influences the rainfall time series. Analyzing the response, the approximation was precise enough to reproduce the behavior of this sub-basin. In Figure 4b, the Nash coefficient in this basin was 0.30, which returns a satisfactory outcome. The main differences were gathered from 1990, when the available records had quality issues, as the frequency of the measurements increased to 4–7 days.

In the rest of the basins, two verifications reviewed the results. We compared monthly contributions for each season, and the water contributions were correlated applying a linear regression of the HBV and SIMGES models. Evaluation through the R^2 coefficient as shown in Figure 5 verifies whether the distribution is similar among HBV and SIMGES volumes. In addition, we checked the Nash-coefficient with each reservoir's computed contributions for all seasons. Table 3 presents the calculated Nash-coefficients.

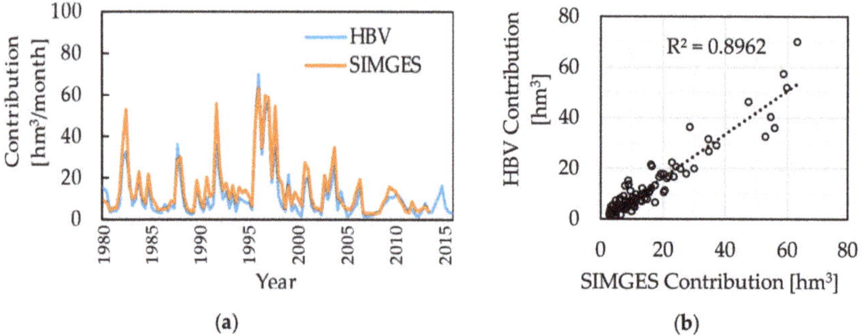

Figure 5. (**a**) La Baells monthly reservoir contribution during winter months, comparison of Hydrologiska Byråns Vattenbalansavdelning (HBV) and SIMGES outputs; (**b**) Comparison analysis of the two contributions. The black line corresponds to a simple linear regression.

Table 3. Nash-coefficient for each reservoir at a monthly time scale, for each season and considering the historical analyzed records (1980–2013).

Nash-Coefficient	Winter	Spring	Summer	Autumn	Total
La Baells	0.85	0.67	0.51	0.89	0.79
La Llosa del Cavall	0.72	0.60	0.44	0.22	0.49
Sant Ponç	0.60	0.10	0.10	0.62	0.33
Llobregat's Contribution	0.85	0.64	0.55	0.87	0.79
Sau	0.82	0.81	0.75	0.69	0.77
Susqueda	0.59	0.31	0.32	0.41	0.42
Ter's Contribution	0.85	0.83	0.77	0.73	0.80

Subsequently, the total contributions from 1980–2013 were analyzed to check if the total volume of water available to serve the demand was the same, regarding the distribution by year at each reservoir unit for the HBV, and the SIMGES model used as a reference, as Figure 6 illustrates.

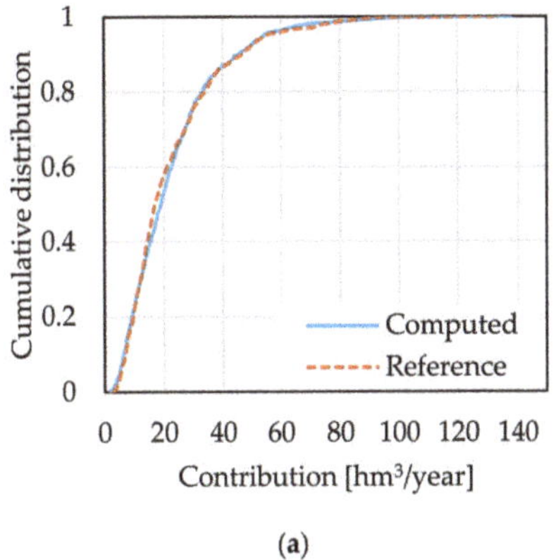
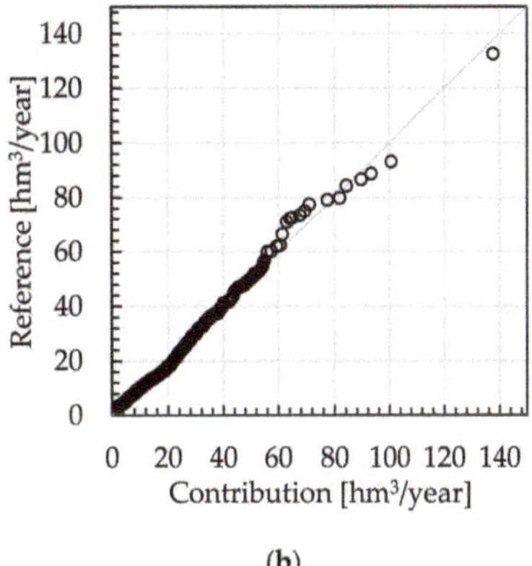

(a)   (b)

Figure 6. Complete Ter–Llobregat system: available water volumes comparison in years 1980–2013: (a) Water contributions data distribution for HBV and SIMGES models. The closer the lines are, the more similar they are. (b) To assess how similar they are, this shows the correlation between the two distributions. As the values are distributed along the reference line (slope 1.00), their correspondence is valid.

The outcomes from both models were similar. The main difference came from the distribution over the year, in particular during summer months, when SIMGES contributions were critical. One of the main reasons to explain this difference may come from the two time-series input models. Recognizing that both datasets are different, the results cannot be the same. Outcomes were similar, ensuring the representation created with the HBV model achieved a reasonable resemblance with the historical dataset.

4.3. Simulations under Future Rainfall Conditions

Reservoir volumes results were from historical data provided by ACA. Results of the projections indicated that not all the climate models were adversely predictive: some of them forecasted rainfall increase (volume), while others estimated a reduction. Figure 7 shows some models forecasting severe drought situations in the RCP8.5 scenario, compared with the RCP4.5 scenario, which showed water scarcity, according to a few models for Ter and Llobregat reservoirs. According to projections, it is expected that in both systems for the RCP8.5 scenario, at least one drought episode is expected with the 20-years return period.

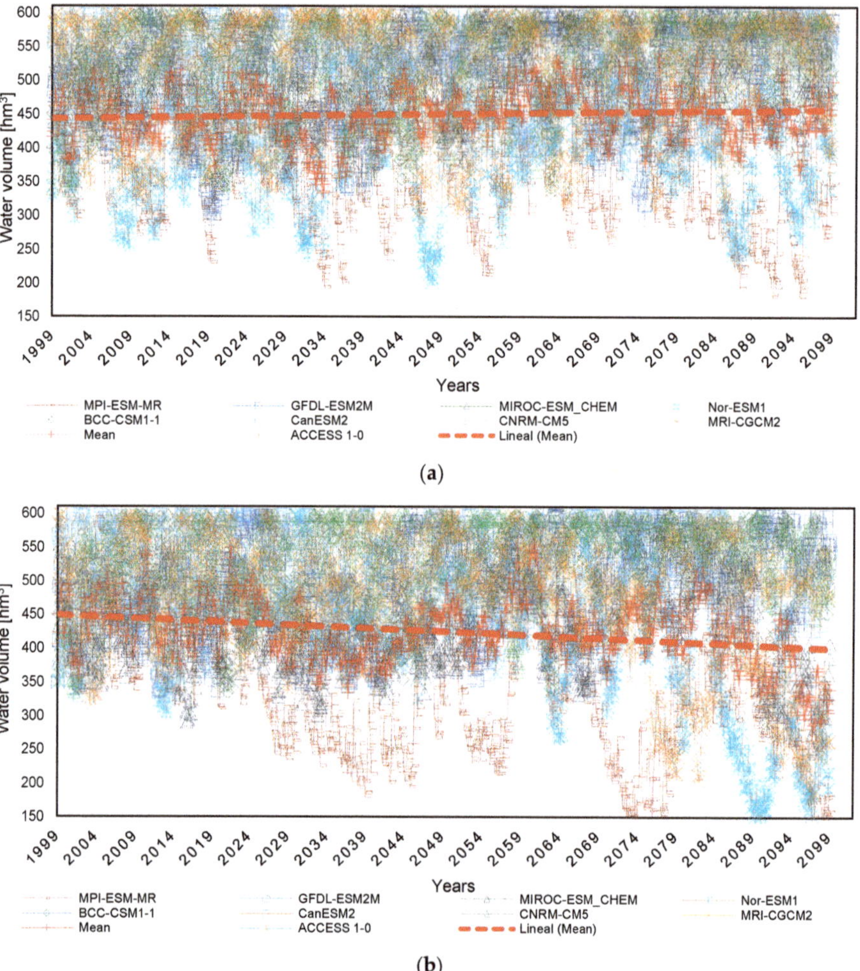

Figure 7. Results summary figure showing range and average water volume outcomes for every representative concentration pathway (RCP) scenario, for the Llobregat and Ter joint system towards the 21st century. The red-spotted line provides a linear trend estimation. (**a**) Results for the RCP8.5 scenario, (**b**) Results for the RCP4.5 scenario.

Model trends and extreme values were plotted to represent the magnitude of future situations and to consider all the possible climate model variables evolution during the 21st century. An implication of the negative trend, as for the RCP8.5 in the reservoir system, is the possibility that a long-term alternative resource will be necessary to preserve reservoir equilibrium. Regression analysis predicts the combined behavior of both systems as a joint water reservoir system. In particular, the analysis of the joint system allows studying the link among the most challenging climate change scenario (RCP8.5) and the predicted water resources availability and bypass any analysis bias.

Figure 7 displays the intercorrelations of the nine performed models, providing a behavior trend-line of the reservoir system water volume. An average trend of these model outputs forecasts an 11.1% decrease in the system water availability, applying the RCP8.5 scenario for the year 2100.

Such water availability variations will have city-scale consequences for social-economic conditions and ecosystems.

5. Discussion

Turning now to the assessment on the applicability of the SIMGES and HBV models as appropriate tools to forecast drought frequency, some factors play a role in determining why the HBV model underestimated the water contribution, as shown in Figure 4, for the first calibration and validation step. The main reason that can be argued to explain this may come from the precipitation records. At high areas, such as this one, convective storms may occur over a localized area not covered by any meteorological station. In any case, the HBV contributions were on the security side. SIMGES model contributions apply in ACA Water Management Plan were closer to HBV computed values. Thus, the response of the HBV model is reasonable.

Analyzing Table 3 outcomes for each season, it is observed how the winter and autumn volume contributions present the best correlation to the SIMGES model. Throughout the spring and summer periods, the dispersion of the results was high. However, the Nash-coefficients were satisfactory for the most significant reservoirs: La Baells, Sau, and Susqueda. La Llosa del Cavall and Sant Ponç reservoirs could not have been simulating with the same precision, notwithstanding, its contribution to the total water reserves was 13% and 4%. Hence, their contribution to the entire system was small in comparison with the other dams.

On the question of understanding and analyzing the water resource availability in Barcelona in the future under different climate change situations, this study found that the water availability would drop in this 21st century for the reservoir system. By 2019–2050, the models average predicts a 9% decrease in surface water volume availability over the reservoir system. However, by 2019–2100, due to precipitation reduction and warming-enhanced evaporation, Climate Change effects, the models average predicts an 11% decrease with a remarkably high consensus among analyzed models for the RCP8.5 scenario, as shown in Figure 7.

The results obtained herein are consistent with Barcelona regional and Barcelona city council results [10], estimating in the year 2050, a 12% surface water resources decrease. Table 4 compares the summary statistics for the water resources availability forecast, comparing other studies for the reservoir system.

Table 4. Comparison of the summary statistics for the forecast of the water resources' availability for the Ter and Llobregat reservoir system.

Study	Mean Expected Reduction by 2050	Mean Expected Reduction by 2100
Climate change impacts study in Barcelona—water cycle [10]	12%	No Data
RESCCUE Project	9%	11%
Water and climate change. Diagnosis of the impacts predicted in Catalonia [17,37]	7%–15% according to diverse scenarios	No Data

These results are consonant with related studies [17], finding that according to valid data control models and detailing low-heterogeneity results, Llobregat river discharges will decrease, in a 2% scale for years 2070–2100.

Our approach in this paper explored the water availability of the principal sources providing about 92% of contemporary water demand. Due to practical constraints, a full discussion of adaptation measures coping with water availability shortage lies beyond the scope of this study. Catalonian Drought Plan documents designate some current adaptation measures [11,37]. They relate an increase in alternative water sources and a decrease in drinking water consumption, such as the planning and implementation of water reclamation and reuse, desalination as a technical option to increase the drinking water availability, to increase groundwater extraction, and to decrease consumption (stronger for agriculture, breeding, and recreative uses).

However, these solutions have significant technical, legal, political, and economic hurdles. In the case of the existing desalination plant, its maximum potential for water production is 60 Hm^3/year, an amount that could meet the current and future water shortages set in this study, nevertheless, the average unit cost of desalinated water production (€ 0.58/m^3) cost four times more than the cost of potable water production in the drinking water treatment plant (€ 0.14/m^3) [24,38].

6. Conclusions

Drought model outcomes cannot predict future conditions for a fact. Still, these results will aid researchers and stakeholders to have an idea concerning the order of magnitude of eventual future drought episodes. The associated uncertainty in climate model variables does not allow them to be used to define a single future path, even though the results can be valuable to design future system improvements and investments.

Overall, these results highlight that under the RCP8.5 scenario, cyclic drought episodes are expected to occur every twenty years with more than one year of drought state persistency. Further research can evaluate climate change impacts, updating models forecast every 5/10 years, to estimate a most reliable behavior forecasting. For the water supply side, renewable water resources are influenced by anthropogenic factors, precipitation, temperature, and other climate variables fluctuations, yet, we dismiss these variations in this study scope.

This study combined hydrological models and the latest greenhouse gas concentration scenarios to synthesize the proposed behavior of water sources in Barcelona. We showed that climate change is likely to affect local and regional water scarcity modestly. Moreover, this research forecasts a water-availability slightly downward trend from the middle of the 21st century. This proposed behavior does not mean that the annual water contributions are ever lower than the current ones. We identified an increase in drought cycle frequency, following a reduction in the average water availability, even in years of hydrological ascent.

By contrast, these reductions lead to a trend (i.e., the one conditioning the water supply system capacity) alleviated by a constant alternative source with the same magnitude and adding specific support when extreme events occur. With regard to the research methods, some limitations need to be acknowledged. After defining how the expected water availability decreases in the system, it is necessary to consider that these analyses do not consider the future growth of water demand or any new planned infrastructure. Likewise, this study does not consider variations in land-use future states. The reservoir watershed has undergone a revegetalization process since 1997, as assorted researches carried out in Catalonia indicates a farmland abandonment process and the resulting increase in forest mass [39]; as a result, we consider these land-use conditions will remain stable in the future.

Further research is expected to have a better understanding of the mechanisms underlying Barcelona reservoirs' management yielding an effective and sustainable water supply scheme, in conjunction with other hydrological involved processes. Adaptation measures studies, which take these variables into account, will need to be undertaken.

Author Contributions: E.F.-O.: Conceptualization, Methodology, Software, Formal analysis, Investigation, Data curation, Writing—original draft preparation. E.M.-G.: Conceptualization, Validation, Resources, Writing—review and editing; Supervision, Project administration. R.M.: Investigation, Methodology, Visualization, Data curation, Resources. All authors have read and agreed to the published version of the manuscript.

Funding: This research was funded by the RESCCUE project, which is sponsored by the European Union's Horizon 2020 research and innovation program under grant agreement No. 700174, whose support is gratefully recognized.

Acknowledgments: The contents of this research are a part of the findings of the project RESCCUE, which has obtained funding from the EU H2020 (grant agreement n. 700174). Re-use of the knowledge enclosed in this paper for commercial and/or non-commercial purposes is allowed and free of charge, on the requirements of compliance by the re-user of the research, not distortion of the original meaning or information of this research and the non-liability of the project RESCCUE partners for any consequence stemming from the re-use. The project RESCCUE partners do not accept any liability for the errors, consequences, or omissions herein contained.

Conflicts of Interest: The authors declare no conflict of interest. The funders had no role in the design of the study; in the collection, analyses, or interpretation of data; in the writing of the manuscript, or in the decision to publish the results.

References

1. Chen, L.; Guo, S. *Copulas and its Application in Hydrology and Water Resources*; Springer: Singapore, 2019; ISBN 9789811305733.
2. Grillakis, M.G. Increase in severe and extreme soil moisture droughts for Europe under climate change. *Sci. Total Environ.* **2019**, *660*, 1245–1255. [CrossRef]
3. Tian, W.; Liu, X.; Liu, C.; Bai, P. Investigation and simulations of changes in the relationship of precipitation-runoff in drought years. *J. Hydrol.* **2018**, *565*, 95–105. [CrossRef]
4. Hao, Z.; Hao, F.; Singh, V.P.; Sun, A.Y.; Xia, Y. Probabilistic prediction of hydrologic drought using a conditional probability approach based on the meta-Gaussian model. *J. Hydrol.* **2016**, *542*, 772–780. [CrossRef]
5. Singh, S.; Sharma, V.K. *Urban Droughts in India: Case Study of Delhi*; Springer: Singapore, 2019; ISBN 9789811089466.
6. Roudier, P.; Andersson, J.C.M.; Donnelly, C.; Feyen, L.; Greuell, W.; Ludwig, F. Projections of future floods and hydrological droughts in Europe under a +2°C global warming. *Clim. Change* **2016**, *135*, 341–355. [CrossRef]
7. Koutroulis, A.G.; Papadimitriou, L.V.; Grillakis, M.G.; Tsanis, I.K.; Wyser, K.; Betts, R.A. Freshwater vulnerability under high end climate change. A pan-European assessment. *Sci. Total Environ.* **2018**, *613–614*, 271–286. [CrossRef] [PubMed]
8. Iglesias, A.; de Garrote, L.; Cancelliere, A.; Cubillo, F.; Wilhite, D. *Coping with Drought Risk in Agriculture and Water Supply Systems Drought Management and Policy Development in the Mediterranean*; Springer: Dordrecht, Netherlands, 2009; ISBN 9781402090448.
9. Anand, J.; Gosain, A.K.; Khosa, R. Optimisation of multipurpose reservoir operation by coupling soil and water assessment tool (SWAT) and genetic algorithm for optimal operating policy (case study: Ganga River Basin). *Sustainability* **2018**, *10*, 1660. [CrossRef]
10. Barcelona Regional; Ajuntament de Barcelona Cicle de L'Aigua - Estudi dels Impactes del Camvi Climàtic a Barcelona. Available online: http://www3.amb.cat/repositori/PSAMB/Canvi_clima.pdf (accessed on 19 November 2019).
11. Generalitat de Catalunya; Agència Catalana de L'Aigua; Molist, J. Pla especial d'actuació en situació d'alerta i eventual sequera. Informe de sostenabilitat ambiental. Available online: http://aca.gencat.cat/web/.content/30_Plans_i_programes/30_Pla_sequera/bloc1/04_pes_Isa_ca.pdf (accessed on 7 November 2019).
12. Martin-Ortega, J.; González-Eguino, M.; Markandya, A. The costs of drought: The 2007/2008 case of Barcelona. *Water Policy* **2012**, *14*, 539–560. [CrossRef]
13. Coll, J.R.; Aguilar, E.; Prohom, M.; Sigró, J. Variabilidad y tendencias a largo plazo de las sequías en Barcelona (1787-2014). *Cuad. Investig. Geogr.* **2016**, *42*, 29–48. [CrossRef]
14. Pascual, D.; Pla, E.; Lopez-Bustins, J.A.; Retana, J.; Terradas, J. Impacts du changement climatique sur les ressources en eau dans le bassin méditerranéen: Une étude de cas en Catalogne, Espagne. *Hydrol. Sci. J.* **2015**, *60*, 2132–2147. [CrossRef]
15. Gallart, F.; Delgado, J.; Beatson, S.J.V.; Posner, H.; Llorens, P.; Marcé, R. Analysing the effect of global change on the historical trends of water resources in the headwaters of the Llobregat and Ter river basins (Catalonia, Spain). *Phys. Chem. Earth* **2011**, *36*, 655–661. [CrossRef]
16. Barcelona Regional. Ajuntament de Barcelona Estratègia Delta del Llobregat. Available online: https://ajuntament.barcelona.cat/economiatreball/sites/default/files/documents/MemoriaESTRATEGIADELTA_A4_completa.pdf (accessed on 18 September 2019).
17. Agència Catalana de l'Aigua Aigua i Canvi Cimàtic. Diagnosi dels Impactes Previstos a Catalunya. Available online: http://www.gencat.cat/mediamb/publicacions/monografies/aigua_canvi_climatic.pdf (accessed on 25 September 2019).
18. Seibert, J.; Vis, M.J.P. Teaching hydrological modeling with a user-friendly catchment-runoff-model software package. *Hydrol. Earth Syst. Sci.* **2012**, *16*, 3315–3325. [CrossRef]

19. Versini, P.A.; Pouget, L.; McEnnis, S.; Custodio, E.; Escaler, I. Climate change impact on water resources availability: Case study of the Llobregat River basin (Spain). *Hydrol. Sci. J.* **2016**, *61*, 2496–2508. [CrossRef]

20. Velasco, M.; Versini, P.A.; Cabello, A.; Barrera-Escoda, A. Assessment of flash floods taking into account climate change scenarios in the Llobregat River basin. *Nat. Hazards Earth Syst. Sci.* **2013**, *13*, 3145–3156. [CrossRef]

21. Andreu, J.; Solera, A. Methodology for the analysis of drought mitigation measures in water resources systems. *Drought Manag. Plan. Water Resour.* **2006**, 133–168.

22. Céspedes, R.; Lacorte, S.; Ginebreda, A.; Barceló, D. Chemical monitoring and occurrence of alkylphenols, alkylphenol ethoxylates, alcohol ethoxylates, phthalates and benzothiazoles in sewage treatment plants and receiving waters along the ter River basin (Catalonia, N.E. Spain). *Anal. Bioanal. Chem.* **2006**, *385*, 992–1000. [CrossRef]

23. Honey-Rosés, J.; Acuña, V.; Bardina, M.; Brozović, N.; Marcé, R.; Munné, A.; Sabater, S.; Termes, M.; Valero, F.; Vega, À.; et al. Examining the demand for ecosystem services: The value of stream restoration for drinking water treatment managers in the Llobregat river, Spain. *Ecol. Econ.* **2013**, *90*, 196–205. [CrossRef]

24. Pouget, L.; Escaler, I.; Guiu, R.; Mc Ennis, S.; Versini, P.A. Global Change adaptation in water resources management: The Water Change project. *Sci. Total Environ.* **2012**, *440*, 186–193. [CrossRef]

25. Climate Research Foundation (FIC). Downscaled climate model outputs of the RESCCUE project. Available online: https://www.ficlima.org/intercambio/indexed/RESCCUE/ (accessed on 19 November 2019).

26. Ribalaygua, J.; Torres, L.; Portoles, J.; Monjo, R.; Gaitan, E.; Pino, M. Description and validation of a two-step analogue/regression downscaling method. *Theor. Appl. Climatol.* **2013**, *114*, 253–269. [CrossRef]

27. Bi, D.; Dix, M.; Marsland, S.J.; O'Farrell, S.; Rashid, H.; Uotila, P.; Hirst, A.C.; Kowalczyk, E.; Golebiewski, M.; Sullivan, A.; et al. The ACCESS coupled model: Description, control climate and evaluation. *Aust. Meteorol. Ocean. J* **2013**, *63*, 41–64. [CrossRef]

28. Xiao-Ge, X.; Tong-Wen, W.; Jie, Z. Introduction of CMIP5 Experiments Carried out with the Climate System Models of Beijing Climate Center. *Adv. Clim. Chang. Res.* **2013**, *4*, 41–49. [CrossRef]

29. Chylek, P.; Li, J.; Dubey, M.K.; Wang, M.; Lesins, G. Observed and model simulated 20th century Arctic temperature variability: Canadian Earth System Model CanESM2. *Atmos. Chem. Phys. Discuss.* **2011**, *11*, 22893–22907. [CrossRef]

30. Voldoire, A.; Sanchez-Gomez, E.; y Mélia, D.; Decharme, B.; Cassou, C.; Sénési, S.; Valcke, S.; Beau, I.; Alias, A.; Chevallier, M.; et al. The CNRM-CM5.1 global climate model: Description and basic evaluation. *Clim. Dyn.* **2013**, *40*, 2091–2121. [CrossRef]

31. Dunne, J. GFDL's ESM2 Global Coupled Climate-Carbon Earth System Models. Part I: Physical Formulation and Baseline Simulation Characteristics. *J. Clim.* **2012**, *25*, 6646–6665. [CrossRef]

32. Watanabe, S.; Hajima, T.; Sudo, K.; Nagashima, T.; Takemura, T.; Okajima, H.; Nozawa, T.; Kawase, H.; Abe, M.; Yokohata, T.; et al. MIROC-ESM 2010: Model description and basic results of CMIP5-20c3m experiments. *Geosci. Model Dev.* **2011**, *4*, 845. [CrossRef]

33. Marsland, S.J.; Haak, H.; Jungclaus, J.H.; Latif, M.; Röske, F. The Max-Planck-Institute global ocean/sea ice model with orthogonal curvilinear coordinates. *Ocean Model.* **2003**, *5*, 91–127. [CrossRef]

34. Yukimoto, S.; Yoshimura, H.; Hosaka, M.; Sakami, T.; Tsujino, H.; Hirabara, M.; Tanaka, T.; Deushi, M.; Obata, A.; Nakano, H.; et al. Meteorological Research Institute-Earth System Model Version 1 (MRI-ESM1)—Model Description. *Tech. Reports Meteorol. Res. Inst.* **2011**, *64*, 1–96.

35. Bentsen, M.; Bethke, I.; Debernard, J.B.; Iversen, T.; Kirkevåg, A.; Seland, Ø.; Drange, H.; Roelandt, C.; Seierstad, I.A.; Hoose, C.; et al. The Norwegian Earth System Model, NorESM1-M – Part 1: Description and basic evaluation of the physical climate. *Geosci. Model Dev.* **2013**, *6*, 687–720. [CrossRef]

36. Iversen, T.; Bentsen, M.; Bethke, I.; Debernard, J.; Kirkevåg, A.; Seland, Ø.; Drange, H.; Kristjansson, J.; Medhaug, I.; Sand, M.; et al. The Norwegian Earth System Model, NorESM1-M - Part 2: Climate response and scenario projections. *Geosci. Model Dev.* **2013**, *6*, 389. [CrossRef]

37. Agència catalana de l'aigua Pla especial d'actuació en situació d'alerta i eventual sequera Districte de conca fluvial de Catalunya. **2009**.

38. Cetaqua Identification of impacts and definition of adaptation measures. Available online: https://climate-adapt.eea.europa.eu/metadata/projects/medium-and-long-term-water-resources-modelling-as-a-tool-for-planning-and-global-change-adaptation-application-to-the-llobregat-basin (accessed on 12 October 2019).

39. Duran, X.; Picó, M.J.; Reales, L. *El Cambio Climático en cataluña. Resumen ejecutivo del Tercer informe sobre el cambio climático en Cataluña*; Institut d'Estudis Catalans y Generalitat de Catalunya: Barcelona, Spain, 2017; ISBN 9788439396208.

 *sustainability*

Article

Investigating the Effects of Pluvial Flooding and Climate Change on Traffic Flows in Barcelona and Bristol

Barry Evans [1,2,*], Albert S. Chen [1], Slobodan Djordjević [1], James Webber [1], Andoni González Gómez [3] and John Stevens [4]

[1] Centre for Water Systems, University of Exeter, Exeter EX4 4QF, UK; A.S.Chen@exeter.ac.uk (A.S.C.); S.Djordjevic@exeter.ac.uk (S.D.); J.Webber2@exeter.ac.uk (J.W.)
[2] School of Built Environment, College of Sciences, Massey University, Auckland 0745, New Zealand
[3] Adjuntament de Barcelona, 08012 Barcelona, Spain; agonzalezgom@bcn.cat
[4] Bristol City Council, Bristol BS3 9FS, UK; john.stevens@bristol.gov.uk
* Correspondence: b.evans@exeter.ac.uk or b.evans2@massey.ac.nz; Tel.: +44-1392-724-075

Received: 31 January 2020; Accepted: 13 March 2020; Published: 17 March 2020

Abstract: This paper outlines the work carried out within the RESCCUE (RESilience to cope with Climate Change in Urban ArEas) project that is, in part, examining the impacts of climate-driven hazards on critical services and infrastructures within cities. In this paper, we examined the methods employed to assess the impacts of pluvial flooding events for varying return periods for present-day (Baseline) and future Climate Change with no adaptation measures applied (Business as Usual) conditions on traffic flows within cities. Two cities were selected, Barcelona and Bristol, with the former using a meso-scale and the latter a micro-scale traffic model. The results show how as the severity of flooding increases the disruption/impacts on traffic flows increase and how the effects of climate change will increase these impacts accordingly.

Keywords: flooding; climate change; traffic modelling; resilience

---

1. Introduction

Within major cities, the transportation network serves as an essential component in its functionality allowing for the movement of goods, services, and the general population, with an estimated 81.7 thousand vehicles per mile of motorway, and 12.2 thousand miles of rural "A" roads per day within the UK [1].

The implications of flooding within the road network can be severe both in terms of risks to human lives both directly as a result of drowning and indirectly due to impacts on the ability of first responders to respond to incidents [2–4] as well as to a region's economy. Flooding in Barcelona 2011 disrupted the transport network both directly as a result of flooded road sections and indirectly as a result of traffic light failures [5]; the traffic disruption alone caused by the Summer Floods of 2007 in the UK cost the UK economy in the range of £22–£174 million (depending on assumptions) [6].

From a climate change perspective, the Department for Transport have stated that the Strategic Road Network (the main roads that connect the country) has been identified as being particularly vulnerable to weather-related flooding [7], with Highways England highlighting that the current drainage systems in place may not be sufficient to deal with the increased rainfalls associated with climate change predictions [8]. The UK Climate Impact Projections Report 2009 (UKCP09) predicts the precipitation across the UK will increase up to 70% in certain locations by the 2080's [9], which could result in more frequent and greater levels of disruption to traffic movements.

Previous works have investigated both the risks and impacts of flooding poses to the transport sector such as the combined interactions of flood depths and flow velocities on vehicular stability [10,11], the relationship between vehicular speed and standing water depths on road surfaces [2,12–14] and the significance of which roads within a network are flooded [15].

This paper investigates the impacts on traffic of pluvial flood events in two European cities (Barcelona and Bristol) via linking flood model outputs with traffic models and examine how the magnitude of these impacts could change in the future with respect to climate change model predictions.

2. Materials and Methods

Previous work by Pyatkova et al. [12,13] demonstrated the use of loosely coupling flood model outputs with micro-simulation traffic model inputs as a means simulating and assessing the impacts of flood events upon traffic flows.

The approach proposed here utilizes maximum flood-depth data derived from flood mapping as the criteria for determining the properties of individual road sections at various timings during the traffic model period to simulate effects of flooding to a transportation network. Figure 1 shows conceptually how the flood model outputs are utilized as a means for preparing the traffic model inputs to simulate the effects of flooding.

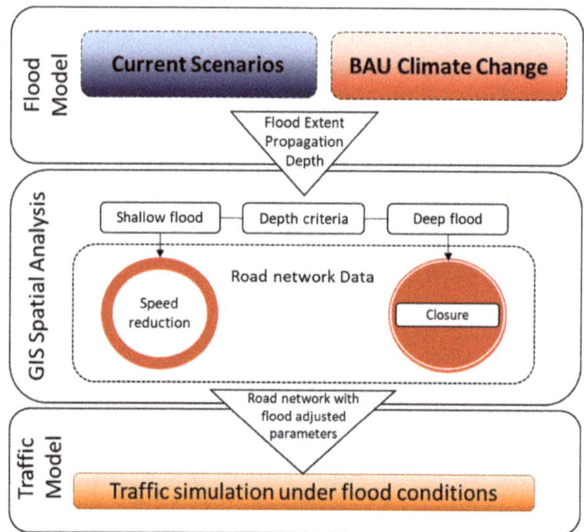

Figure 1. Loosely coupling flood model outputs with traffic model input parameters (based on methodology outlined by Pyatkova, 2018 [16]).

2.1. Barcelona Research Site

For the Barcelona Case Study, a 1D/2D-coupled flood model that utilizes the dual drainage concept [17] using Infoworks ICM (Intergrated Catchment Modelling) [18] has been employed. This hydrodynamic model was used to provide outputs of both water depths and velocities for different return periods under the present (Baseline) and Future Climate Change scenarios whereby a Business As Usual (BAU) policy is assumed (i.e., no adaptation measures applied within the city). For the climate change scenarios in Barcelona, a Representative Concentration Pathway of 8.5 (RCP8.5) has been considered. Table 1 shows the comparison of the maximum rainfall intensities for the Baseline and BAU scenarios from the synthetic rainfall events generated via the Foundation for Climate Research (FIC) used within the flood model.

Table 1. Comparison between maximum 5 intensities of Current Scenario (Baseline) and Future Climate Change Scenario (BAU) of synthetic rainfall events used within Barcelona Case Study (Deliverable 2.3 RESCCUE).

Return Period	Current Scenario (Baseline)		Future Climate Change Scenario (BAU)	
	Peak Rainfall Intensity (mm/h)	Rainfall Depth (mm)	Peak Rainfall Intensity (mm/h)	Rainfall Depth (mm)
1	63.6	22.2	73.8	24.1
10	177.2	83.7	195.7	88.4
50	217.2	104.2	234.6	112.9
100	239.6	115.8	256.4	127.3
500	291.7	143.4	312.1	157.2

2.1.1. Barcelona Traffic Model

Within the city of Barcelona, the Departament d'Estratègia de la Mobilitat has developed/provided a pre-existing meso-scale traffic model using the commercial software package TransCAD® [19]. For this meso-scale traffic model, the road network was divided up into road sections referred to links and each link had a wealth of properties relating both the physical characteristics and imposed rules parameters of the road including, but not limited to speed restrictions, number of parking maneuvers per hour, and lane capacity. Upon addition of all the required parameters within TransCAD®, the model can evaluate how the road network performs. Figure 2 shows the extent of the Barcelona traffic model and the relative speed reductions that are calculated by the model under normal operating (dry weather) conditions. In this figure the %Speed Difference (Equation (1)) refers to the calculated/modelled speed of the vehicles derived from the TransCAD® software, relative to the "Free Flow Speed" (the speed at which a vehicle could move along the section unimpeded by other vehicles). The figure highlights the high levels of congestion within the heart of the city and through the major roadways along the outer perimeter in the modelling result.

$$\%Speed\ Reduction = \left(\frac{Speed_{Modelled} - Speed_{FreeFlow}}{Speed_{FreeFlow}} \right) \times 100,$$ (1)

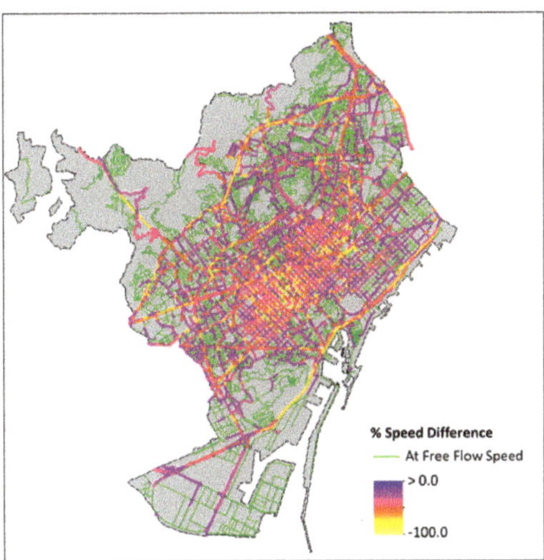

Figure 2. Percentage speed reduction of traffic in meso-scale model with respect to free-flow speed.

2.2. Bristol Research Site

For the Bristol research site, the Infoworks ICM was also employed at a city-wide level. In contrast to the FIC Climate data used in the Barcelona case, the Bristol case study's climate data was derived from UKCP09 [9] predictions. Table 2 shows a comparison of the associated rainfall depths of a 1 in 100-year, 60-min duration event for BAU in Bristol based on UKCP09. From this analysis, for a comparative study of pluvial flooding, the upper end (high emissions scenario) and upper epoch (furthest future projection: 2071-2100) were selected for the climate change scenarios as they show a comparative climate change uplift (highlighted in Table 2).

Table 2. Synthetic rainfall depths from a 1 in 100-year, 60-min duration event derived from UKCP09 climate change projections.

Climate Change Scenario	Year/Epoch	Climate Change Uplift	Rainfall Depth (mm)
Current Scenario (Baseline)	Present day	N/A	44
FIC	2041–2070	30%	57.2
	2071–2100	40%	61.6
UKCP09–Central	2041–2070	10%	48.4
	2071-2100	20%	52.8
UKCP09–Upper end	2041–2070	20%	52.8
	2071–2100	40%	61.6

For the Bristol research site, we were limited with the number of return periods available from the flood model outputs. For the Baseline scenarios, we used 1 in 10, 30, and 100-year return periods and for the Future Climate Change scenario, we selected the 1 in 10, 20, and 100-year scenario with the 1 in 20 Year being deemed to be the closest available approximation to the Baseline 1 in 30 year event with climate change uplift applied.

Bristol Traffic Model

Unlike the Barcelona case study, Bristol did not have a pre-existing traffic model available for testing. Due to this, we looked to develop a micro-scale traffic model using the Open Source "Simulating Urban Mobility" (SUMO) software [20]. In contrast to the meso-scale model, the micro-scale model used for Bristol in this analysis simulates the movement of each individual vehicle separately as shown in Figure 3.

Figure 3. Example of SUMO micro-simulation.

For the road network the Bristol model has been built using OpenStreetMap (OSM) data [21]. Using SUMO's "netconvert" tool, OSM data was converted into a network file suitable for use within SUMO that contains road property information including but not limited to, the number of lanes, junctions, and traffic light locations. In the absence of traffic data, the traffic flows within the network were derived via generating Origin-Destination (OD) matrix database using data from the National-Receptor-Database

(NRD) [22]. For this process, we assumed vehicles start from either Residential locations or from the boundary of the road network extent (Network Entry Points) (assuming from outside the city) and that the journeys terminate either at a place of work or they leave the network at a boundary (Network Exit Points). Table 3 shows the composition of the origin-destination points with Table 4 showing the percentage distribution of the Origin and Destination locations accordingly. For the 'School' classification, some vehicles can use the school as a mid-way point in their journey to simulate school drop-offs during the morning.

Table 3. Origin Destination points within the road network.

Origin Location	Count	Destination Location	Count
Residential	666,445	Office	3054
		Retail	3911
		Industry	1051
Network Entry Points	17	Warehouse	910
		School	49
		Network Exit Points	17

Table 4. Origin Destination percentage distribution.

Origin	Percentage	Destination	Percentage
Home	90	Office	49
		Retail	11
		Industry	12
Enters Network	10	Warehouse	8
		School	*10*
		Leaves network	10

Using the spatial information of land-use points from the NRD, SUMO's 'Duarouter' tool was used to generate the OD catalogue of vehicular journeys within the network. An additional rule applied states that each journey must have a journey length equal to or greater than 1 km.

To simulate morning rush hour flows, a sigmoid style curve was used in determining the number of vehicles that were added to the network over time during the simulation. Figure 4 shows the number of vehicles being added to the network over time for a 5000 and 10,100 vehicle scenario respectively using the same curve function.

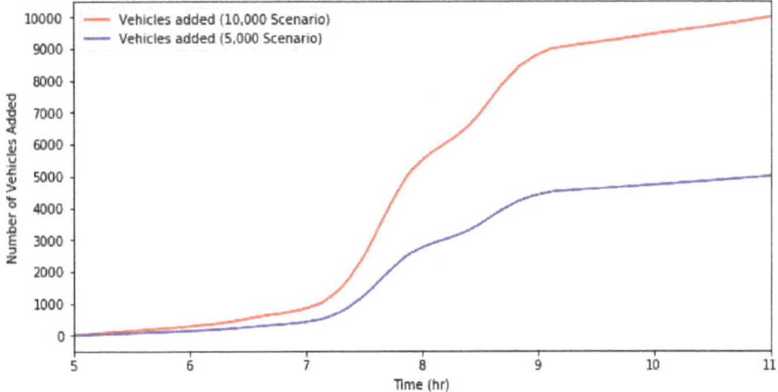

Figure 4. Number of vehicles added to the network over time for the 5000 and 10,000 traffic volume scenarios.

Figure 5 shows a comparison of the SUMO model outputs for dry weather scenarios, in which the different volumes of traffic present within the network over time whereby vehicles are only being added to the network (starting their journeys) between the hours of 5 am and 11 am. The result demonstrates that a doubling of vehicle journeys from 5000 to 10,000 vehicles within that period results in a seven-fold increase of the number of vehicles present within the network at its peak and a subsequent long tail section as the vehicles leave the network. For both scenarios, between the hours of 7 and 9 am there is a large increase in the number of vehicles being added to the network (approximately 4000 and 8000, respectively). In each scenario, the vehicles added to the network are subsequently removed from the network upon completion of their respective journeys.

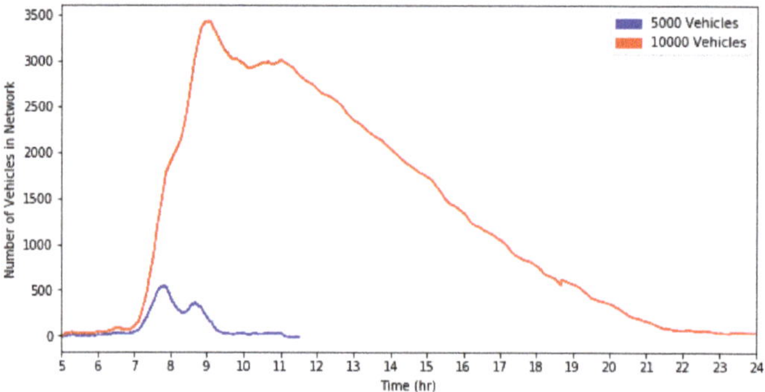

Figure 5. Comparison of Traffic Volumes in the Network for 5000 journey and 10,000 journey scenarios.

The reason behind this substantial difference in the number of vehicles within the network is a result of traffic jam formation leading to the delay in the completion of vehicle journeys. Figure 6 highlights the variations in the Average Journey Speed (Equation (2)) of vehicles for both the 5000 and 10,000-journey scenarios. The majority of vehicles within the network during the 10,000-journey scenario are travelling at relatively low speeds (less than 10 km/h) whereas the average journey speed for traffic in the 5000-vehicle scenario is around 30 km/h.

$$Average\ Jounrey\ Speed_{Vehicle} = \frac{Total\ Journey\ Distance_{Vehicle}}{Total\ Journey\ Time_{Vehicle}} \tag{2}$$

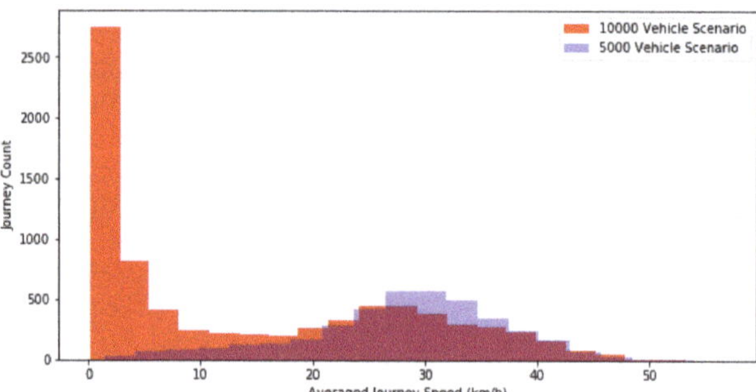

Figure 6. Average Journey Speed of Vehicles within the Network during the hours of 7 am and 9 am for 5000 and 10,000 journey scenarios.

Figure 7 shows a section within in the network at the peak times (determined in Figure 5) for both journey cases, where the 10,000 journey scenario presents a considerable worse congestion. Because of the congestion both here and in other sections of the network the time for the traffic to clear the network (complete their respective journeys) becomes dependent on the interval timing of the traffic lights and the settings in place within the model "time-to-teleport" to handle these obstruction issues.

Traffic signal timing data is often unavailable and has a dominant influence on intersection capacity and network performance [23]. If the timings of these traffics lights are not configured correctly, under high volumes of traffic a stalemate scenario can occur whereby traffic can neither enter nor exit an area thus resulting in severe gridlock. To minimise gridlock scenarios within the traffic model (due to imperfections in the network design) and to deal with instances of vehicles becoming an obstruction, the teleportation rule is applied. In the examples shown in Figure 5, if a vehicle is stationary 40 min (flood duration of 30 min plus and arbitrary 10-min window), it is deemed to be assumed to be erroneously stuck and is teleported to the next edge within its route. Note that it is important that the teleportation rule has a time-limit set to be equal or greater than the duration of a flood event to prevent traffic teleporting past the blocked roads under flooded conditions.

(A) Peak time 7:48 am   **(B) Peak time 9:03 am**

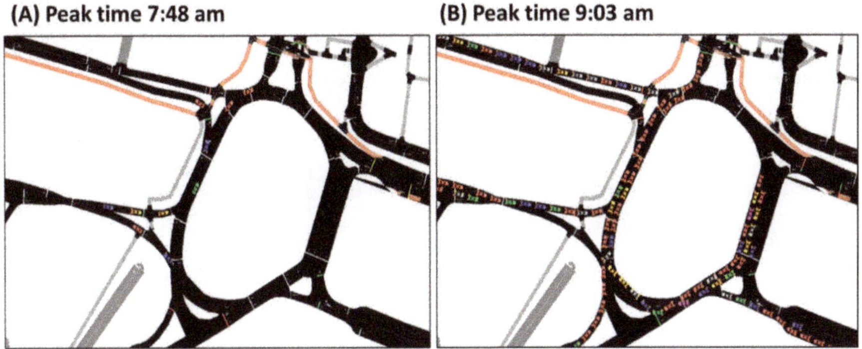

Figure 7. Comparison of Traffic Volumes at peak times for (**A**) 5000 journeys and (**B**) 10,000 journey scenarios.

For the purpose of the study within this paper, we selected 5000 vehicle traffic scenario as a means of analysing the effects of flooding on obstructing or causing diversions to vehicles within the network and to minimise the implications of imperfections in the network configuration itself causing disruptions to traffic flows. Ten 6-h duration traffic scenarios were generated, where each scenario contains 5000 randomly selected journeys from the OD catalog whereby the Origin and Destination's match the percentage distribution outlined in Table 4. The synthetic sigmoid curve, as shown in Figure 4, was applied to stagger the start times of the vehicles during the simulation to generate a pseudo morning rush scenario. Figure 8 shows the variation/range of the number of vehicles present in the network over time across the ten scenarios, highlighting the two temporal peaks in traffic volumes within the network during the morning rush hours between 7 am and 9 am. Figure 9 shows the extent of the Bristol traffic model and the percentage route distribution of the ten scenarios (10 × 5000 journeys) whereby the higher percentage values correspond to the road sections where vehicles have traversed the most within the 10 scenarios. Here, we observe that within the modelled scenarios, there is a preference for vehicles to traverse the river section (that bisects the city) across the bridges.

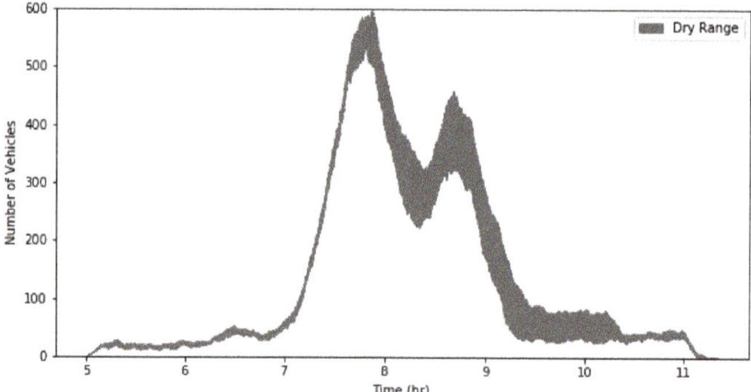

Figure 8. Range of vehicle distributions from 10 scenarios under dry weather conditions in Bristol road network.

Figure 9. Percentage Route Distributions under dry conditions.

2.3. Translating Flood Hazards into Traffic Models

To simulate the impacts of flooding within the traffic network the relationship between maximum allowable speeds of vehicles with respect to flood-depths is required. Previous work by Pyatkova et al. [12,13] discretised flood hazards via the relationship of maximum permitted speed limits along road sections with respect to flood depths along those sections. Table 5 shows the discrete ranges for the flood depths, their hazard classification and the subsequent speed reductions along the road sections where these flood depths are present.

Table 5. Flood hazards' effect on traffic flows.

Flood Depth Range (m)	Hazard Classification	Maximum Vehicle Speed (kmh^{-1})
Depth < 0.1	Low	Road Speed Limit
$0.1 \leq$ Depth < 0.3	Medium	20
0.3 < Depth	High	0 (Road Closed)

Through an intersect analysis, analyzing the depths of water on road surfaces, we modify the input parameters (speed limit) of links within the traffic model. Figure 10 shows an example of the links affected by flooding within Barcelona when analyzing maximum flood-depths of a 1 in 10 year BAU climate change scenario.

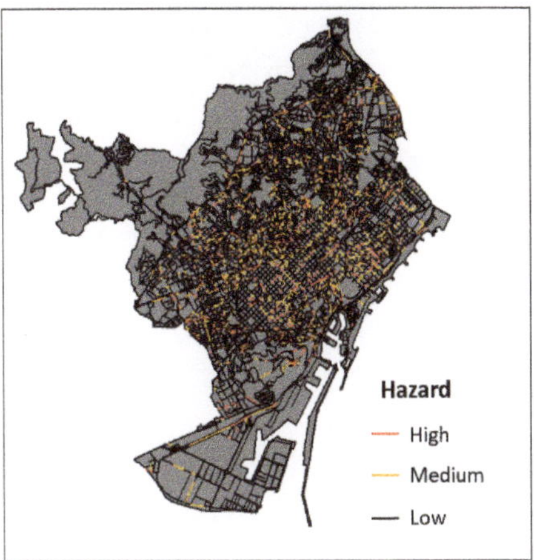

Figure 10. Example of hazard classification for a 1 in 10 Year Future Climate Change Scenario Event.

2.4. Quantifying Impacts from Traffic Model

The impacts of flooding on traffic flows can be quantified in a number of ways including but not limited to, lost time, fuel consumption and pollution levels. Based on the Multi-Coloured Manual (MCM) [24], we estimated the costs accrued to a vehicle (in GBP) over time and distance in relation to its speed. Table 6 shows a breakdown of costs in pence per unit of speed for five vehicle classes.

Table 6. Total Costs of travel as a function of speed (pence) [23].

	Speed km/h								
Vehicle	**1**	**2**	**5**	**10**	**20**	**40**	**50**	**80**	**100**
Car	1023	515	210	109	57	31	25	17	15
LGV	1181	596	245	128	68	37	32	23	20
OGV_1	1241	634	268	144	79	44	37	29	27
OGV_2	1454	746	320	175	98	57	50	40	37
PSV	7406	3742	1514	774	403	216	178	124	106

For simplicity in our analysis, we assumed all the vehicles to be of generic petrol driven cars. Figure 11 shows the derived relationship between the estimated costs incurred per car per hour in relation to its speed whereby the line of best fit is described by the function in Equation (3).

$$Cost = 9.6275 \times Speed^{-0.925} \tag{3}$$

For the TransCAD® model the accumulated costs were derived from analysing the model outputs at a link level where the total accumulated cost is derived using Equation (4).

$$Total\ Cost = \sum_{Links} (Traffic\ Volume_{Link} \times Length_{Link} \times SFn_{Link}) \tag{4}$$

where

$Traffic\ Volume_{Link}$ = Number of Vehicles per km per hour per link.
$Length_{Link}$ = Length of link in km.
SFn_{Link} = Calculated cost value with respect to average speed derived via Equation (3).

An estimate of monetary impacts of a flood event with respect to changes in vehicular speed is thus derived via comparing the costs under flooded conditions to the costs under dry weather conditions (Equation (5)).

$$Monetary\ Impact = Costs_{Flooded\ Conditions} - Costs_{Dry\ Conditons} \tag{5}$$

As the Barcelona model has no temporal information applied to traffic flows, the associated costs were derived with respect to incurred costs per hour of disruption. For the Bristol case study, as we adopted a micro-scale traffic model with a temporal component, the incurred costs were assessed during the period of time the network was deemed to be impacted.

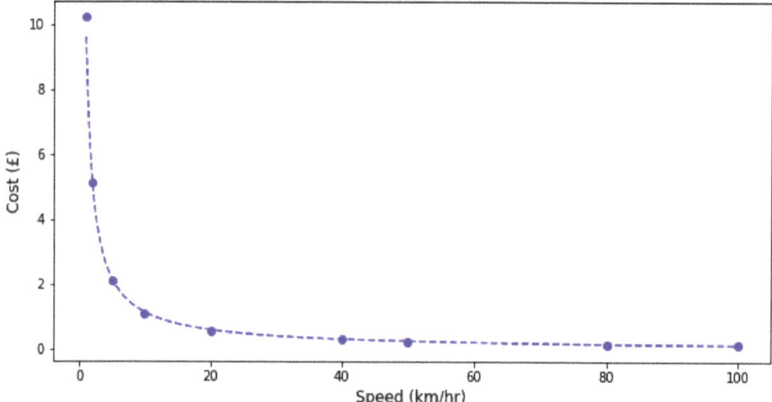

Figure 11. Derived Cost to Speed relationship Cars based on MCM data.

For the Bristol model, data at the link level were also exported but in contrast to the Barcelona model, these outputs vary over time, therefore, Equations (3)–(5) were applied accordingly during the identified period of traffic disruption.

3. Results

3.1. Hazard Analsysis

3.1.1. Barcelona Road Hazards

Figure 12 shows the percentage of road sections that were affected by flood-depths at various return periods for the Baseline (a) and Climate Change (b) scenarios using the rules as described in Table 5. As the severity of the events increases, the number of deep-flooded roads continues to grow whereas the percentage of shallow flooded roads affected begins to level out around 12%. This "levelling out" of the number of shallow flooded sections is due to the transition of road hazard classifications whereby as the severity of the event increases, previously shallow ponding areas upon the surface continue to accumulate flood waters thus moving their flood depths from below 30 cm to 30 cm+ thus transitioning to deep-flooded road status.

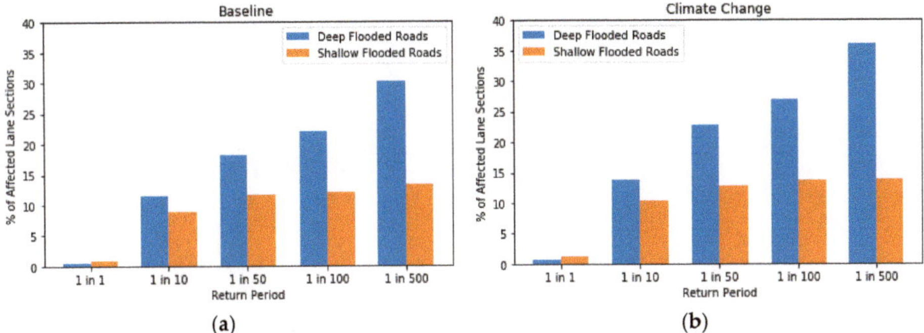

Figure 12. Affected road sections in Barcelona case study area for (**a**) Baseline Scenarios, (**b**) Climate Change Scenarios.

3.1.2. Bristol Road Hazards

Figure 13 shows percentage of affected lane sections for the Bristol case study with respect to the severity of the pluvial flooding events for Baseline (a) and Climate Change (b) scenarios. Within the Bristol case study, an additional criteria specifying the minimum length of a flooded road was included to reduce the number of "flood zones" in the traffic model. This reduction of flood zones was implemented in order improve model performance. In this instance, the minimum length for a flooded road was set to 10 m; therefore, if the length of the section of road that is flooded is less than 10 m, the road will be regarded as not flooded, thereby bringing the overall percentages down. Like that of the Barcelona case study, there is a positive correlation between the severity of the event and the percentage of affected links. For both the Baseline and Climate Change scenarios, we see climbing numbers of affected links and again the transition of shallow flooded areas to deep flooded as the severity of rainfall events increase.

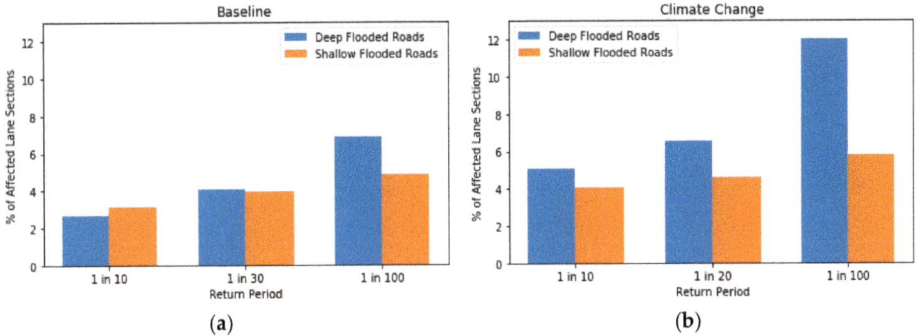

Figure 13. Affected road sections in the Bristol case study area for (**a**) Baseline Scenarios, (**b**) Climate Change Scenarios.

3.2. Impact Assessment

3.2.1. Impacts on Traffic in Barcelona

Upon changing the respective speed limit parameters within TransCAD®, the model was reapplied to assess how the traffic flows within the city have changed. Figure 14 shows speed difference maps calculated from the TransCAD® model runs for a 1 in 100 year event for the Baseline and Climate Change scenarios. Overall appearance shows a similar spatial distribution of speed reductions with slight increases (higher negative values) in reductions for the Climate Change Scenario.

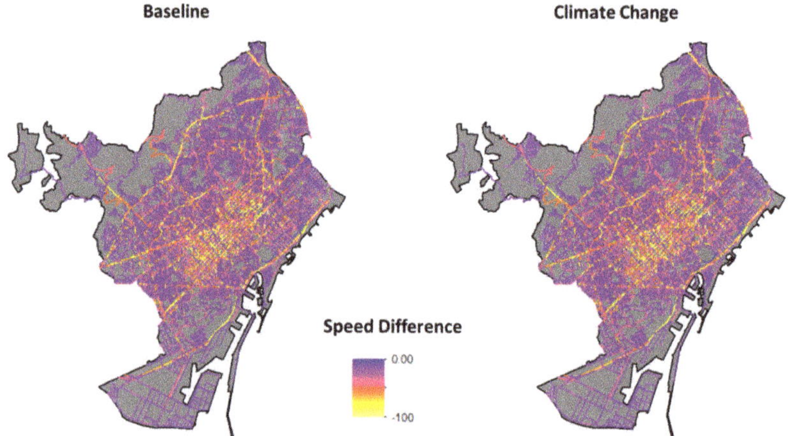

Figure 14. Comparison of speed differences for 1 in 100-year event for Baseline and Climate Change Scenarios.

Figure 15 shows a comparison of the derived Exceedance Probability (EP) Curves for the Baseline and Future Climate Change prediction scenarios based on summation of all the link cost values. This curve was derived via plotting the total accumulated losses for a pluvial flood event across the network with respect to the probability of occurrence (1/Return Period) of the event. Here, we see that under future climate change conditions the predicted monetary losses/impacts when traffic disruption increases with respect to the severity of the flood event. Table 7 shows a comparison of the loss values with an average increase of monetary losses due to the vehicular speed in the network was reduced by 10% or more under future climate change scenarios.

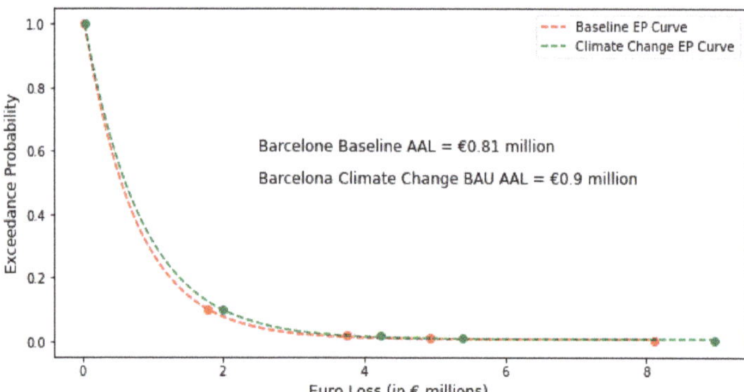

Figure 15. Flood Impact on Traffic EP Curves for Barcelona.

Table 7. Monetary losses as a result of traffic speed disruption.

Return Period (Years)	Baseline Monetary Losses (€)	Climate Change Monetary Losses (€)	% Difference of Monetary Losses
1 in 1	26,236	28,664	9.3
1 in 10	1,805,005	1,999,440	10.8
1 in 50	3,751,575	4,231,176	12.8
1 in 100	4,938,234	5,393,834	9.2
1 in 500	8,123,795	8,955,965	10.24

3.2.2. Impacts on Traffic in Bristol

Within the Bristol model, as we also considered the temporal aspect of the flooding in the analyses such that we can specify the time and duration of the flood event. In this example we have specified that flooding occurs as 7 am and lasts for 30 min. During this period, the maximum permitted speeds along hazard affected road sections are temporarily modified and will return to normal at 7:30 am after the flood event has ended.

As the Bristol model is micro-scale, if some journey starts within a flooded region they cannot begin and therefore, are not added to the network this can lead to reduced traffic number on the road during and after the flood event and needs to be considered as part of the impact assessment. Table 8 shows the percentage of journeys whose start-times begin during a flood event and are unable to begin their journey as lie within a closed road.

Table 8. Journeys unable to begin due to flooding.

(Baseline) Return Period	Journeys Lost (%)	(Climate Change) Return Period	Journeys Lost (%)
1 in 10	0.26	1 in 10	0.55
1 in 30	0.40	1 in 20	0.68
1 in 100	0.75	1 in 100	1.1

One of the additional indicators used to assess the impacts of traffic flows through a micro-scale simulation is to examine the number of vehicles within the network at any given time under flooded conditions and compare this distribution against dry weather conditions. Under flooded conditions, as some road sections will temporarily have their maximum allowable speeds reduced and some sections are temporarily closed. The journey times for vehicles that usually traverse these sections along their assigned routes will increase as vehicles are forced to either move at a reduced speed through shallow water or are diverted onto alternative routes if their original route is obstructed. Figure 15 shows a comparison of the number of vehicles within the road network over time for the different severities of flood events for baseline and future climate change conditions. Within this figure, the "Dry Count Range" represents the range (minimum and maximum) of vehicle counts across the 10 generated OD matrix routing scenarios, and the "Flooded Traffic Count Range" shows the ranges with respect to the network during the 30 min flood simulations. The "Average Flood Traffic Count" represents the average number of vehicles within the network during the respective flood scenarios. The figures highlight that even though a percentage of journeys are lost/unable to start during the flood event, the vehicular saturation of the network both during and immediately after the flood event surpasses the dry weather conditions with the network (on average) recovering by 9 am for all scenarios. The Flooded Traffic Count Range further highlights the different effects flooding has across the 10 generated route scenarios. Figure 16 further shows that even after the flood event has finished the road network still takes time to recover as previously impeded vehicles are continuing to complete their journeys and their remaining presence within the network effects other vehicles that travelling.

Figure 17 shows the comparison of the relative EP curves for the Baseline and Future Climate Change scenarios utilising the same cost to speed relationship applied to the Barcelona case study. Here the points for the Baseline and Climate Change scenarios represent the average calculated losses derived from the simulations with the curves interpolated from these points respectively. Within this example, we are examining the relative cost increases between the hours of 7 am and 9 am that corresponds to the period of disruption shown in Figure 16. In contrast to the Barcelona case study the calculated loss values depicted for Bristol simulations are considerably less. There are a number reasons for this including, but not limited to, the case study area examined within the city of Bristol (24 km^2) is considerably smaller than that assessed within Barcelona (102 km^2). A second reason relates to the limited number of vehicles used in the duration of the model. With Bristol having a population of approximately 463,500 [25] and 41% of the population driving a car to work [26] the simulated

5000 vehicles over a 6 h period could be a significant under estimate of the traffic volumes/journeys undertaken within the network during this period. The results, therefore, merely serve to show how the effects of climate change can result in observed increases in disruption to traffic flows and potential losses within the traffic network.

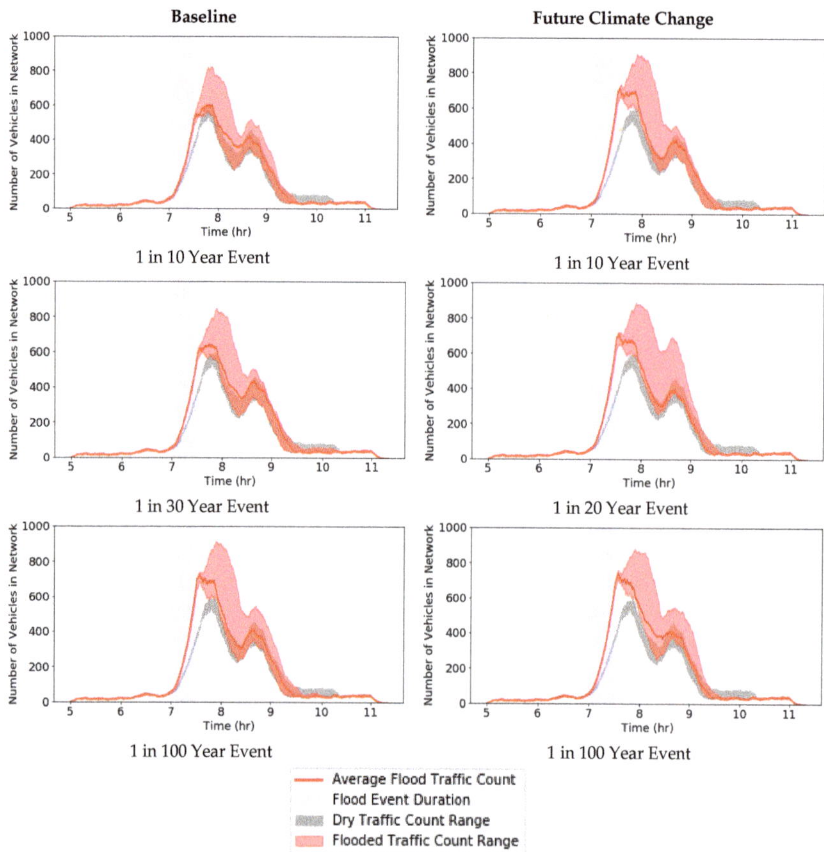

Figure 16. Comparison of Traffic Network Recovery Times.

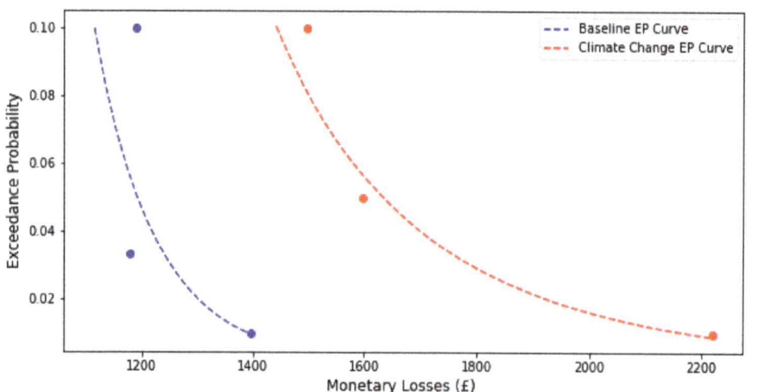

Figure 17. Flood Impact on Traffic Relative EP Curves for Bristol.

4. Discussion

4.1. Limitations and Assumptions

This paper shows the application of two distinct traffic-modelling approaches with different levels of data availability.

The meso-scale traffic model used in Barcelona case study had very detailed traffic information across the city but the example does not represent the temporal aspect of a flood event. The impact assessment therefore only considers the effects during a flood event and not the recovery period after the event. In addition, within the Barcelona traffic model the criteria for determining hazard classification did not consider the length of the portion of the flooded road, which was an additional restriction applied to the Bristol case study, such that there could be an overestimate of flooded roads. Further assessment could investigate the reduction in perceived flood hazards with the length restriction included.

In contrast to Barcelona's traffic model, the Bristol model lacked both real traffic-count data and a pre-existing traffic model. Due to these challenges, the traffic model was built from the ground up using freely available Open Source software and data and deriving traffic flows from land-use classifications. As highlighted in Section 2.1.1 there were limitations with this approach when dealing with large traffic volumes and as such the simulations within the paper utilise relatively low traffic counts. For future assessment, a more detailed analysis of the performance of the network could be carried out with the aim of improving/optimising the network under standard dry weather conditions.

For further work and improvements, it would be of interest to see how the micro-scale model performs with real traffic-count data to determine both the volume of traffic over time and the routes/journeys taken by vehicles within the network and determine loss estimates and recovery times under flooded conditions accordingly. Moreover, where such data available, it would be of further interest to evaluate the effectiveness of land-use data for determining route distribution in comparison to real traffic data.

Additionally, in this paper, for the micro-scale simulation we have only considered the flood event occurring with a fixed duration (30 min) starting at one specific time (7 am) which is at the beginning of the rush hour scenario. The degree of disruption to traffic flows within the network would however be dependent upon both when the event occurs and for how long, therefore future work could examine the effects/sensitivity of the time of occurrence and duration parameters.

4.2. Verification of Results

The costs of disruption to traffic flows to cities is generally quite high and within the UK, and has been shown to be within a range of 3–7% of the total accumulated estimated losses from flood events [27]. The flash floods and landslides that occurred as a result of high and prolonged precipitation in Catalonia in September 2006 resulted in the Consorcio de Compensaci´on de Seguros (CCS), the national insurance company paying out €55.9 million and resulted in bringing traffic to a standstill in Barcelona due to jams [28]. In the region of Co. Galway Ireland, the 2015/2016 floods were thought to have losses of €3.8 million of losses through traffic disruption [29]. The calculated losses for these events however are not limited solely to disruption of traffic as a result of standing water but also consider traffic light failures as in the case of Barcelona 2011 [5] and also the road closures due to potential risk to like from flooding. For example, the summer floods of 2007 resulted in the closure of the M1 in the UK for 40 h between junctions 31 and 41 due to the risk of a dam breach and the cost of this disruption alone was estimated to be £2.3 million [6].

The work outlined in this paper shows the potential of combining climate change data with flood mapping and traffic models as a means of assessing the possible implications change may have. For the Barcelona case study, the estimated losses with respect to the measured return periods seem to portray values within the orders of magnitude of similar climate events (as shown in Co. Galway flood event that was a 1 in 100-year event). As the data used for traffic model in the Bristol case study was assumed based on NRD data and with low traffic counts the estimated losses serve more as a benchmark/guide relative to the severity of input flood events to highlight the potential implications of climate change on traffic.

The model, however, highlights the impact of flooding is not limited to the period of time where there is standing water upon the roads surface but also post flood event as the network takes time to recover.

5. Conclusions

Two case studies each with different traffic modelling approaches were presented within this paper. Both cases have differing qualities and sources of input data where one has been derived from traffic counts/surveys within the city and the other approximated from land-use classifications. The two approaches both demonstrate the feasibility of loosely coupling traffic models to flood mapping as a basis of assessing the potential impacts to traffic flows within the city. The Barcelona case study illustrates how changing parameters within the model input data can serve to approximate the effects of flooding within the model. The Bristol case study shows that even with limited data, we can begin to create a traffic model for basic impact assessment that can be built upon within the future. In addition, the micro-scale approach used within the Bristol case study shows the effect of flooding is not solely limited to the duration of the flood and that the impact assessment needs to consider the recovery time of the network.

Author Contributions: The main author B.E. worked on each component of the paper including the conceptualization, methodology, software, modelling, analysis and validation, and original draft. A.G.G. facilitated with the modelling and validation with respect to the Barcelona case study and the review and editing of the paper. S.D., A.S.C., J.W., and J.S. were each involved with the validation processes of the research and the review and editing of the paper. All authors have read and agreed to the published version of the manuscript.

Funding: This research was funded by the European Union's Horizon 2020 Research and Innovation Programme, RESCCUE project grant number 700174.

Acknowledgments: The author would like to thank Katya Pyatkova for her support with the development of the micro-scale traffic model for the city of Bristol.

Conflicts of Interest: The authors declare no conflict of interest.

References

1. Havaei-Ahary, B. Road Traffic Estimates: Great Britain 2018. 2019. Available online: https://assets.publishing.service.gov.uk/government/uploads/system/uploads/attachment_data/file/808555/road-traffic-estimates-in-great-britain-2018.pdf (accessed on 11 March 2020).
2. Arrighi, C.; Pregnolato, M.; Dawson, R.J.; Castelli, F. Preparedness against mobility disruption by floods. *Sci. Total Environ.* **2019**, *654*, 1010–1022. [CrossRef] [PubMed]
3. Green, D.; Yu, D.; Pattison, I.; Wilby, R.; Bosher, L.; Patel, R.; Thompson, P.; Trowell, K.; Draycon, J.; Halse, M.; et al. City-Scale Accessibility of Emergency Responders Operating During Flood Events. 2016. Available online: https://repository.lboro.ac.uk/articles/City-scale_accessibility_of_emergency_responders_operating_during_flood_events/9483023 (accessed on 11 March 2020).
4. Coles, D.; Yu, D.; Wilby, R.L.; Green, D.; Herring, Z. Beyond 'flood hotspots': Modelling emergency service accessibility during flooding in York, UK. *J. Hydrol.* **2017**, *546*, 419–436. [CrossRef]
5. Travieso, J. Una Gran Tormenta Derriba 50 Árboles en Barcelona y Causa Graves Inundaciones. 2011. Available online: https://www.20minutos.es/noticia/1124029/0/inundaciones/barcelona/emergencias/ (accessed on 11 March 2020).
6. Chatterton, J.; Viavattene, C.; Morris, J.; Edmund Penning-Rowsell, S.T. The Costs of the Summer 2007 Floods in England. 2010. Available online: http://www.publications.parliament.uk/pa/cm201012/cmselect/cmtran/794/794.pdf (accessed on 11 March 2020).

7. Department for Transport. Government Response to the Transport Resilience Review. 2014. Available online: https://assets.publishing.service.gov.uk/government/uploads/system/uploads/attachment_data/file/380211/cm-8968-print.pdf (accessed on 11 March 2020).

8. Highways England. *Climate Adaptation Risk Assessment Progress Update—2016*; Highways England: London, UK, 2016.

9. Murphy, J.M.; Sexton, D.M.H.; Jenkins, G.J.; Boorman, P.M.; Booth, B.B.B.; Brown, C.C.; Clark, R.T.; Collins, M.; Harris, G.R.; Kendon, E.J.; et al. *UK Climate Projections Science Report: Climate Change Projections*; Met Office Hadley Centre: Exeter, UK, 2009.

10. Martínez-Gomariz, E.; Gómez, M.; Russo, B.; Djordjević, S. Stability criteria for flooded vehicles: A state-of-the-art review. *J. Flood Risk Manag.* **2018**, *11*, S817–S826. [CrossRef]

11. Martínez-Gomariz, E.; Gómez, M.; Russo, B.; Djordjević, S. A new experiments-based methodology to define the stability threshold for any vehicle exposed to flooding. *Urban Water J.* **2017**, *14*, 930–939. [CrossRef]

12. Pyatkova, K.; Chen, A.S.; Butler, D.; Vojinović, Z.; Djordjević, S. Assessing the knock-on effects of flooding on road transportation. *J. Environ. Manag.* **2019**, *244*, 48–60. [CrossRef] [PubMed]

13. Pyatkova, K.; Chen, A.S.; Djordjević, S.; Butler, D.; Vojinović, Z.; Abebe, Y.A.; Hammond, M. Flood impacts on road transportation using microscopic traffic modelling techniques. In *Simulating Urban Traffic Scenarios*; Behrisch, M., Weber, M., Eds.; Springer International Publishing: Cham, Switzerland, 2019; pp. 115–126.

14. Pregnolato, M.; Ford, A.; Wilkinson, S.M.; Dawson, R.J. The impact of flooding on road transport: A depth-disruption function. *Transp. Res. Part D Transp. Environ.* **2017**, *55*, 67–81. [CrossRef]

15. Sohn, J. Evaluating the significance of highway network links under the flood damage: An accessibility approach. *Transp. Res. Part A Policy Pract.* **2006**, *40*, 491–506. [CrossRef]

16. Pyatkova, K. Flood Impacts on Road Transportation. Ph.D. Dissertation, University of Exeter, Exeter, UK, 2018. Available online: https://ore.exeter.ac.uk/repository/handle/10871/37346 (accessed on 11 March 2020).

17. Djordjević, S.; Prodanović, D.; Maksimović, Č. An approach to simulation of dual drainage. *Water Sci. Technol.* **1999**, *39*, 95–103. [CrossRef]

18. Innovyze. *User Manual References, InfoWorks ICM (Integrated Catchment Modeling) v.3.5*; Innovyze: Newbury, UK, 2013.

19. TransCAD Transportation Planning Software. Available online: https://www.caliper.com/tcovu.htm (accessed on 20 December 2019).

20. Lopez, P.A.; Behrisch, M.; Bieker-Walz, L.; Erdmann, J.; Flötteröd, Y.; Hilbrich, R.; Lücken, L.; Rummel, J.; Wagner, P.; WieBner, E. Microscopic traffic simulation using SUMO. In Proceedings of the 2018 21st International Conference on Intelligent Transportation Systems (ITSC), Maui, HI, USA, 4–7 November 2018; pp. 2575–2582. [CrossRef]

21. OpenStreetMap. Available online: https://www.openstreetmap.org/ (accessed on 21 September 2019).

22. Find Open Data. Available online: https://data.gov.uk/ (accessed on 21 December 2019).

23. Flötteröd, Y.-P.; Behrisch, M. Improving SUMO's signal control programs by introducing route information. In Proceedings of the SUMO 2018—Simulating Autonomous and Intermodal Transport Systems, Berlin, Germany, 14–16 May 2018; Volume 2, pp. 162–172. [CrossRef]

24. Penning-Rowsell, E.; Viavattene, C.; Pardoe, J.; Chatterton, J.; Parker, D.; Morris, J. *The Benefits of Flood and Coastal Risk Management: A Handbook of Assessment Techniques*; Flood Hazard Research Centre, Middlesex University: London, UK, 2010.

25. Bristol City Council "Population of Bristol". Available online: https://www.bristol.gov.uk/statistics-census-information/the-population-of-bristol (accessed on 31 January 2020).

26. Bristol Is Open "How Do Bristolians Go to Work". Available online: https://www.bristolisopen.com/how-does-bristol-go-to-work/ (accessed on 31 January 2020).

27. Environment Agency. Estimating the Economic Costs of the 2015 to 2016 Winter Floods. 2018. Available online: https://assets.publishing.service.gov.uk/government/uploads/system/uploads/attachment_data/file/672087/Estimating_the_economic_costs_of_the_winter_floods_2015_to_2016.pdf (accessed on 11 March 2020).

28. Llasat, M.; López, L.; Barnolas, M.; Llasat-Botija, M. Flash-floods in Catalonia: The social perception in a context of changing vulnerability. *Adv. Geosci.* **2008**, *17*. [CrossRef]

29. McDermot, T.; Kilgarriff, P.; Vega, A.; O'donoghue, C.; Morrisey, M. The indirect economic costs of flooding: Evidence from transport disruptions during Storm Desmond. In Proceedings of the Irish Economic Association Annual Conference 2017, Dublin, Ireland, 4–5 May 2017; Available online: https://iea2017.exordo.com/files/papers/120/initial_draft/commuting-discruption-floods-v5-IEA-submission.pdf (accessed on 11 March 2020).

Article

Flood Depth-Damage Curves for Spanish Urban Areas

Eduardo Martínez-Gomariz [1,2,*], Edwar Forero-Ortiz [1,2], María Guerrero-Hidalga [1], Salvador Castán [3] and Manuel Gómez [1,2]

[1] Cetaqua, Water Technology Centre, Carretera d'Esplugues, 75, 08940 Cornellà de Llobregat, Spain; eaforero@cetaqua.com (E.F.-O.); maria.guerrero@cetaqua.com (M.G.-H.); manuel.gomez@upc.edu (M.G.)

[2] Flumen Research Institute, Universitat Politècnica de Catalunya, Jordi Girona 1-3, 08034 Barcelona, Spain

[3] Agencia Pericial (AGPERICIAL). Calle de Vista Alegre, 6 Bjs, 08940 Cornellà de Llobregat, Spain; castan@agpericial.es

* Correspondence: eduardo.martinez@cetaqua.com; Tel.: +34-933-124-899

Received: 22 February 2020; Accepted: 25 March 2020; Published: 27 March 2020

Abstract: Depth-damage curves, also known as vulnerability curves, are an essential element of many flood damage models. A relevant characteristic of these curves is their applicability limitations in space and time. The reader will find firstly in this paper a review of different damage models and depth-damage curve developments in the world, particularly in Spain. In the framework of the EU-funded RESCCUE project, site-specific depth-damage curves for 14 types of property uses have been developed for Barcelona. An expert flood surveyor's opinion was essential, as the occasional lack of data was made up for by his expertise. In addition, given the lack of national standardization regarding the applicability of depth-damage curves for flood damage assessments in Spanish urban areas, regional adjustment indices have been derived for transferring the Barcelona curves to other municipalities. Temporal adjustment indices have been performed in order to modify the depth-damage curves for the damage estimation of future flood events, too. This study attempts to provide nationwide applicability in flood damage reduction studies.

Keywords: depth-damage curves; urban floods; properties; claims; flood expert surveyor

---

1. Introduction

According to the European Environment Agency (EAA) [1], the total reported economic losses in Europe caused by weather and climate-related extremes over the period 1980–2017 amounted to approximately EUR 453 billion; the losses in Spain amounted to EUR 37 billion. In the words of the Spanish Insurance Compensation Consortium (CCS, for its acronym in Spanish), it is estimated that around 50% of the damage was covered by insurance in Spain. For example, the total losses from the Lorca earthquake in 2011 were estimated at EUR 1 billion, EUR 0.5 billion of which was insured and thus compensated [2]. Nowadays, economic losses from flood events at the urban level are increasingly relevant, in line with socioeconomic changes such as population growth and the expansion of infrastructure density in cities around the world [3]. Floods are the most damaging natural hazard in Europe, with around two-thirds of the total damage costs. Moreover, rising temperatures are expected to intensify the hydrological cycle, thus leading to more frequent and intense floods in many regions, together with a corresponding increase in economic losses. Nevertheless, it has to be noted that increases in costs from flooding in recent decades can be partly attributed to more people living in flood-prone areas [4].

The types of damage caused by floods are numerous and can be classified as tangible and intangible; these, in turn, can be categorized as direct or indirect [5]. Traditionally, economic flood damage assessment (i.e., direct damage) concerning flood impacts has been studied in more depth.

Particularly in urban areas, the focus has been on damage caused to flooded properties. Thus, a variety of methodologies, which still need more development, have led to important advancements.

Gilbert F. White (1945) [6] was a pioneer in considering the damage to properties. Among other aspects, White [6] defined in greater detail the types of losses in urban areas when a flood occurs, such as those related to properties and shops. His study addressed losses that can occur in residential areas, such as to the foundations and structure of dwellings and other buildings, garages, and vehicles. Also, the loss of property rental income (i.e., indirect damage) was considered. A relevant statement in White's [6] work was that water depth and velocity variables established the degree of severity of damage to the foundation and structure of dwellings. Water depth was stated to be the most limiting factor for such losses. Although White [6] did not distinguish directly between direct and indirect damage, both categories were addressed in his study.

In Spain, the Directorate General for Civil Defense and Emergencies, and the Spanish Insurance Compensation Consortium (Consorcio de Compensación de Seguros, CCS) have reported that flooding has caused the death of 312 people over the last 20 years, and economic damage amounting to EUR 800 million per annum [7]. In this context, European Directive 2007/60/CE on the assessment and management of flood risks [8] was published, and enacted by the Royal Decree 903/2010 on flood risk evaluation and management [9] in the Spanish legislation. It requires the Member States to develop, adopt, and implement flood risk management plans. These plans encompass a number of measures that involve land management and urban planning, civil protection, insurance, early warning, and improving the condition of rivers and coastal areas. One of the measures included in these plans was the development of guidelines to reduce the vulnerability of properties exposed to floods [7]. The main aim of these guidelines was to improve knowledge of flood consequences and foster citizens' commitment to risk reduction, focusing on the vulnerability of people and assets and enhancing the resilience of high risk properties.

Although measures to increase buildings' resilience have been proposed, these plans are focused on riverine floods, which indeed involve important risks that must be dealt with, but only for those urban areas located in flood-prone areas. However, sewer flooding should not be underestimated since all cities are prone to this type of flooding once a drainage system exceeds its design capacity, regardless of the distance to rivers. This type of flood is also expected to become more frequent due to the effects of climate change [10], increasing risk, damage, and disruptions to citizens.

In the framework of the RESCCUE project, tailored depth-damage curves for Barcelona have been developed. These curves encompass 14 different types of properties that are usually found in highly urbanized areas. There is no standardization at a national level regarding the employment of depth-damage curves for flood damage assessment. Therefore, this study attempts to bridge the gap in the inability to compare flood damage reduction studies from different Spanish regions. Ultimately, it expects to provide a tool to carry out homogeneously nationwide flood damage assessments. To do so, regional adjustment indices have been derived for transferring the Barcelona curves to other Spanish municipalities. Moreover, temporal adjustment indices have been performed to modify the depth-damage curves for the damage estimation of future flood events.

This paper offers in Section 2 a review of a variety of flood damage models and depth-damage curves that can be found within the literature, grouped according to their geographical application: a) worldwide or b) Spain-specific. Section 3 presents the particular context of pluvial floods in Barcelona, the role of the Spanish public insurance company (CCS), and the methodology applied in this study. The data used, its analysis and the processes to create semi-empirical depth-damage curves tailored to Barcelona are described. The procedure to transfer them in space and time is presented in this section. In Section 4, the relative depth-damage curves are presented together with their monetization for Barcelona city. Moreover, depth-damage curves for the most damaged municipalities in Spain due to flooding (pluvial and fluvial) for the 2020 reference year are presented. Finally, Section 5 recaps the main messages of this study, and the usefulness and adequacy of the proposed depth-damage curves for Spanish cities are argued.

2. Literature Review and Analysis

Even though there is currently no universally trusted system for assessing flood damage in urban areas, most damage models rely on depth-damage curves (also known as stage-damage functions) for simplicity [11]. In order to apply damage models to assess the economic impact of flooding over urban areas, the required floodwater depths across the inundated area are usually obtained from 2D simulations [12]. In this section, some of the most relevant damage models worldwide are presented together with a variety of depth-damage curves developed for regions across the world and those specifically developed for Spanish regions.

2.1. Flood Damage Models

Different approaches have been adopted worldwide in order to develop models to assess damage due to flooding. However, all share a common purpose: evaluating the cost-effectiveness of projects designed to alleviate flood impacts.

In the USA, two well-known models are currently being used to assess the damage caused by floods and other natural hazards. The first is HAZUS-MH [13], which is a multihazard estimation model developed by the U.S. Federal Emergency Management Agency (FEMA) that assesses the impacts of earthquakes, wind, and floods. It was mainly developed by the U.S. National Flood Insurance Program (NFIP), because the insurance industry plays a key role when it comes to natural hazards. The outputs of the damage module are area-weighted estimates of damage as a percentage of replacement cost, at the Census Block or for a given building. Since the U.S. NFIP pays claims based on depreciated value, the model considers depreciation as opposed to cost of repair as the general measure of economic loss. The damage assessment model includes a library of more than 900 damage curves estimating damage to various types of buildings and infrastructure. Some drawbacks of this model are the complexity of the data input process and the U.S. regional-based stage-damage curves. The second well-known model is HEC-FDA [14], developed by the U.S. Army Corps of Engineers (USACE), which is a freely downloadable software provided together with the rest of the Hydrologic Engineering Center (HEC) resources, and includes extensive documentation. Among its several features, it stores hydrologic and economic data necessary for analysis, provides tools to visualize data and results, and computes expected annual damage. Generic depth-damage relationships are provided to be utilized for a flood damage study conducted in the USA, in the absence of regionally developed relationships.

A comprehensive study conducted by Jongman et al. [15] compares seven different damage models developed for a variety of regions across Europe and the United States: FLEMO (Germany), Damage Scanner (Netherlands), the Rhine Atlas (Rhine Basin), the Flemish Model (Belgium), Multi-Coloured Manual (MCM) (United Kingdom), HAZUS-MH (United States), and the JRC (Germany, European Commission/HKV). The fact that five out of the seven models are based on aggregated land use data (e.g. CORINE) rather than individual objects (HAZUS-HM and MCM) indicates that the scale of work is an essential matter when either selecting or developing a damage model. Moreover, it should be noted that only two out of the seven models are based on individual objects, which indicates the complexity of developing such detailed damage models. While the object-based models can control for varying building density in areas with same CORINE land use, the area-based models can be applied for rapid calculations over larger areas. However, object-based models such as HAZUS-MH and MCM use a large number of object types and corresponding flood damage characteristics [16]. FLEMO, HAZUS-MH, and the Rhine Atlas models are empirically based and could be more accurate when applied to similar case studies. The others are mainly synthetic with the intrinsic issue of their unreliable application to another region or country. An essential improvement in these recent damage models is their GIS-based characteristic; however, the complexity of the data input process, together with the inherent regional (USA) dependency of depth-damage curves, may be considered as important drawbacks.

As Jongman et al. [15] noted, the use of depth-damage curves involves great uncertainty, which makes the models very sensitive. The need to adjust asset values to the regional economic situation and property characteristics when using aggregated land use data was also highlighted by Jongman et al. [15]. In addition, the actual damage to a property is not only due to floodwater depth, but also to factors such as the time of the year the flood occurs, flood duration, water velocity, suspended debris, or warning time. Therefore, there is an intrinsic uncertainty to depth-damage approaches and a known influence of these other factors on the extent and severity of flood damage to buildings and their contents. However, it is a general practice to accept the water level as a fundamental criterion for estimating the damage caused by these events. Lately, other factors beyond the water level have been incorporated into so-called multiparameter damage models; nonetheless, such models require more complex and extensive datasets [17].

Next, we will present descriptions of a variety of depth-damage curves developed for different parts of the world, and a more in-depth analysis of those performed for Spanish regions.

2.2. Depth-Damage Curves in the World

Within damage models, depth-damage curves, also known as vulnerability curves, are used to represent the vulnerability of elements at risk. These can be found in either a relative or an absolute form, by considering percentages of the total property value or damage expressed in monetary terms, respectively. While relative curves are more transferable in space and time, since they do not depend on the market value of assets, absolute curves need a periodic recalibration to incorporate depreciation and inflation. In addition, depth-damage curves can be classified by their development process, namely analytical, empirical, and synthetic, and a combination of these could also be possible. The first group (analytical) are laboratory-based curves, where the effect of flood variables, such as depth, velocity, or duration, is assessed through monitoring. Empirical curves are developed through the collection of properties' damage data by means of survey campaigns. Synthetic functions are derived from the study of a theoretical standard property, assuming that all properties within the studied area are similar. This last type of curve is proposed when no actual data are available. Within the literature, other approaches have been found in terms of classification of the curves, such as by land use, building structure type, building contents or inventory, social status (income level), duration of flood, and warning time. Moreover, curves for buildings (i.e., structures) and contents are usually provided separately.

An important collection of damage curves is provided by the HAZUS-MH model, which offers to users the Federal Insurance Administration's (FIA) "credibility weighted" depth-damage curves and selected curves developed by various districts of the U.S. Army Corps of Engineers (USACE). The latter group is included in the HEC-FDA model and described in the Economic Guidance Memorandum (EGM) 04-01 developed by the USACE [18], with the purpose of providing guidance for the use of generic depth-damage curves in their flood damage reduction studies. These are developed based on actual flood losses in various parts of the United States (from 1996 to 2001) in the framework of the Flood Damage Data Collection Program, aiming at providing Corps districts' offices with standardized relationships for estimating flood damage. These cover one-story homes, two- or more story homes, and split-level homes, all of them with or without basements.

More recently, Huizinga et al. [19], as one of the JRC technical reports, carried out work to provide a methodology and a database of depth-damage functions for a variety of assets and land use classes (i.e., residential, commercial, industrial, transport, infrastructure, and agriculture). This work is based on an extensive literature survey to normalize damage curves for each continent at a national scale. The purpose of this work was trying to bridge the gap on the inability to compare flood damage assessments from different countries. The variety of depth-damage curves developments across countries is displayed in Table 1, at a national level, and Table 2 at a regional level. These tables summarize the main characteristics and sources of depth-damage curves found in the literature.

Table 1. Relative depth-damage curves at a national level.

Region	Year	Property Types	Classification (Development)	Source
Australia	2016	Residential	Empirical	[20–22]
	2017	Residential, Industrial, Roads	Synthetic	[23]
Belgium	2006	Residential, Industrial, Vehicles, Recreation, Agriculture, Railways, Wind turbines	Analytical-Synthetic	[24]
	2018	Residential, Commerce, Industrial, Roads, Agriculture	Synthetic	[17]
Canada	2019	Residential	Synthetic	[25]
Czech Republic	2018	Residential, Commerce, Industrial, Roads, Agriculture	Synthetic	[17]
	2013	Residential	Empirical	[26]
Germany	2017	Residential and Commerce	Empirical	[23]
	2018	Residential, Commerce, Industrial, Roads, Agriculture	Synthetic	[17]
Italy	2017	Residential	Empirical	[27]
New Zealand	2010	Residential, Commerce, Industrial	Empirical-Synthetic	[28]
	2005	Residential, Industrial, Vehicles, Recreation, Agriculture, Railways	Synthetic	[29]
The Netherlands	2007	Residential, Industrial, Vehicles, Recreation, Agriculture, Nature	Analytical - Synthetic	[30]
	2018	Residential, Commerce, Industrial, Roads, Agriculture	Synthetic	[17]
United Kingdom	2010 – Updated on 2013	Residential, Non-residential	Empirical-Synthetic	[31]
	2018	Residential, Commerce, Industrial, Roads, Agriculture	Synthetic	[17]
United States of America	2019	Residential, Essential Facilities, Transportation systems, Commerce, Industrial, Vehicles, Agriculture	Analytical-Empirical	[32–34]
Switzerland	2018	Residential, Commerce, Industrial, Roads, Agriculture	Synthetic	[17]
A variety of countries from Europe, Africa, Asia, and South America	2017	Residential, Commerce, Industrial, Roads, Agriculture	Analytical-Empirical	[35]

Table 2. Relative depth-damage curves at a regional level.

Region	Year	Property Types	Classification (Development)	Source
Lombardy (Italy)	2009	Residential	Analytical	[36]
Palermo (Italy)	2010, 2014	Residential	Analytical	[37,38]
Chenab River (Pakistan)	2014	Residential, Commerce, Industrial, Roads, Agriculture, Nature	Analytical	[39]
Jakarta (Indonesia) Manila (Philippines) Ho Chi Minh (Vietnam) Bangkok (Thailand)	2015	Residential, Commerce, Industrial	Empirical-Synthetic Synthetic Empirical-Synthetic Empirical-Synthetic	[40]

For comparison purposes, a variety of European residential relative depth-damage curves [19] have been represented together in the graph in Figure 1.

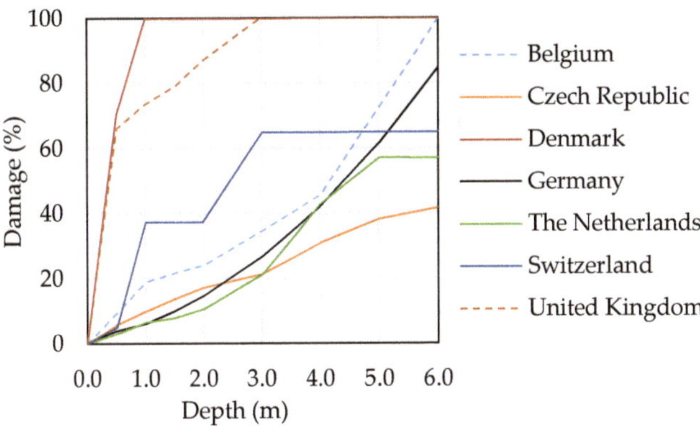

Figure 1. Residential depth-damage curves for European countries proposed by Huizinga et al. [19].

As can be observed, there is no correlation between the economic development and the estimated damage percentage of the countries. For instance, the curves developed for the Netherlands present the lowest estimated damage percentage, while it ranks third on GDP per capita of the countries listed [41]. This could be attributed to the level of investment in adaptation measures against flood events of in Netherlands [42], causing lower potential flood event damage compared to other countries. Such dispersion of damage among curves in different European regions was also highlighted by Velasco et al. [5], where seven different types of curves for residential and commercial uses were compared.

2.3. Depth-Damage Curves in Spain

Only a few recent depth-damage curves' developments have been found for Spain, almost all of them addressing specific regions and only one providing national coverage. In 2013, in the framework of a Flood Defense Master Plan for the Spanish region of Marina Baja (Alicante), relative depth-damage curves for the different CORINE land uses in the region studied [43] were developed. The curves were validated based on actual damage data (i.e., claims paid) provided by the Spanish Insurance Compensation Company, CCS hereafter, and the City Councils of the region studied. Only one month later, in July 2013, the Ministry of Agriculture, Fisheries and Food published the report "Propuesta de mínimos para la metodología de realización de los mapas de riesgo de inundación" [44] (proposal for a common methodology of flood risk maps development). The purpose of this document was to offer a common framework to implement the Directive 2007/60/EC on the assessment and management of flood risks in Spain, and a nationwide relative depth-damage curve was proposed. It was based on the flood damage caused in the Spanish Ebro basin. Only four water levels were considered, assuming total damage (i.e., 100%) when the water depth exceeds 2 m, and 20% for depths lower than 30 cm. The single curve did not distinguish among land uses, although a table to monetize ($€/m^2$) it according to different land uses was provided.

In 2015, an updated version of the former Flood Risk Prevention Territorial Action Plan for the Spanish Autonomous Community of Valencia (PATRICOVA) [45] was published. Again, a single relative depth-damage curve was provided, not distinguishing by type of land uses either. It stated that, regardless of the type of land use flooded, below 80 cm of water depth low damage were expected; however, once the water level exceeds this value, the damage increases rapidly up to a depth of 1.2 m, when the increase levels off. Finally, the River Basin Management Plan produced by the Eastern Cantabrian Spanish River Basin District [46], part of the requirements of the Directive 2007/60/EC, has been reviewed. A single relative depth-damage curve is found in this Plan, which is the result of averaging others developed for other countries, such as the USA (FEMA) and the United Kingdom (Flood Hazard Research Centre). The monetization of these curves is conducted by providing the maximum estimated value per land use. Unlike previous developments, in this case building and contents are differentiated. It considers a percentage from the building damage to constructing the contents' depth-damage curve. All reviewed depth-damage curves for Spanish regions have been jointly graphed and are presented in Figure 2.

Ritter et al. [47] conducted a flood damage assessment for the Spanish municipality of Agramunt, in which the depth-damage curves for Spain were selected from the database provided by Huizinga et al. [19]. It was concluded that the total computed damage for an actual riverine flood event was clearly overestimated after being compared with the claims paid by the Spanish Insurance Compensation Consortium (CCS). This may indicate that the depth-damage curves employed for the study could overestimate individual assets.

Although there might be other curves developed regionally in Spain, the review conducted has been considered enough to identify the great variety of approaches taken. As observed in Figure 2, where residential and general depth-damage curves are represented together, the damage for the same water depth varies significantly. For instance, a water depth of 2 m would suggest total damage if the nationwide curve is applied, and only 52% of total damage for the ones developed for the region of Marina Baja. Low water depths would also provide a very different level of damage depending on the curve selection. While the nationwide curve starts from 20% damage, the Marina Baja region curve does not rise above zero until a 10 cm water depth. This is assumed to be the average height of pavement curbs. Moreover, some of them aggregate building and contents (i.e., furniture and household furnishings) damage, while others consider them separately.

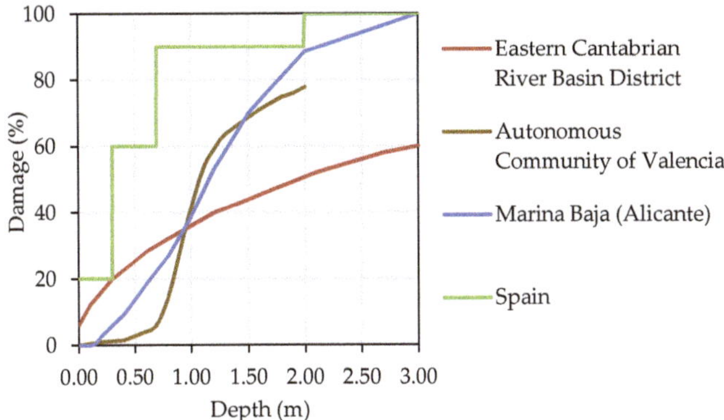

Figure 2. Depth-damage curves for Spanish regions.

3. Materials and Methods

3.1. Context of Pluvial Floods in Barcelona

From 1996 to 2018, pluvial floods in Barcelona alone have required more than EUR 34 million in compensation for industries, offices, dwellings, vehicles, and civil works, according to the Spanish insurance company CCS (Figure 3). The CCS is a state-owned enterprise attached to the Ministry of Economy Affairs and Digital Transformation that performs several complementary functions in the Spanish Insurance Industry, enhancing its stability and protecting the insured. In 2018, four heavy rainfalls hit Barcelona, which caused extraordinary damage (Figures 3 and 4). In 1999 and 2002 alone, the total amount of compensatory damage exceeded that of 2018. It has to be noted, however, that since 2002 several improvements in the drainage network have been carried out. The analysis of these insurance payouts according to the CCS classification (Figure 3b) shows that in 2018 almost 75% of the total payouts were due to damage to commercial buildings, warehouses, and other types. This pattern is common in the last 22 years, representing more than 50% of the total payouts per event, which indicates the vulnerability to pluvial floods of commercial properties in urban areas.

These figures clearly indicate the relevance of the ever more frequent pluvial floods, highlighting the need to provide tools to estimate the damage that future flood events may cause in urban areas. The existing methodologies to assess flood damage in urban areas are usually based on the use of so-called depth-damage curves. As described in previous sections, these are merely the mathematical relationship between the floodwater depth reached in the property and the economic damage caused [12,25,48–50]. When it comes to urban pluvial floods, very detailed depth-damage curves are required to provide for the heterogeneity of building uses within an urban area. Therefore, the scale of the study concerning the damage assessment is essential when either selecting or developing depth-damage curves.

3.2. The Role of the Spanish Insurance Company, Insurance Compensation Consortium (Consorcio de Compensación de Seguros, CCS)

The high loss potential from natural hazards and the need to make more generalized insurance cover viable has led many countries to involve the state in specific cover schemes, collaborating to varying degrees with the private market. The CCS is a government institution attached to the current Ministry of Economy Affairs and Digital Transformation. This institution has its legal personality and full capacity to act, and is subject to private law. This means that, although it is a government institution, it is subject to the rules contained in the legislation establishing the legal regulation and supervision of private insurance.

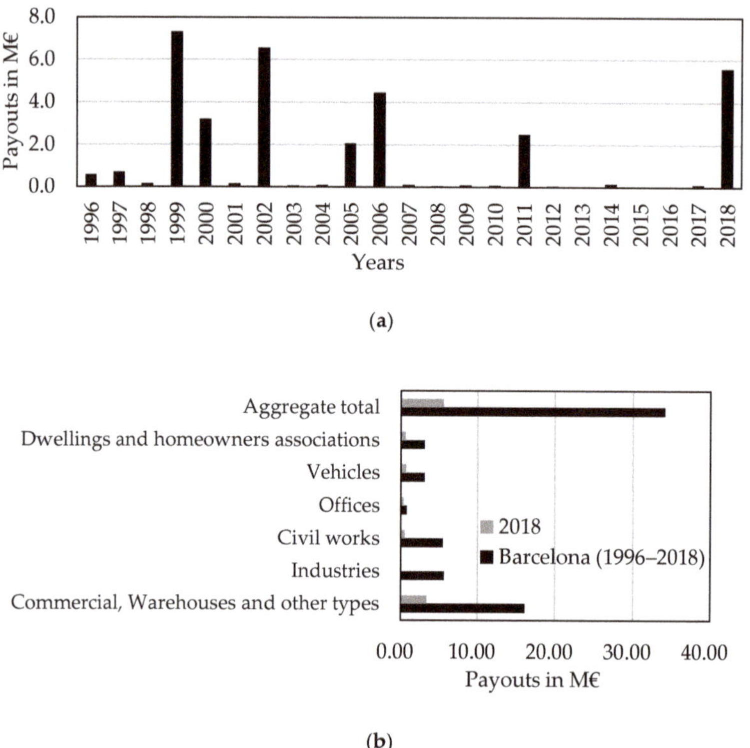

Figure 3. CCS's payouts due to damages caused by pluvial floods in Barcelona: (**a**) historical total annual amounts (1996 to 2018); (**b**) total historic (22 years) amounts grouped into diverse types of properties, according to CCS classification.

(**a**)   (**b**)

Figure 4. Consequences of pluvial floods that occurred in Barcelona on (**a**) 9 October 2018, and (**b**) 15 November 2018. Sources: a) https://www.elperiodico.com, and b) https://www.telecinco.es.

The CCS covers so-called extraordinary risks, which include both natural hazards and those of a political or social nature, not expressly assumed by the insurance company issuing the standard policy. Its coverage encompasses losses derived from direct damage caused to people or to property, as well as business interruption due to the alteration of normal outcomes of production or business processes concerning such activity. The coverage of extraordinary risks is compulsory and is necessarily linked to underwriting (an insurance policy) in certain branches of insurance related to property (vehicles, home, etc.) and persons (life, accidents, etc.). All insurance policies rates include a surcharge to the endowment of a common fund, under the principle of solidarity. Claims must be lodged within seven days following the damage event to the private insurer when it expressly covers the extraordinary event that caused the damage. Otherwise, the CCS will be responsible for the payouts, and could receive the claim directly from the policyholder or through the private insurer. A policyholder will be entitled to compensation after the damage assessment of an expert surveyor designated by the CCS.

When an extraordinary risk occurs, such as the pluvial flood events that hit Barcelona on 9 October and 15 November 2018 (Figure 4), the CCS sends one or more expert surveyors to provide an estimation of the extent of the damage. According to discussions with the CCS, these estimations are extremely close to the final payouts, which denotes the remarkable know-how of such experts. For these reasons, the depth-damage curves developed in this study are mainly based on the knowledge of a flood expert surveyor. In the same line, the curves developed by USACE [51] were based on the expert opinion method described in the Handbook of Forecasting Techniques (IWR Contract Report 75-7, December 1975) [52] and the Handbook of Forecasting Techniques, Part II, Description of 31 Techniques (Supplement to IWR Contract Report 75-7, August 1977) [53].

3.3. Methodology

3.3.1. Data

A comprehensive analysis of 378 records of properties damaged by floods at a national level was carried out. These records come from the actual damage assessed by a flood expert surveyor and include the CCS compensation and the floodwater depth inside the property when available. The source of this data is from a number of floods that occurred during the period 2012 to 2018 and affected Spanish municipalities of different economic levels located across the Mediterranean (pluvial floods) and Cantabrian (fluvial floods) areas. Records from pluvial floods include damage caused by medium and lower water depths (up to 50 cm for ground floors), while those from fluvial floods contain losses originating from high floodwater depths (up to 100 cm) inside the properties. Among the available information was the type of asset damaged: a) building, b) furniture and household furnishings, and c) inventory. It has to be noted that not all insurance policies cover both the structure (building) and the contents (furniture and household furnishings, and inventory). For instance, the tenant may pay only for the coverage of furniture and household furnishings, and there could be no inventory to insure. Therefore, not all 378 records provide data for the three types of assets that are "potentially insurable." In this way, the useful data can be classified into: 354 records for buildings, 242 records for furniture and household furnishings, and 98 records for the inventory. Regarding the floodwater depth inside the property, 52 different depths from 1 cm to 280 cm were available. Water depths were distributed as follows: 43% (163) up to 10 cm, 46% (147) between 10 and 50 cm, 8% (30) between 50 and 100 cm, and 3% (11) corresponding to depths higher to 100 cm. Table 3 presents a summary of the records available, grouped by type of asset and property.

3.3.2. Analysis

It is assumed in this study that buildings cannot collapse. Variables such as the erosion of the terrain over which the building is located could be responsible for a possible structural collapse, rather than the water level itself. In addition, this type of failure is very unusual, thus its consideration could undermine the curve profile for the frequent cases (i.e., no collapse). Therefore, the maximum relative damage established is limited to the percentage that represents the building components over the construction costs (i.e., floors, carpentry, electrical installation, air conditioning, plastering, cladding, painting, etc.). In order to set this maximum loss, construction price records [54] have been consulted, and for each construction stage we considered the relative damage that flooding can cause. As an example, the maximum loss resulted in 34% for dwellings, 30% for industries, 15% for car parks, and 36% for offices. On the other hand, furniture and household furnishings, such as crockery, metal shelves, or pallet trucks, are not generally ruined by flooding. Thus, a maximum relative damage value has been set for this type of asset, between 90% and 97% depending on the type of property.

Fourteen types of properties have been proposed, and the available records have been classified accordingly. For each type of property and asset (i.e., building, furniture and household furnishings, and inventory) the correlation between economic damage and water depth inside the property has been analyzed. It should be noted that economic damage refers to the actual damage assessed by the flood surveyor rather than the compensation paid by the CCS (usually a lower amount).

Table 3. Summary of available records per type of property.

Type of Property	Building			Furniture and Household Furnishings			Inventory		
	N of Records	*N* of Different Depths	Average Relative Damage	*N* of Records	*N* of Different Depths	Average Relative Damage	*N* of Records	*N* of Different Depths	Average Relative Damage
Dwelling	100	22	4.15%	97	20	20.38%			
Workshop	15	11	13.84%	15	11	27.10%	16	8	40.33%
Health	8	5	4.45%	8	5	2.75%			
Office	14	10	7.42%	5	3	39.21%			
Industry	12	10	2.74%	11	9	9.03%	12	10	17.09%
Hotel	2	2	16.24%	1	1	100.00%			
Education	14	8	3.32%	8	5	2.75%			
Sport	6	4	5.96%						
Homeowners association	44	14	0.56%						
General trade	67	23	5.07%	67	23	19.81%	52	16	26.45%
Restaurant	15	10	11.14%	14	9	18.77%	0	3	41.41%
Car park	39	11	0.38%						
Warehouse	18	11	1.60%	16	10	14.70%	18	11	23.85%
Churches and singular buildings	A single record was available for this type of property. Its corresponding depth-damage curve has been developed under the criterion of the flood expert surveyor.								
Total	354	141		242	96		98	48	

To standardize the diversity of geographical locations, economic level of construction, and type of property, the first action was to develop relative depth-damage curves by determining the ratio between economic damage and total property value. To do this, we set the value of each asset according to the availability of either the assessment of the flood surveyor or the insured amount, when a prior evaluation of the assets was not done. The asset value set divided by the total square meters of the entire building (i.e., whether flooded or not) results in the cost per square meter ($€/m^2$). In turn, the value set by the flood surveyor divided by the flooded floor area results in the damage per square meter ($€/m^2$). It has to be noted that buildings may have different numbers of upper floors, which usually results in a single flooded floor, and thus the total floor area is not flooded. The ratio between cost and damage per square meter provides the relative damage value, which has been averaged among all records from the same type of property, as indicated in Table 4 for a commercial property. Thus, the 67 records classified as general trade and corresponding to building assets are grouped into 23 different water depths inside the property.

We analyzed the correlation between relative damage and water depth for each type of property and asset. A great variety of coefficients of determination have been obtained, offering a good fit in some cases but a poor one in others. For industrial use and building assets, an accurate correlation was observed ($R^2 = 0.81$); however, in the case of general trade, a very poor value was obtained ($R^2 = 0.0022$) (Figure 5).

Table 4. Records from commercial use (building), grouped by water depth inside the property.

Depth (cm)	N of Records	Average Value (€/m^2)	Average Damage (€/m^2)	Relative Average Damage (%)
1	1	1052.63	95.00	9.02
2	8	434.33	8.56	2.25
3	11	502.28	16.33	24.85
4	2	416.94	29.36	7.04
5	6	411.15	12.98	2.28
6	2	367.71	5.33	1.45
7	1	735.46	15.45	2.10
10	8	415.81	4.38	1.22
12	1	742.41	1.14	0.15
15	3	494.94	2.07	0.42
18	2	741.07	4.65	0.46
20	8	753.99	22.61	12.32
30	1	753.41	7.73	1.03
32	1	493.75	18.11	3.67
35	1	539.77	15.64	2.90
40	3	370.84	5.84	1.60
45	1	453.72	6.61	1.46
48	2	520.17	18.92	3.70
60	1	598.09	6.89	1.15
74	1	842.39	4.95	0.59
82	1	467.29	36.78	7.87
85	1	200.07	21.28	10.63
100	1	750.00	137.63	18.35
TOTAL	67	567.75	21.66	5.07

For the scatter plot of some types of properties and assets, some outliers were identified for which low water depths caused unexpectedly high damage values. Some explanations in this regard may be given: 1) the heterogeneity of construction elements, 2) different furniture quality and costs, 3) stowage conditions, and 4) the existence of cold stores. As an example, the red dot in Figure 5 (general trade and inventory) indicates that very high relative damage occurred to the inventory (i.e., 70%) of a general trade when the property was only flooded by 1 cm.

Overall, some other inconsistencies may be discussed:

- The general trade category includes a variety of trades. They range from those that are more flood-resilient, such as outlets established in an industrial warehouse, to those more vulnerable to floods, such as fashion boutiques with parquet floors, cladding, and wood furniture.
- Damage to the inventory of chilled food trade occurs in a cascade. When cold stores are flooded with even a low floodwater depth, the damage could be total. However, in the case of a trade of construction materials, even when part of the inventory is flooded, it is possible to salvage the inventory placed on upper floors.

Particularly in terms of building assets, the linear correlations present an accurate fit in some cases but not in others, highlighting that, overall, the phenomenon is not well explained for low depths. It has been observed that for buildings from the majority of property types, the maximum relative damage is reached at 180 cm of water depth. This is the depth fixed for all buildings to reach maximum damage.

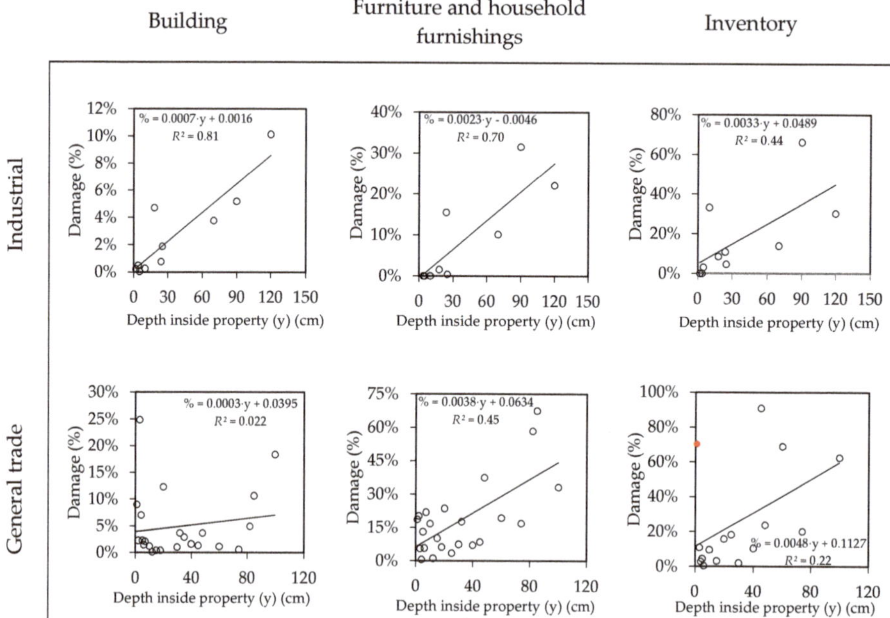

Figure 5. Relationship of water depth inside the property and relative damage to building, furniture and household furnishings, and inventory, for industrial and general trade property uses.

3.3.3. Depth-Damage Curves' Development for Barcelona

Based on the analysis of the available data and applying corrections according to the expert opinion, nationwide relative depth-damage curves are initially developed. As stated by Van Vloten [55], in situations where there are no previous damage data or when the elements at risk are not comparable, consulting expert opinions on the matter is a good choice. This involves asking their opinion on the percentage of damage they expect for each structural type and for each hazard intensity. Expert opinions are also sought in the U.S. HAZUS methodology for the assessment of the impacts of flooding, for which depth-damage functions were derived from expert opinion and historical data [56], as has been done in the present study. The report of the Gulf Engineers & Consultants (GEC) [51], developed for the USACE, explicitly highlighted the importance of insurance experts as a primary source for obtaining depth-damage relationships. Also, in the work done by Bedford et al. [57], a variety of depth-damage curves were proposed based on expert opinions and damage claims in 1993 and 1995.

To make the depth-damage curves applicable to a specific municipality (i.e., Barcelona), these must be monetized by converting relative damage into economic damage per square meter (€/m²). In order to do this, the economic level of the target city is included. We must stress the importance of monetizing the curves of each type of asset separately and aggregating them afterwards into a single curve per type of property. In doing so, the disparity of prices for each type of asset in different municipalities may be taken into account. For instance, the cost of a building could be the same between two different municipalities, while the furniture, household furnishings, and inventory prices could be significantly different.

3.3.4. Regional Transferability to Other Spanish Urban Areas

Departing from the semi-empirical depth-damage curves developed for Barcelona in the project RESCCUE, the present research goes further, proposing a methodology to transfer them to other

Spanish municipalities. This allows for the use of depth-damage curves for flood damage assessments at a national level through curves obtained under a standard methodology.

The regional transfer of Barcelona's curves to a large number of the 8131 Spanish municipalities considers demographic, economic, and geographical factors, as they substantially influence the prices of goods and services across the country [16]. Regional adjustment indices have been obtained, taking as a reference Barcelona, based on indicators that are used as proxies of the expected regional price variability for the different assets' curves (Figure 10). The original 14 types of property uses were grouped into three general sectors: commercial, industrial, and residential and others. These have been classified by type of asset in order to obtain three indicators (i.e., building, furniture and household furnishings, and inventory) (Table 5). For instance, the prices of buildings used as warehouses are assumed to vary, as commercial buildings do, but warehouses' furniture, household furnishings, and inventory are more closely related to the price variability of the industrial sector. Relevant economic or market data at the municipal level are scarce, which was a limiting factor when developing the curves. Thus, when necessary, assumptions were made regarding the price or value variability of similar structures within a municipality.

Table 5. Relationship between property uses and general sectors for assets.

Type of property use	Residential & Others		Commercial			Industrial		
	Building	Furniture & household furnishing	Building	Furniture & household furnishing	Inventory	Building	Furniture & household furnishing	Inventory
Warehouse			•				•	•
Car park	•							
Restaurant			•	•	•			
General trading			•	•	•			
Homeowners association	•							
Sport	•	•						
Education	•	•						
Hotel	•	•			•			
Industries						•	•	•
Office			•	•				
Health	•	•			•			
Workshop						•	•	•
Dwelling	•	•						
Churches & singular building	•	•						

Buildings, for the residential and others sector, represent the physical structure of the living space and the indicator selected to define their relative value per municipality is the average tax value per square meter for all properties' transactions during the reference year 2020. These were obtained from an online real estate agent (www.idealista.com). For municipalities with no data available, the lowest value of their corresponding autonomous region has been considered as a proxy of the value, as those not represented are small, low-income towns. The baseline assumption is that the differences at a municipal level of the costs of damage reconstruction are comparable to the differences in property value. Continuing with the residential and others sector, the cost of damage to furniture and household furnishings is assumed to be aligned with the average disposable income per municipality. Hence, the indicator to compute the variation in content damage curves for the residential sector among municipalities is obtained through the statistics published by three sources: 1) the National Tax Agency (www.agenciatributaria.es), 2) the regional tax agency of the Basque Country (www.eustat.eus), and 3) the statistics agency of Navarra (www.navarra.es). Data were limited by the information provided by tax agencies that display small municipalities' results in groups; thus, it was not possible to include municipalities with under 3000 inhabitants.

The residential and others sector does not consider inventory. Not all types of property uses include all three asset types. For instance, while car parks only consider building assets, offices contain the three types of assets.

Regarding the commercial and industrial sectors, the price variability among municipalities of the furniture (there are no household furnishings related to these sectors) and inventory has been explained through two indicators ($n = 2$): a) average revenues of each of the two sectors at the autonomous region scale, and b) the number of businesses per sector at a municipality level. The Sauerbeck index [58] (Equation (1)) was applied, defined as the arithmetic average of two or more reference prices to the rest of the municipalities' relative values, on the basis of Barcelona prices. This was found to be the most appropriate way to introduce two related datasets that were at different geographical scales.

$$\text{Index}_{s,mi} = \frac{1}{n} \times \frac{AV_{s,mi}}{AV_{s,m0}} + \frac{NC_{s,mi}}{NC_{s,m0}}, \tag{1}$$

where n is the number of indicators ($n = 2$), s is the sector represented (i.e., commercial or industrial), m_i denotes municipality i, and m_0 is the reference municipality (i.e., Barcelona); $AV_{s,mi}$ represents the average economic value for the sectors of the autonomous region municipality i belongs to; $NC_{s,mi}$ stands for the number of companies of sector s registered in municipality i; and $AV_{s,m0}$ and $NC_{s,m0}$ represent the same values for the reference city of Barcelona.

In the furniture and household furnishings asset category, the variable AV takes the average investments in tangible assets per autonomous region for all companies registered under commercial and services sectors on the one hand, and the industrial sector on the other hand. For the inventory, the variable AV considers the average business revenue per autonomous region and sector. The number of commercial (commercial and services registered companies) and industrial companies per municipality (NC) comes from the national statistics office (www.ine.es). The lack of economic data at a local scale was an obstacle to including more precise values, and the two datasets alone do not provide any relevant measure. However, when combined they provide the local average revenue of the businesses belonging to each of the sectors displayed.

In summary, the spatial variability of the damage costs for the furniture and household furnishings can be explained by the differences in the average investment in tangible assets per municipality. In this sense, a city where the average investment to improve their assets is higher than the reference city (i.e., Barcelona), the damage caused to its business would be higher. The average local revenues are proposed to explain the inventory costs' variability. This follows the rationale that, the higher the revenue in a municipality, the higher the inventory would be stocked in local businesses. Hence, the damage would be higher in the case of a flooding event. The spatial costs variability for buildings was established based on the property values for commercial, industrial, and residential sectors. The lack of data at a municipal level of commercial and industrial property values limited the range of action. However, the official property values ($€/m^2$) of the three sectors from the Spanish Registrar Chartered Institute (www.registradores.org) were used to obtain the value variation at the autonomous regional level, which was then applied to the municipal property values. The final regional adjustment indices (RI) are obtained as decimal fractions referring to Barcelona (i.e., the unit).

3.3.5. Temporal Transferability

Regarding the price variability over time, a method to transfer damage curves to the future has been applied to the original depth-damage curves developed for the year 2020 in Barcelona. The time horizon has been set to 2060, defined by the availability of the economic forecast. Long-term economic forecasting is a projection based on an assessment of the economic climate in individual countries and the world economy using both econometric models outputs and expert judgement [59]. Therefore, they can be characterized by uncertainty and complexity [60]. Considering this, and the scarcity of long-term projections, a temporal indicator has been developed using the OECD real GDP long-term forecast for Spain [61]. This is the most reliable source of information of its kind. Using the year 2020 as a reference, an indicator of the potential economic growth has been included up to 2060 in order to transfer present damage costs to future estimates. The final temporal adjustment indices (TI) are obtained as decimal fractions referring to 2020. Consequently, in order to obtain the depth-damage curve of a certain municipality for a specific year, the total adjustment index (TAI) (Equation (2)) will be applied to each monetized asset curve of Barcelona (Table 7 and Figure 12):

$$TAI = RI * TI,\tag{2}$$

where TAI is the total adjustment index, RI is the regional index, and TI is the temporal index.

Figure 6 presents the expected economic trend until 2060, thus the current depth-damage curves can be updated accordingly by multiplying the monetized aggregated curves (i.e., including building and contents) by the temporal index for a specific year obtained through this function.

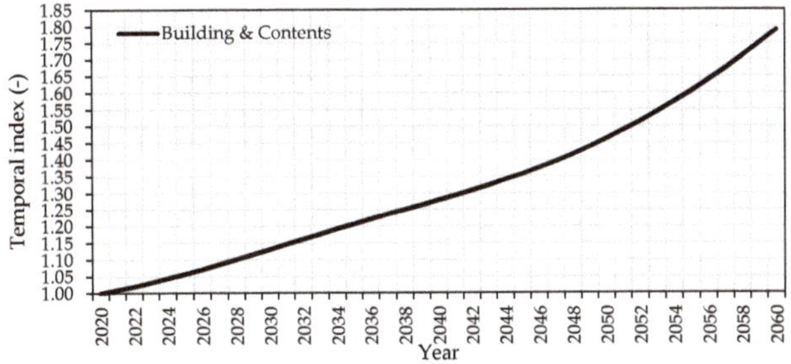

Figure 6. Temporal adjustment indices to transfer present damage costs to future estimates for buildings and contents until the year 2060. Values based on the OECD real GDP long-term forecast for Spain [61].

3.3.6. Graphical Overview of the Proposed Methodology

Figure 7 gives an overview of the research process. It summarizes the key steps in the development of nationwide depth-damage curves.

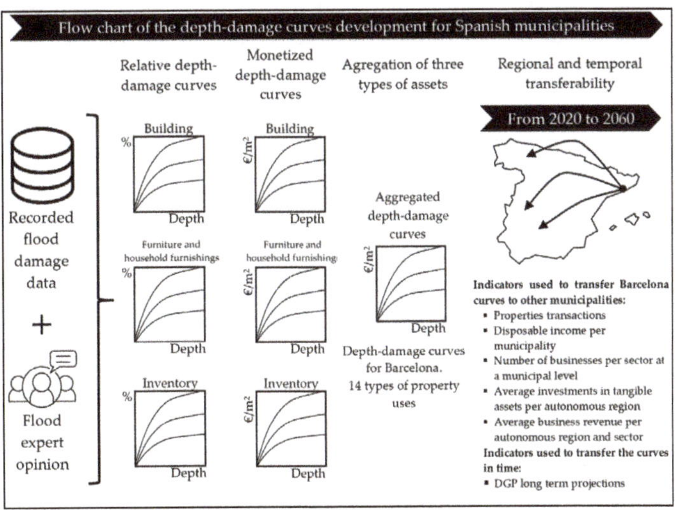

Figure 7. Flowchart of the development of depth-damage curves for Spanish municipalities.

4. Results

4.1. Relative Depth-Damage Curves Development

The relative damage corresponding to intermediate water depths (between 10 cm and 180 cm), when the coefficient of determination is acceptable, has been set as the range for the linear regression. Nevertheless, relative damage for low depths (<10 cm) has been adapted according to the opinion of the flood expert surveyor. Regarding the damage related to high water depths (>180 cm), the maximum value has been set. Therefore, only in those situations where the correlation was not accurate enough has the curve been adjusted according to expert opinion. Figure 8a showcases an example of a relative depth-damage curve proposed for the building of an industrial property. Inventory for commercial uses, though, except for those elements that require cold stores, tends to be at the same height, and accordingly its damage is also evenly distributed, as shown in Figure 8b. For the inventory, a maximum 100% relative damage has been considered, assuming that those elements that can be saved after flooding (e.g., plastic, construction materials, etc.) are not representative.

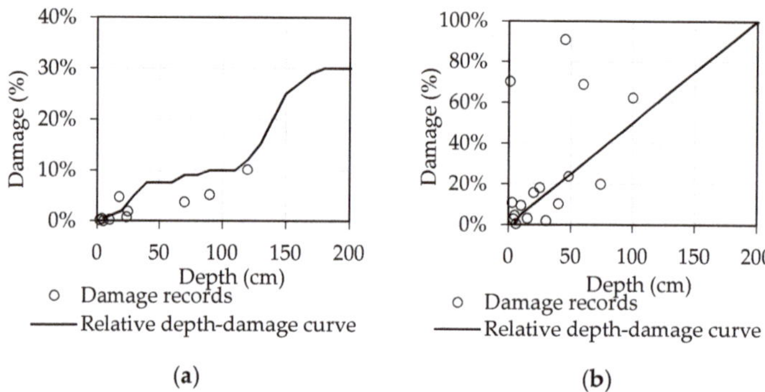

Figure 8. Proposed relative depth-damage curves and actual damage recorded for type of asset and property: (**a**) industrial buildings and (**b**) inventory, general trade.

Overall, the best correlations (linear) (i.e., high R^2) were found for furniture and household furnishings, with values of $R^2 = 0.74$ for restaurants and $R^2 = 0.82$ for workshops. In these cases, damage related to low depths was set according to the expert opinion, and the maximum damage was set between 90% and 97% according to the type of property. Figure 9 shows, as an example, the relative depth-damage curves proposed for furniture and household furnishings in restaurants and workshops.

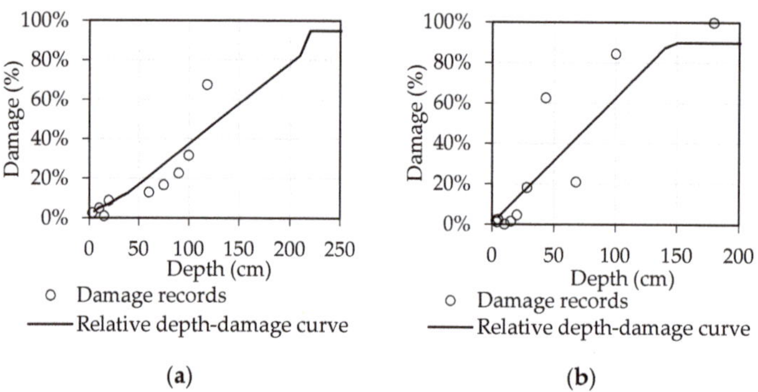

Figure 9. Proposed relative depth-damage curves and actual damage recorded for furniture and household furnishings and type of property: (**a**) restaurants and (**b**) workshops.

Figure 10 shows the proposed relative depth-damage curves for each asset, building, furniture and household furnishings, and inventory, according to 14 different types of property uses. These semi-empirical curves are developed to be applied in any Spanish municipality. Therefore, the curves' shape are invariable in both space (in Spain) and time.

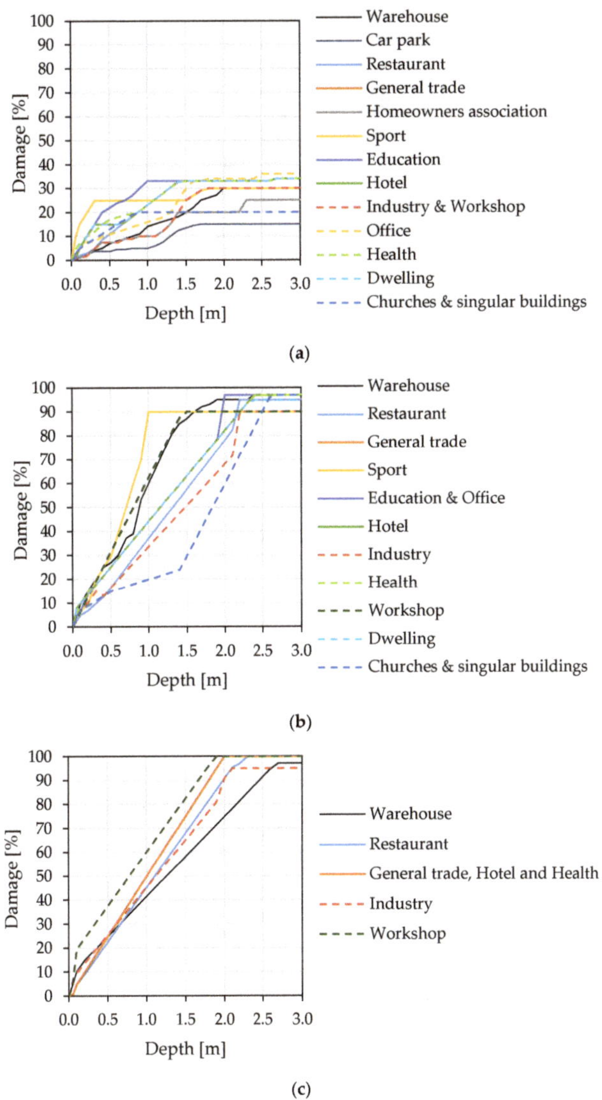

Figure 10. Relative depth-damage curves for Spanish cities: (**a**) buildings, (**b**) furniture and household furnishings, and (**c**) inventory.

4.2. Monetization of Relative Damage for Barcelona

To monetize the curves for Barcelona, initially the deciles (cost(€)/m^2) from the distribution of the data sample for each type of property and asset have been determined. According to expert opinion, a specific decile has been established for each type of property based on the target city. In some cases, the expert suggested providing an average value or making a direct estimation when the deciles did not match with his criterion, as Table 6 presents.

Table 6. Proposed cost (€/m²) for assets (i.e., building, furniture and household furnishings, and inventory) for each type of property in Barcelona.

	Building (€/m²)	Decile (Di)	Furniture and Household Furnishings (€/m²)	Decile (Di)	Inventory (€/m²)	Decile (Di)
Dwelling	999.89	D9	227.51	D9	-	-
Workshop	539.00	D8	419.03	D8	190.88	D8
Health	1227.75	D9	1871.91	D8	250.00	Direct estimation
Office	1500.00	D9	401.05	D9	-	-
Industry	568.16	D9	1827.36	D8	404.14	D8
Hotel	1443.00	D8	208.25	D8	50.00	Direct estimation
Education	1521.23	D6	151.14	D8	-	-
Sport	1811.85	90% D9	86.68	D6	-	-
Homeowners association	1629.62	D8	-	-	-	-
General trade	743.93	D8	338.85	D8	394.84	D8
Restaurant	1050.74	D9	470.78	D9	60.93	D9
Car park	1064.59	90% D6	-	-	-	-
Warehouse	733.43	D8	446.07	D8	853.86	Average D6 to D9
Churches and singular buildings	906.00	Average D6 to D9	250.00	Direct estimation	-	-

Once an adequate cost per square meter was associated with each property and asset according to the expert opinion for Barcelona, depth-damage curves were constructed for each asset. The following aggregation of the three types of assets provided the total depth-damage curves (Figure 11).

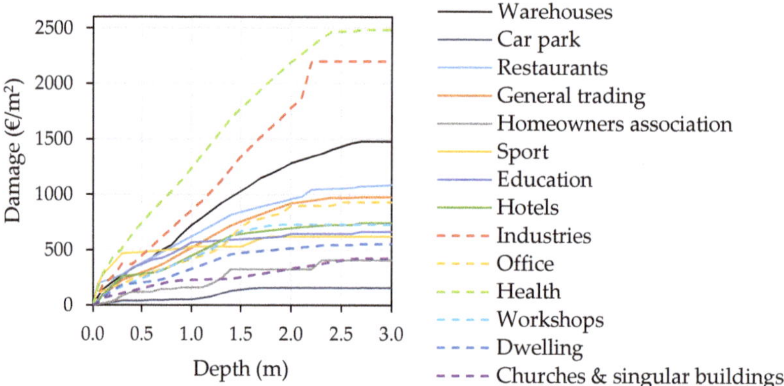

Figure 11. Semi-empirical depth-damage curves for Barcelona.

For those type of properties and assets with data scarcity or those that presented a low coefficient of determination, the expert used his criterion to adjust a curve from other property uses expected to be similar.

4.3. Depth-Damage Curves for Other Spanish Municipalities

As an example, Table 7 presents the regional indexes (RI) corresponding to some of the most damaged municipalities in Spain (Figure A1) due to flooding, which allows for constructing their own depth-damage curves (Figure 12).

Table 7. Regional indexes by assets in general use for the 10 most damaged municipalities in Spain, 1995–2019.

General sector	Asset	Barcelona	Orihuela	Los Alcázares	Vera	Murcia	San Sebastian	Málaga	Valencia	San Javier	Zaragoza	Lorca
Residential &	Build.	1.00	0.42	0.30	0.29	0.26	1.11	0.50	0.44	0.30	0.39	0.23
Others	Furnit.	1.00	0.59	0.59	0.61	0.75	0.70	0.73	0.82	0.67	0.78	0.59
	Build.	1.00	0.39	0.31	0.30	0.28	0.75	0.51	0.40	0.31	0.34	0.25
Commercial	Furnit.	1.00	0.37	0.40	0.32	0.55	0.69	0.44	0.55	0.40	0.74	0.43
	Invent.	1.00	0.41	0.37	0.33	0.45	0.49	0.44	0.57	0.37	0.49	0.38
	Build.	1.00	0.47	0.25	0.31	0.22	1.13	0.53	0.48	0.25	0.31	0.20
Industrial	Furnit.	1.00	0.37	0.35	0.27	0.51	0.73	0.39	0.55	0.36	0.81	0.39
	Invent.	1.00	0.37	0.40	0.32	0.55	0.69	0.44	0.55	0.40	0.74	0.43

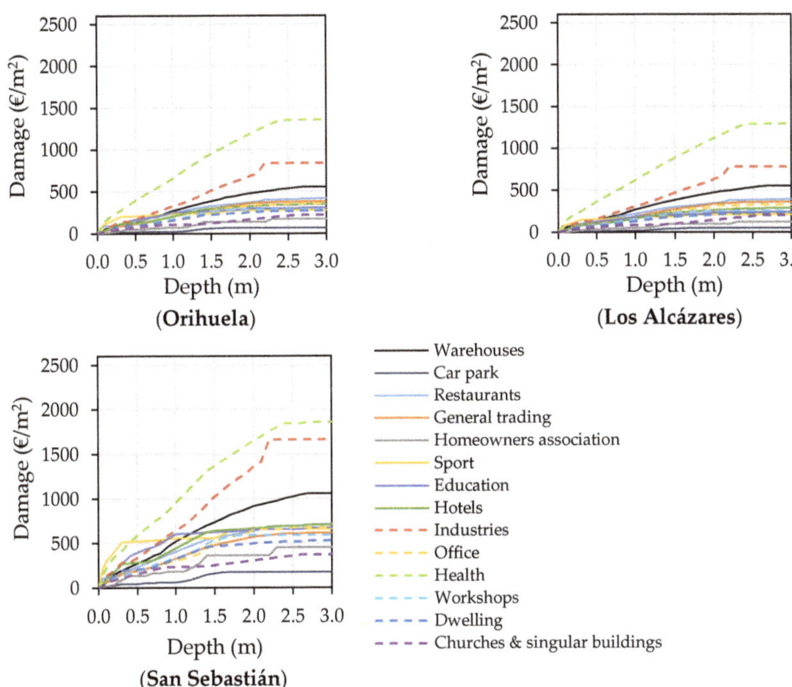

Figure 12. Depth-damage curves for some of the most damaged municipalities in Spain due to flooding (pluvial and fluvial) for the 2020 reference year.

The literature is scarce on the spatial transferability of damage curves, but a similar study by the IBI Group for Alberta (Canada) [16] uses the results of a local consumer price index survey to transfer damage to all economic sectors. Therefore, although limitations still exist, the present study represents an improvement on the current state of the field.

5. Conclusions

The use of depth-damage curves is globally accepted, even acknowledging the omission of other relevant variables, such as the water velocity or the floodwater residence time. A number of flood damage models are based on the use of these water-damage relationships. One of their major limitations is their site-specific nature, which means they cannot be applied in other regions. Moreover, the need of price updating may be considered a limitation, too. These issues are discussed in the literature and there is broad agreement about the better performance of relative depth-damage functions that remain static at least in time. Their shape, though, in terms of regional transferability, is more dependent on the style and typology of construction, which could be assumed to be fairly uniform at a national level. Maybe, for these reasons, standardizing a methodology of curves at a European level does not seem feasible yet, as some authors indicate discrepancies among computed and actual damage when using these continent-wide curves. Besides, countries such as the United Kingdom (Multi Colour Manual) and the United States of America (HAZUS-MH) apply their own nationwide depth-damage curves, and they have been well accepted for many years.

Barcelona is one of the case studies of the EU-funded RESCCUE project, whereby a comprehensive analysis of its climate resilience has been carried out. The impact of increasingly frequent pluvial floods has been analyzed. Namely, a detailed flood damage assessment has been conducted for the entire city, focusing primarily on properties. Due to the lack of existing detailed depth-damage curves for Barcelona, a tailored approach has been used, considering the 14 types of properties commonly found in highly urbanized cities. These developments were based on a sample of actual damage records; when insufficient or when the correlation between the damage and water depth was poor, the contribution of a flood expert surveyor was essential. Therefore, as in many other previous studies, expert opinion was included, resulting in the construction of semi-empirical depth-damage curves for Barcelona. In addition, this paper offers a methodology to obtain the depth-damage curves for any Spanish municipality, which provides standardization of depth-damage curves at a national level.

The methodology to standardize the construction of depth-damage curves in Spain presented here will contribute to enhancing cost-benefit studies in flood damage assessments at a micro scale and will allow for the comparison of results between different regions of the country. This could take on special relevance for future reviews and updates of flood risk management plans in Spain in the framework of the European Directive 2007/60/CE, included in the Spanish legislation through the Royal Decree 903/2010.

Author Contributions: Conceptualization, E.M.-G. and M.G.-H.; methodology, E.M.-G., M.G.-H., E.F.-O. and S.C.; validation, E.M.-G. and M.G.; formal analysis, E.M.-G.; investigation, E.M.-G., E.F.-O. and M.G.-H.; resources, S.C.; data curation, S.C. and E.M.-G.; writing—original draft preparation, E.M.-G.; writing—review and editing, M.G.; visualization, E.M.-G.; supervision, M.G.; project administration, E.M.-G.; funding acquisition, E.M.-G. and M.G.-H. All authors have read and agreed to the published version of the manuscript.

Funding: This research was funded by Horizon 2020 Framework Programme, Grant Agreement No. 700174.

Acknowledgments: The authors thank the Spanish insurance company Consorcio de Compensación de Seguros (CCS) for its important role in this research. Without its collaboration by providing claims data the damage model would have not been calibrated properly.

Conflicts of Interest: The authors declare no conflict of interest.

Appendix A

Figure A1 indicates the 20 municipalities in Spain most damaged by flooding in 1995–2019, which considers both pluvial and fluvial floods, but only damage to properties. The values indicated in the graph correspond to the compensation paid by the CCS.

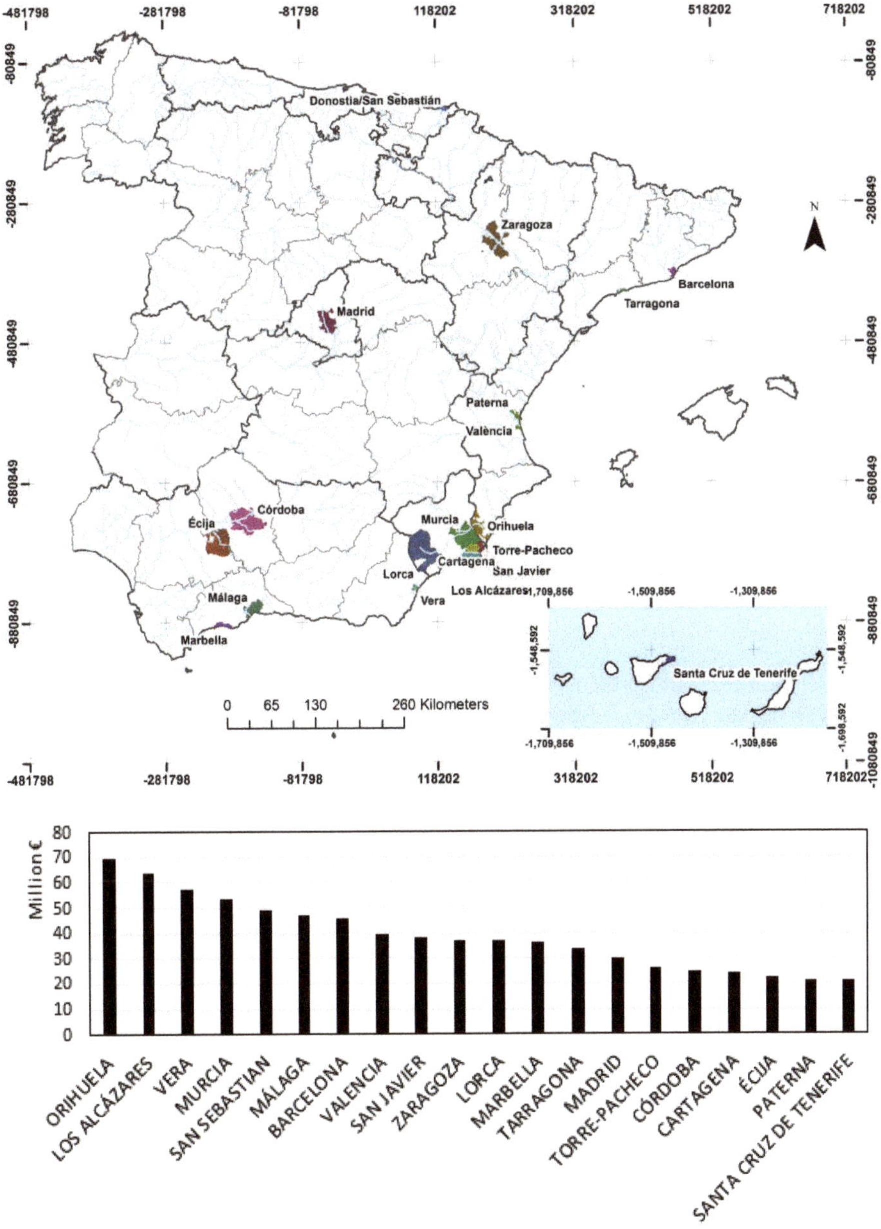

Figure A1. The 20 most damaged municipalities in Spain due to flooding (pluvial and fluvial) according to the compensations paid by the CCS.

References

1. European Environment Agency. *Economic Losses from Climate-Related Extremes in Europe*; Copenhagen, Denmark, 2019; p. 28. Available online: https://www.eea.europa.eu/data-and-maps/indicators/direct-losses-from-weather-disasters-3/assessment-2 (accessed on 20 January 2020).

2. Improving Damage Assessments to Enhance Cost-Benefit Analyses (IDEA) Project Earthquake of Lorca in 2011. Available online: http://www.ideaproject.polimi.it/?portfolio=floods-in-vall-daran-and-pyrenees (accessed on 22 February 2020).

3. United Nations, Department of Economic and Social Affairs, Population Division. *World Urbanization Prospects: The 2018 Revision (ST/ESA/SER.A/420)*; New York, NY, USA, 2018; p. 126. Available online: https://population.un.org/wup/Publications/Files/WUP2018-Report.pdf (accessed on 20 January 2020).

4. European Environmental Agency (EEA). *Flood Risk in Europe: the Long-Term Outlook*; European Environmental Agency (EEA): Copenhague, Denmark, 2016.

5. Velasco, M.; Cabello, À.; Russo, B. Flood damage assessment in urban areas. Application to the Raval district of Barcelona using synthetic depth damage curves. *Urban Water J.* **2016**, *13*, 426–440. [CrossRef]

6. White, G.F. Human Adjustment to Floods: A Geographical Approach to the Flood Problem in the United States. Ph.D. Dissertation, University of Chicago, Chicago, MA, USA, 1945.

7. Manrique, A.; Nájera, A.; Escartín, C.; Moreno, C.; Martínez, E.; Espejo, F.; Sánchez, F.J.; Aparicio, M.; Cordero, S.; González, S. *Guía Para la Reducción de la Vulnerabilidad de Los Edificios Frente a Las Inundaciones*. *Consorcio de Compensación de Seguros (CCS)*; Madrid, Spain, 2017; p. 106. Available online: https://www.consorseguros.es/web/documents/10184/48069/guia_inundaciones_completa_22jun.pdf (accessed on 14 March 2018).

8. *The European Parliament and the Council of the European Union Directive 2007/60/EC of the European Parliament and of the Council of 23 October 2007 on the Assessment and Management of Flood Risks (Text with EEA Relevance)*. 2007, p. 8. Available online: https://eur-lex.europa.eu/legal-content/EN/TXT/PDF/?uri=CELEX:32007L0060&from=EN (accessed on 11 February 2020).

9. Ministerio de la Presidencia Real Decreto 903/2010, de 9 de julio, de evaluación y gestión de riesgos de inundación. 2010, p. 14. Available online: https://www.boe.es/buscar/doc.php?id=BOE-A-2010-11184. (accessed on 11 February 2020).

10. Arnbjerg-Nielsen, K.; Willems, P.; Olsson, J.; Beecham, S.; Pathirana, A.; Bülow Gregersen, I.; Madsen, H.; Nguyen, V.-T.-V. Impacts of climate change on rainfall extremes and urban drainage systems: A review. *Water Sci. Technol.* **2013**, *68*, 16–28. [CrossRef] [PubMed]

11. Nafari, R.H.; Ngo, T. Predictive applications of australian flood loss models after a temporal and spatial transfer. *Geomat. Nat. Hazards Risk* **2018**, *9*, 416–430. [CrossRef]

12. Amadio, M.; Rita Scorzini, A.; Carisi, F.; Essenfelder, H.A.; Domeneghetti, A.; Mysiak, J.; Castellarin, A. Testing empirical and synthetic flood damage models: The case of Italy. *Nat. Hazards Earth Syst. Sci.* **2019**, *19*, 661–678. [CrossRef]

13. Federal Emergency Management Agency (FEMA). *Department of Homeland Security. Mitigation Division Multi-hazard Loss Estimation Methodology. Flood Model. Hazus-HM MR5 Technical Manual*; Federal Emergency Management Agency (FEMA): Washington, DC, USA, 2015; Volume 499.

14. U.S. Army Corps of Engineers (USACE). *HEC-FDA User's Manual. Flood Damage Reduction Analysis*; U.S. Army Corps of Engineers (USACE): Washington, DC, USA, 2016; p. 392.

15. Jongman, B.; Kreibich, H.; Apel, H.; Barredo, J.I.; Bates, P.D.; Feyen, L.; Gericke, A.; Neal, J.; Aerts, J.C.J.H.; Ward, P.J. Comparative flood damage model assessment: Towards a European approach. *Nat. Hazards Earth Syst. Sci.* **2012**, *12*, 3733–3752. [CrossRef]

16. IBI Group. *Provincial Flood Damage Assessment Study. Prepared for Government of Alberta*; IBI Group: Calgary, AB, Canada, 2015.

17. Carisi, F.; Schröter, K.; Domeneghetti, A.; Kreibich, H.; Castellarin, A. Development and assessment of uni- and multivariable flood loss models for Emilia-Romagna (Italy). *Nat. Hazards Earth Syst. Sci.* **2018**, *18*, 2057–2079. [CrossRef]

18. U.S. Army Corps of Engineers (USACE). *Economic Guidance Memorandum (EGM) 04-01, Generic Depth-Damage Relationships for Residential Structures with Basements*; U.S. Army Corps of Engineers (USACE): Washington, DC, USA, 2003.

19. Jan, H.; Hans, D.M.; Wojciech, S. *Global Flood Depth-Damage Functions: Methodology and the Database with Guidelines | EU Science Hub*; Joint Research Centre: Ispra, Italy, 2017.

20. Nafari, R.H.; Ngo, T.; Mendis, P. An assessment of the effectiveness of tree-based models for multi-variate flood damage assessment in Australia. *Water* **2016**, *8*, 282. [CrossRef]

21. Hasanzadeh Nafari, R.; Ngo, T.; Lehman, W. Development and evaluation of FLFAcs—A new Flood Loss Function for Australian commercial structures. *Int. J. Disaster Risk Reduct.* **2016**, *17*, 13–23. [CrossRef]

22. Hasanzadeh Nafari, R.; Ngo, T.; Lehman, W. Calibration and validation of FLFArs-A new flood loss function for Australian residential structures. *Nat. Hazards Earth Syst. Sci.* **2016**, *16*, 15–27. [CrossRef]

23. Olesen, L.; Löwe, R.; Arnbjerg-Nielsen, K. *Flood Damage Assessment Literature Review and Recommended Procedure*; Cooperative Research Centre for Water Sensitive Cities: Melbourne, Australia, 2017; Volume 4, ISBN 9781921912399.

24. Vanneuville, W.; Maddens, R.; Collard, C.; Bogaert, P.; De Maeyer, P.; Antrop, M. *Impact op mens en economie t.g.v. overstromingen bekeken in het licht van wijzigende hydraulische condities, omgevingsfactoren en klimatologische omstandigheden* **2006**, *2*.

25. McGrath, H.; Abo El Ezz, A.; Nastev, M. Probabilistic depth–damage curves for assessment of flood-induced building losses. *Nat. Hazards* **2019**, *97*, 1–14. [CrossRef]

26. Cammerer, H.; Thieken, A.H.; Lammel, J. Adaptability and transferability of flood loss functions in residential areas. *Nat. Hazards Earth Syst. Sci.* **2013**, *13*, 3063–3081. [CrossRef]

27. Hasanzadeh Nafari, R.; Amadio, M.; Ngo, T.; Mysiak, J.; Nafari, R.H.; Amadio, M.; Ngo, T.; Mysiak, J. Flood loss modelling with FLF-IT: A new flood loss function for Italian residential structures. *Nat. Hazards Earth Syst. Sci.* **2017**, *17*, 1047–1059. [CrossRef]

28. Reese, S.; Ramsay, D. *RiskScape: Flood fragility methodology. Technical Report: WLG2010-45*; New Zealand Climate Change Research Institute, Victoria University of Wellington: Kilbirnie, New Zealand; Wellington, New Zealand, 2010.

29. Huizinga, H.J.; Dijkman, M.; Barendregt, A.; Waterman, R. *HIS - Schade en Slachtoffer Module Versie 2.1*; Ministry of Transport and Water Management: The Hague, The Netherlands, 2005.

30. Milieu- en Natuurplanbureau (MNP). *Delft Hydraulics Overstromingsrisico's in Nederland in een Veranderend Klimaat*; Milieu- en Natuurplanbureau (MNP): Delft, The Netherlands, 2007.

31. Penning-Rowsell, E.; Viavattene, C.; Parode, J.; Chatterton, J.; Parker, D.; Morris, J. *The Benefits of Flood and Coastal Risk Management: A Handbook of Assessment Techniques-2010 (Multi-Coloured Manual)*; book+ CD-ROM with damage data; FHRC: London, UK, 2010.

32. Federal Emergency Management Agency (FEMA). *Hazus Flood Model User Guidance*; Federal Emergency Management Agency (FEMA): Washington, DC, USA, 2018.

33. Karamouz, M.; Fereshtehpour, M.; Ahmadvand, F.; Zahmatkesh, Z. Coastal flood damage estimator: An alternative to FEMA's HAZUS platform. *J. Irrig. Drain. Eng.* **2016**, *142*, 1–12. [CrossRef]

34. Cutrell, A.K.; Rozelle, J.; Hines, S.H. *FEMA Standard Operating Procedure for Hazus Flood Level 2 Analysis Hazus Flood Model*; Federal Emergency Management Agency (FEMA): Washington, DC, USA, 2018.

35. Huizinga, J.; De Moel, H.; Szewczyk, W. *Global Flood Depth-Damage Functions: Methodology and the Database with Guidelines*; Joint Research Centre: Sevilla, Spain, 2017.

36. Luino, F.; Cirio, C.G.; Biddoccu, M.; Agangi, A.; Giulietto, W.; Godone, F.; Nigrelli, G. Application of a model to the evaluation of flood damage. *Geoinformatica* **2009**, *13*, 339–353. [CrossRef]

37. Freni, G.; La Loggia, G.; Notaro, V. Uncertainty in urban flood damage assessment due to urban drainage modelling and depth-damage curve estimation. *Water Sci. Technol.* **2010**, *61*, 2979–2993. [CrossRef]

38. Pistrika, A.; Tsakiris, G.; Nalbantis, I. Flood depth-damage functions for built environment. *Environ. Process.* **2014**, *1*, 553–572. [CrossRef]

39. Tariq, M.A.U.R.; Hoes, O.A.C.; Van de Giesen, N.C. Development of a risk-based framework to integrate flood insurance. *J. Flood Risk Manag.* **2014**, *7*, 291–307. [CrossRef]

40. Budiyono, Y.; Aerts, J.; Brinkman, J.J.; Marfai, M.A.; Ward, P. Flood risk assessment for delta mega-cities: A case study of Jakarta. *Nat. Hazards* **2015**, *75*, 389–413. [CrossRef]

41. Central Intelligence Agency The World Factbook. Available online: https://www.cia.gov/library/publications/resources/the-world-factbook/index.html (accessed on 11 February 2020).

42. Aerts, J.C.J.H.; Botzen, W.J.W. Climate adaptation cost for flood risk management in the Netherlands. In Proceedings of the conference Storm Surge Barriers to Protect New York City Against The Deluge, Brooklyn, NY, USA, 30–31 March 2009; pp. 99–113.

43. Bussi, G.; Ortiz, E.; Francés, F.; Pujol, L.; De Ingeniería, I.; De València, U.P.; De Vera, C.; Hidrográfica, C.; Ibañez, A.B.; Bellver, P.; et al. Modelación hidráulica y análisis del riesgo de inundación según las líneas guía de la Directiva Marco del Agua. El caso de la Marina Alta y la Marina Baja (Alicante). *II Jornadas sobre Ing. del Agua.* 2007, 10. Available online: http://lluvia.dihma.upv.es/es/publi/congres/050_JIA2011_PRESENTACION_GB_articulo.pdf (accessed on 10 January 2020).

44. Ministry of Agriculture, Fisheries and Food (MAGRAMA). *Propuesta de Mínimos Para la Metodología de Realización de los Mapas de Riesgo de Inundación*; Ministry of Agriculture, Fisheries and Food (MAGRAMA): Madrid, Spain, 2013.

45. Valenciana, C. *Plan de Acción Sobre Prevención del Riesgo de Inundación en la Comunitat Valenciana (PATRICOVA)*; Generalitat Valenciana: Valencia, Spain, 2015.

46. Demarcación Hidrográfica del Cantábrico Oriental. *Plan de Gestión del Riesgo de Inundación 2015-2021*; Demarcación Hidrográfica del Cantábrico Oriental: San Sebastián, Spain, 2016.

47. Ritter, J.; Berenguer, M.; Corral, C.; Park, S.; Sempere-Torres, D. ReAFFIRM: Real-time Assessment of Flash Flood Impacts – a Regional high-resolution Method. *Environ. Int.* **2020**, *136*, 105375. [CrossRef]

48. Van Ootegem, L.; Van Herck, K.; Creten, T.; Verhofstadt, E.; Foresti, L.; Goudenhoofdt, E.; Reyniers, M.; Delobbe, L.; Murla Tuyls, D.; Willems, P. Exploring the potential of multivariate depth-damage and rainfall-damage models. *J. Flood Risk Manag.* **2018**, *11*, S916–S929. [CrossRef]

49. Wagenaar, D.; Lüdtke, S.; Schröter, K.; Bouwer, L.M.; Kreibich, H. Regional and temporal transferability of multivariable flood damage models. *Water Resour. Res.* **2018**, *54*, 3688–3703. [CrossRef]

50. Jamali, B.; Löwe, R.; Bach, P.M.; Urich, C.; Arnbjerg-Nielsen, K.; Deletic, A. A rapid urban flood inundation and damage assessment model. *J. Hydrol.* **2018**, *564*, 1085–1098. [CrossRef]

51. Gulf Engineers & Consultants (GEC). *Depth-Damage Relationships for Structures, Contents, and Vehicles and Content-to-Structure Value Ratios (CSVR) in Support of the Donaldsonville to the Gulf, Luisiana, Feasibility Study*; Gulf Engineers & Consultants (GEC): New Orleans, LA, USA, 2006.

52. Mitchell, A.; Dodge, B.; Kruzic, P.; Miller, D.; Schwartz, P. *Handbook of Forecasting Techniques*; National Technical Information Service U. S. Department of Commerce: Springfield, VA, USA, 1975.

53. Mitchell, A.; Dodge, B.H. *Handbook of Forecasting Techniques. Part 2. Description of 31 Techniques*; National Technical Information Service U. S. Department of Commerce: Springfield, VA, USA, 1977.

54. Boletín Económico de la Construcción (BEC) Revista de Información Económica (Precios Unitarios y Descompuestos) Dirigida a Profesionales del Sector de la Construcción. Nº 309. Available online: https://becsl.es (accessed on 1 July 2019).

55. Van Vloten, S.O. Vulnerability and Flood Risk in Urban Areas. Bachelor's Thesis, Universitat Politècnica de Catalunya, Barcelona, Spain, 2014.

56. Scawthorn, C.; Flores, P.; Blais, N.; Seligson, H.; Tate, E.; Chang, S.; Mifflin, E.; Thomas, W.; Murphy, J.; Jones, C.; et al. HAZUS-MH flood loss estimation methodology. II. Damage and loss assessment. *Nat. Hazards Rev.* **2006**, *7*, 72–81. [CrossRef]

57. Bedford, T.; van Gelder, P.; Guedes Soares, C.; van Manen, S.E.; Brinkhuis, M. Quantitative flood risk assessment for Polders. *Reliab. Eng. Syst. Saf.* **2005**, *90*, 229–237.

58. Statistical Office of the European Communities. *Export and Import Price Index Manual: Theory and Practice*; International Monetary Fund: Washington, DC, USA, 2009; ISBN 9264085416.

59. OECD "Long-term baseline projections, No. 103", OECD. *Econ. Outlook Stat. Proj.* 2020. Available online: https://doi:10.1787/68465614-en (accessed on 18 February 2020).

60. D'Amico, S.; Orphanides, A. *Uncertainty and Disagreement in Economic Forecasting*; Finance and Economics Discussion Series; Divisions of Research & Statistics and Monetary Affairs, Federal Reserve Board: Washington, DC, USA, 2008.
61. OECD Real GDP Long-Term Forecast (Indicator). Available online: https://doi.org/10.1787/d927bc18-en (accessed on 18 February 2020).

Article

Interlinking Bristol Based Models to Build Resilience to Climate Change

John Stevens [1,*], Rob Henderson [2], James Webber [3], Barry Evans [3], Albert Chen [3], Slobodan Djordjević [3], Daniel Sánchez-Muñoz [4] and José Domínguez-García [4]**

[1] Bristol City Council, Strategic City Transport, Flood Risk Management Team, 100 Temple Street, Bristol BS3 9FS, UK

[2] Wessex Water, Engineering & Construction, Claverton Down Rd, Bath BA2 7WW, UK; rob.henderson@wessexwater.co.uk

[3] Centre for Water Systems, University of Exeter, Exeter EX4 4QF, UK; J.Webber2@exeter.ac.uk (J.W.); B.Evans@exeter.ac.uk (B.E.); A.S.Chen@exeter.ac.uk (A.C.); S.Djordjevic@exeter.ac.uk (S.D.)

[4] IREC, Power Systems Department, Jardins de les Dones de Negre, 1, 2ª pl., Sant Adrià de Besòs, 08930 Barcelona, Spain; dsanchezm@irec.cat (D.S.-M.); jldominguez@irec.cat (J.D.-G.)

*   Correspondence: john.stevens@bristol.gov.uk

Received: 31 January 2020; Accepted: 9 April 2020; Published: 16 April 2020

Abstract: Expanding populations and increased urbanisation are causing a strain on cities worldwide as they become more frequently and more severely affected by extreme weather conditions. Critical services and infrastructure are feeling increasing pressure to be maintained in a sustainable way under these combined stresses. Methods to better cope with these demanding factors are greatly needed now, and with the predicted impacts of climate change, further adaptation will become essential for the future. All cities comprise a complex of interdependent systems representing critical operations that cannot function properly independently, or be fully understood in isolation of one another. The consequences of localised flooding can become much more widespread due to the inter-relation of these connected systems. Due to reliance upon one another and this connectedness, an all-encompassing assessment is appropriate. Different model representations are available for different services and integrating these enables consideration of these cascading effects. In the case study city of Bristol, 1D and 2D hydraulic modelling predicting the location and severity of flooding has been used in conjunction with modelling of road traffic and energy supply by linking models established for these respective sectors. This enables identification of key vulnerabilities to prioritise resources and enhance city resilience against future sea-level rise and the more intense rainfall conditions anticipated.

Keywords: fluvial; pluvial; tidal; sewer; flood; risk; climate change; modelling; cascading effects

---

1. Introduction

The challenge facing cities includes the need for urban expansion to accommodate rising populations. This puts a heightened demand on existing infrastructure and this is particularly apparent in many older cities that were not designed for the modern-day population and climate, making them unable to cope with such pressures [1,2]. The effects of climate change are likely to make this impact much worse in the future [3]. More severe storms and prolonged wet or dry periods all increase the risk and likelihood of pluvial flooding problems being encountered [4–7]. Drier landscapes from longer periods of drought in summer will produce surfaces more susceptible to rapid run-off and the increased storms and likelihood of thunderstorms exacerbates this risk [8]. Warmer and wetter winters mean more prolonged periods of wet weather, raised groundwater tables and higher river baseflows [9]. Rising sea-levels and heightened river flows are anticipated to create a more substantial threat from tidal and fluvial flooding sources, especially given the increased storm surge component associated with this [9].

In response to this, as well as many other well-known implications associated with climate change, Bristol City Council (BCC, Bristol, UK) has declared a climate emergency and has issued a Climate Emergency Action Plan as well as the Bristol Resilience Strategy to try and counteract and reduce these factors where possible [10,11]. Change in the climate is inevitable and is already being experienced to some extent; evaluating ways to adapt to this change is therefore essential. Work conducted on the EU RESilience to cope with Climate Change in Urban arEas (RESCCUE) [12] project with BCC and other key partners such as Wessex Water, the University of Exeter and IREC (Catalonia Institute for Energy Research, Barcelona, Spain) has made efforts to devise ways of assessing and managing increased flood-related climate risk and these will be elaborated upon below.

This article responds to increasing hazards by evaluating interdependencies in critical infrastructure and services functioning in the city of Bristol. In particular, the work focuses on the key elements of the existing drainage infrastructure, electricity supply system and road network. Roads also represent a significant conveyance mechanism for urban surface water (Fewtrell et al., 2011) [13]. During intense rainfall, these are likely to act as channels for exceedance from the sewer network. Similarly, the energy distribution network can be disrupted during flooding, leading to cascading damages and service interruption across many sectors of a city. Previous research has typically evaluated these systems independently (Pyatkova et al., 2018), however, it is apparent that safe and effective management of cities requires full consideration of interdependencies between complex and highly connected urban systems [14]. The aim of the work is to identify where the main vulnerabilities lie in areas of central Bristol and its immediate surrounds that are more prone to flooding through interlinked modelling, in order to develop adaptation plans to counteract this risk. The paper is structured through initially setting the case study city background and weather-related climatic threats it is faced with now and that which are anticipated in the future. It then goes on to define how these risks have been modelled and assessed and interprets the findings based on implications posed from the various sources of flooding to certain city services and specific areas of the city. Adaptation measures proposed to counteract this risk and attempt to relieve some of the effects of these impacts are then given further consideration.

Bristol Case Study

Bristol is located in the South of England, UK within the Severn River Basin District (see Figure 1) and is particularly vulnerable to tidal/fluvial flood risk, being subjected to the second-highest tidal range in the world from the Severn Estuary which influences the tidal River Avon as well as having significant surface water flood risks [15]. The River Avon shown in Figure 2 passes through Bristol from East to West, with a portion of the flow entering the Floating Harbour in the central area, which has a regulated water level and has complex interactions between incoming tides and river flows. The majority of river flow is diverted along the River Avon New Cut where it continues westwards and discharges into the Severn Estuary at Avonmouth. The "New Cut" is a man-made channel of the River Avon and was constructed in the early 19th century to allow the creation of the Floating Harbour which provides a permanent dock facility isolated from extreme tidal effects. Many tributaries of the Avon within Bristol are tidally influenced near their outfalls.

The city is also rapidly expanding; in recent years it has seen the second-largest rate of population growth in the UK, outside of London. Urban expansion and the threat from intense downpours, which are expected to become more frequent and of greater severity in the future, combined with sea-level rise will impact on critical drainage infrastructure and land drainage functions in Bristol. Improving urban resilience in the city can be achieved by the capability to anticipate, prepare for, respond to, and recover from these significant multi-hazard threats with minimum damage.

In order to achieve the above aims, adaptation plans (on all scales from strategic, operational to community-based including societal and economic impacts) have all been duly considered. The underlying objective was to find ways in which to adapt to this shift in weather patterns and account for what is "the new norm" through a range of proactive and reactive responses. Analysis of the impact that high tides combined with heavy rainfall have through direct flood damage on riverside areas adjacent to the tidal River Avon (and its tributaries) fulfilled part of this assessment. As well as direct impacts, indirect impacts (on the operation of urban drainage systems for instance) were evaluated. This included analysis of flooding issues linked to tidally influenced sections of the sewer network, for example. Wider impacts, in respect of the cascading effects on critical city services, were also considered as a follow-on consequence of flooding. The way in which this was quantified and an overview of some of the outcomes is described in the following sections.

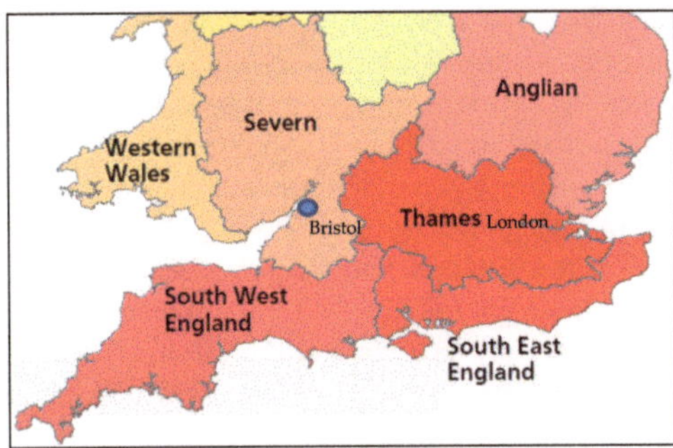

Figure 1. Location of Bristol, in the Severn River Basin District, shown on the South England and South Wales, UK River Basin District map.

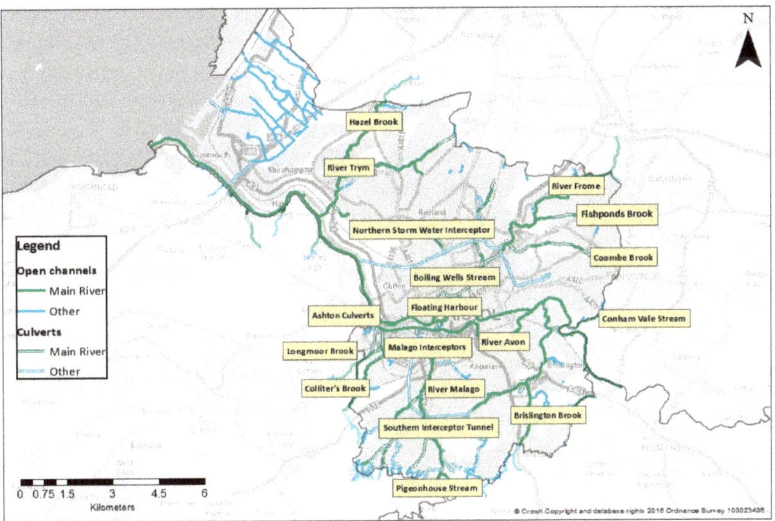

Figure 2. Map of main rivers, streams and surface water interceptor tunnels in the Bristol City Council Local Authority administrative area.

2. Materials and Methods

2.1. Flood Modelling

The city of Bristol is now quite comprehensively modelled as far as sewers, watercourses and the tidal River Avon is concerned. The Bristol Surface Water Management Plan [16] model, developed in conjunction with Wessex Water, covers the entire city and incorporates much of the underground piped sewer network. The tidal River Avon is modelled throughout its expanse within the BCC area providing coverage of the whole watercourse within the city as modelled in the Central Area Flood Risk Assessment (CAFRA) [17]. Tributaries of the River Avon also have detailed flood mapping in the lower reaches of their catchments in the CAFRA [17]. At Avonmouth, in West Bristol, the effect of tidal and fluvial flooding from the Severn Estuary and Avonmouth rhyne network (drainage ditches that serve the area) is mapped through the Avonmouth/Severnside Level 2 Strategic Flood Risk Assessment [18]. Purely fluvial flood extents for other watercourses, not tidally influenced, that appear throughout the remainder of the city are covered by Environment Agency Flood Mapping [19]. By analysing the exposure of urban services and critical city infrastructure to flooding and the vulnerability of key services such as the electricity supply, road network and drainage infrastructure, the impacts and risk can be assessed over time and the necessary adaptation measures considered to ensure their continued functioning. The flood models for pluvial flooding as well as combined fluvial and tidal flood events that exist for Bristol can provide flood extents, levels, depths and hazard ratings inclusive of uplift for climate change. The models used of the sewerage systems for drainage aspects are coupled with models in the flooding sector, most notably with integrated modelling of drainage systems and watercourses. A hazard assessment for current and future (climate change) scenarios exists and this utilized detailed models and software tools including *Infoworks ICM* 1D/2D urban drainage modelling [20]. This analysis was based on models built during the preparation of the City Drainage Master Plans and Surface Water Management Plans. The tidal and fluvial model involved *ISIS* [21] and *TuFLOW* [22] for the joint probability modelling on the River Avon.

The estimated change in future climate parameters was informed by the Met Office and UK guidance and sensitivity checked by the Madrid-based Climate Research Foundation (FIC) as part of the RESCCUE project [12]. Models of the tidal and fluvial system in Bristol have been completed through the CAFRA [17] study conducted by BCC. CAFRA analyzed combination events of tidal floods and fluvial flood flows (i.e., the joint probability of the two flooding sources occurring simultaneously) of varying magnitudes. This was to establish the predominant risk and threat to the city centre, both now and into the future including the predicted impacts of climate change. The conclusion was that the high tidal element causes the greatest flooding risk, far outweighing the fluvial component. The CAFRA study included a large-scale hydraulic model of the tidal and fluvial systems in central Bristol. The model itself was completed using a combined 1D and 2D model built using *ISIS-TuFLOW* software packages. The majority of the river networks in central Bristol were simulated using the 1D *ISIS* software, with topography and ground surface represented using 2D *TuFLOW*. The model was updated as part of the ongoing River Avon Flood Risk Management Strategy.

The modelling has allowed flood depths, velocities and extents to be determined, allowing comprehensive flood hazard mapping for the city in accordance with the UK DeFRA standards [23,24]. Observed and predicted tidal flood levels for the Severn Estuary and tidal River Avon have provided some verification of the model outputs and a series of particularly high tides experienced in 2014 assisted with this, during a stage of the 19-year lunar cycle that caused exceptional astronomical tide levels. In order to predict and assess the likely impacts of climate change, the National Planning Policy Framework (NPPF) [25] and UKCP09 [26] derived uplifts have been applied to the CAFRA tidal/fluvial model. A damages assessment conducted in line with the "Multicoloured Handbook" (MCM) [27] methodology has allowed for the quantification of the flood damages incurred in monetary value in the present day and the future increase in damage value due to the effects of climate change. This involved land-use data acquired via the UK's National Receptor Database. The land-use includes MCM Codes that correspond to depth-damage curves for specified land-use types. Tangible damages on the economic sector were estimated by utilizing the SWMP pluvial model and CAFRA tidal/fluvial model compared against the land-use area distribution through the MCM approach.

Present-day Bristol faces a significant risk of flooding from multiple sources. With the application of a climate uplift factor applied to tide levels, fluvial river flows and rainfall intensity, this gives a resulting increase in flood extents, depths, heightened flow velocities and subsequently an increase in flood hazard risk rating. The flood modelling for the present day has climate change allowances applied to it in line with UK Government guidance [28] to estimate the projected future flood risk. This was based on the data and recommendations available for the Severn River basin district to account for an anticipated increased peak river flow. Upper-end peak rainfall intensity increases were applied, applicable to all areas in England. For sea-level rise, the rate of increase (in mm per year–see Table 1) is reflected in accordance with the advice for the Severn River basin district to use the South West River basin allowances. The increases in flooding over time causes threats directly to land susceptible to this risk but the effects of this are also felt beyond immediate high flood-risk areas, as will be explained in this section. The quantification of damages, identification of key criticalities and vulnerabilities has highlighted the most vulnerable areas.

Table 1. Reflecting the predicted future sea-level rise from a *UK Government website* [28].

Epoch	1990 to 2025	2026 to 2055	2056 to 2085	2086 to 2115	Cumulative Rise 1990 to 2115 (m)
Rate of rise (mm/yr)	3.5	8	11.5	14.5	1.11
Cumulative rise in epoch (compared to 1990) (mm)	123	232	334	421	

The astonishing estimated increase in the tide levels over time is reflected in Table 1 which states national values recommended for planning purposes.

Table 2 defines the DeFRA/Environment Agency Flood Hazard rating and the danger posed to people, including the emergency services which could be called upon in times of flooding disruption caused to critical city services [23]. This has been used to assess the flood hazard posed in the Bristol case study areas.

Table 2. DeFRA/Environment Agency Flood Hazard rating and the danger posed to people for different combinations of flood depth and velocity [23].

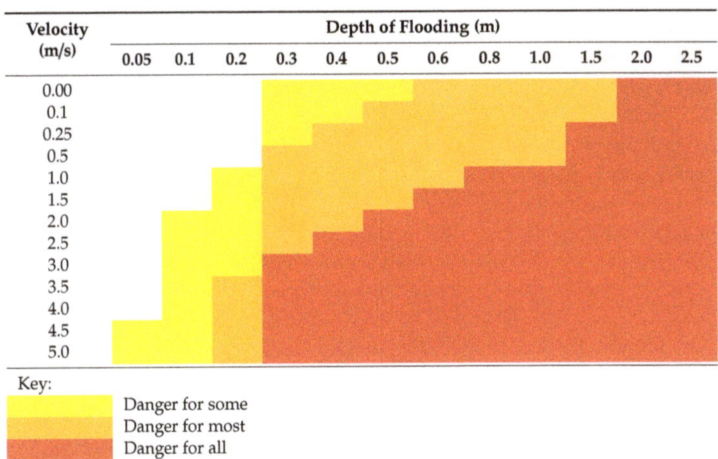

2.2. Integrated Modelling

As outlined in Figure 3, not only were the various sources of flooding combined but these were interfaced with models used to manage other city functions, such as traffic management and power supply by overlapping models that exist for these sectors, with the output of one model providing input to another. An integrated flooding-traffic model applied traffic simulations and flood impact modelling carried out using the Simulation of Urban MObility (*SUMO*) [29] micro-scale traffic software package. The SUMO system can accommodate large road networks, appropriate for that modelled in Bristol, and provides a continuous traffic simulation using open source data [30–33].

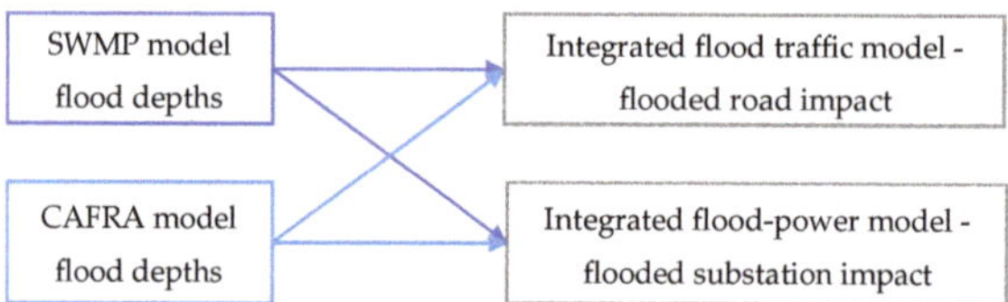

Figure 3. Schematic diagram showing the linkage between the integrated modelling.

Flooding impacts on traffic flow in several ways including: redirection of traffic, reduction in travel speeds, increases in journey times, congestion and increased pollution levels. The impacts may be localized or widespread as drivers try to avoid a problem area and, in doing so, cause congestion elsewhere [34,35].

The hazards posed to traffic flows are represented in relation to predicted flood depths along individual road segments/links in the *SUMO* model. Flood depths define whether a link (section of road) is closed (severe flooding) or if the maximum allowable speed along the link is to be reduced (moderate flooding). This shows which roads would be closed and which would suffer congestion and reduced speed. Any link that experiences flood depths of 0 to 0.10 m is determined as a non-affected road, 0.10–0.30 m is described as a reduced speed road link and road links with flooding of over 0.30 m are classified as closed. An indication of where the anticipated road closures would occur during a flood scenario can be inferred. In 100 years and with the potential effects of sea-level, the future effects of flooding can be surmised, allowing prediction of how the city will suffer in these areas in response to this. Network management plans can be devised in response to this in the current day or longer-term strategic solutions, and improved flood defences can be scoped out for the future.

The outputs of both the pluvial/sewer model and the tidal/fluvial models were also used to analyse the impact on the electricity supply system serving the central area of Bristol (8 km^2). The electrical modelling created a sampling layer through the use of the open-source software *QGIS* [36] using infrastructure location and attribute data provided by Western Power Distribution (WPD) [37], including critical flood-depth thresholds (where known) for individual stations. This integrated flooding-electrical model (*IFEM*), allows the generation of a GIS layer showing fully-affected, partially-affected and non-affected substations and their areas of influence [38]. Knowledge of flooding extents and depths allow the impact on urban services to be assessed in detail, thereby informing the planning of remedial and mitigation measures contributing to the development of a Resilience Action Plan (RAP).

By combining flood mapping and electrical modelling of the power network (using data derived from WPD), complications and cascading effects can be predicted. As the tidal cycle and future astronomical tide levels are forecast well in advance, high spring tides that may combine with adverse prevailing weather conditions can be foreseen with greater warning time ahead of a preceding tidal flood event. Low atmospheric pressure systems and westerly winds raise the tidal storm surge component in Bristol. Knowledge of these factors can then help in tidal flood preparations and electricity substations within the potential flood area can be identified and actions taken to mitigate or eliminate the flooding risk. Sewer flood maps and tidal/fluvial flood maps highlight how many substations could potentially be affected with an increased magnitude of flood events if there is no protection around the substations. The greater vulnerabilities and particular areas of concern can then be demonstrated from this and used to inform the selection of effective protection measures [39]. Impacts from the electricity supply system resulting in power outages further afield can be yet another implication and cascading effect felt by other city services reliant upon this facility.

Two particular high-risk areas within Bristol were then focused on to provide an in-depth detailed analysis at significant locations. The problematic areas were analysed to formulate a RAP to cope better with this and to enhance future sustainability. The impact of flooding on urban services helped quantify more of the overall risk faced. The two areas where analysis of traffic and energy disruption caused by flooding has been conducted are (a) St Phillips Marsh and (b) Ashton.

2.3. Focus Areas

St Phillips Marsh, located east of central Bristol tidal/ fluvial flooding is the principal risk to this mainly commercial and light industrial area. This area is heavily trafficked since it feeds in and out of the main central business hub of the city and is already inundated with commuters.

Ashton, located in South-West Bristol where flooding occurs from the main watercourse (Colliters Brook) and sewer systems (both combined and surface water); flooding is also significantly affected by the water level in the tidal River Avon due to backing up of drainage systems in this low-lying part of the city. Detailed *Infoworks* ICM-2D modelling was undertaken within this subcatchment.

3. Results

3.1. Sources of Flooding Modelled

3.1.1. Tidal Fluvial Flooding

In the present day, there are currently 1000 properties shown as "at risk" to an extreme (1 in 200-year Return Period (RP) or 0.5% Annual Exceedance Probability (AEP)) tidal event in Bristol. The future number of properties at risk rises to 3700 in the eventuality of an extreme (1 in 200-year RP or 0.5% AEP) tidal event in Bristol becoming apparent in the future (2115) when considering the predicted effects of sea-level rise.

The increase in predicted flood extents over the coming decades is illustrated in Figures 4–6. The flow of the River Avon through central Bristol is from East to West, discharging in the Severn Estuary. The tidal influences already cause flooding from the River Avon and Bristol Floating Harbour during exceptional high spring tides and sea-level rise will exacerbate this problem in the future.

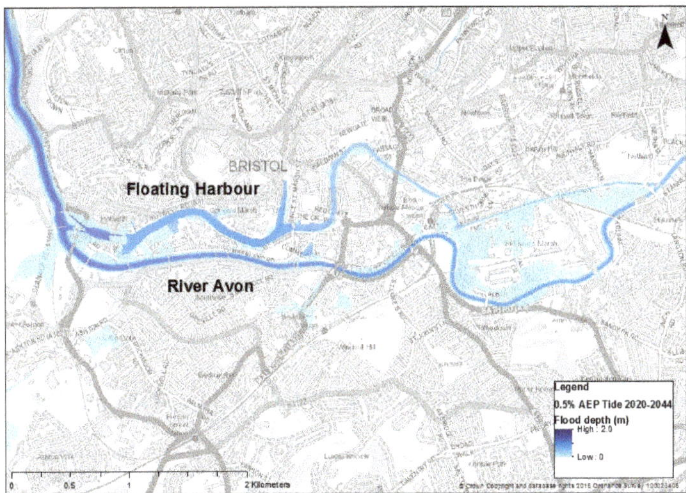

Figure 4. Flood depths and extent for a 0.5% AEP tidal flood event for the 2010–2044 scenario.

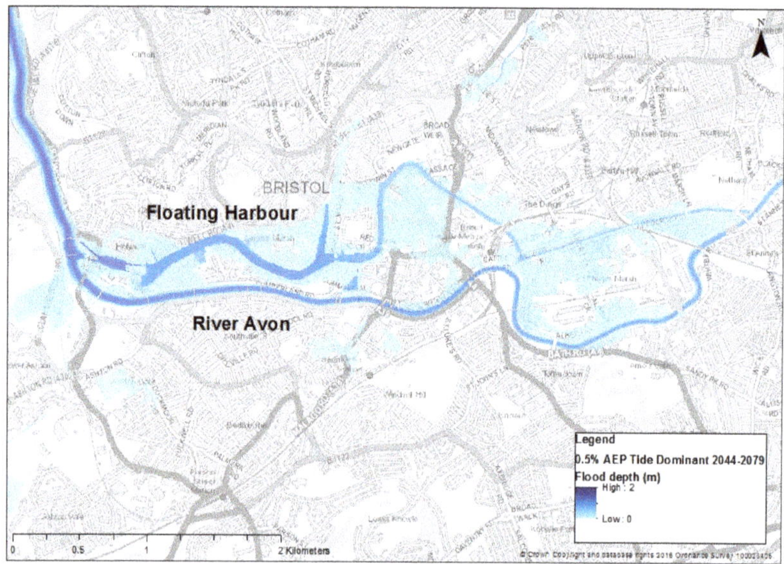

Figure 5. Flood depths and extent for a 0.5% AEP tidal flood event for the 2044–2079 scenario.

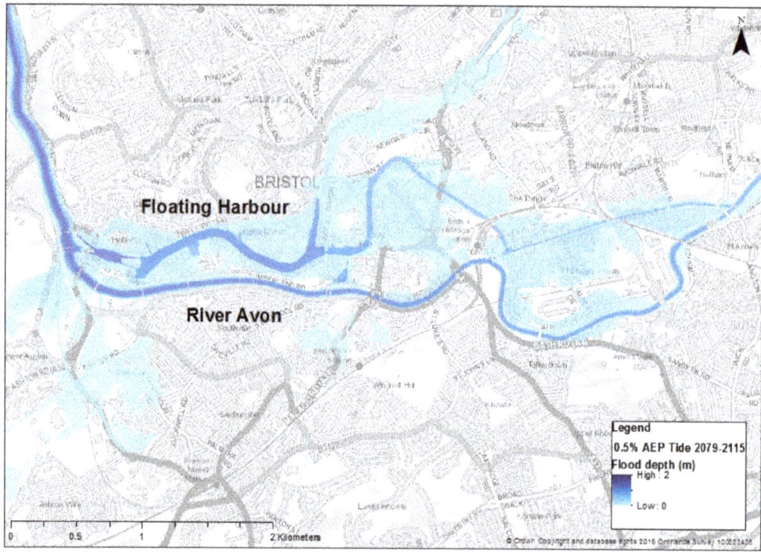

Figure 6. Flood depths and extent for a 0.5% AEP tidal flood event for the 2079–2115 scenario.

In respect of flood hazard mapping, it is evident that the potential impact of climate change (primarily sea-level rise) will have huge implications for properties at risk and for the continuity of city services. Figure 7a,b displays this.

The DeFRA methodology for assessing flood hazard is accepted for application in the UK, applicable to Bristol. Flood hazards posing a danger to people have also been assessed, however, through other means in the works conducted by Martinez-Gomariz E et al. (2016), Chanson H and Brown R (2015), Russo B et al. (2013), Arrighi C et al. (2017) that could be more applicable at other localities [40–43].

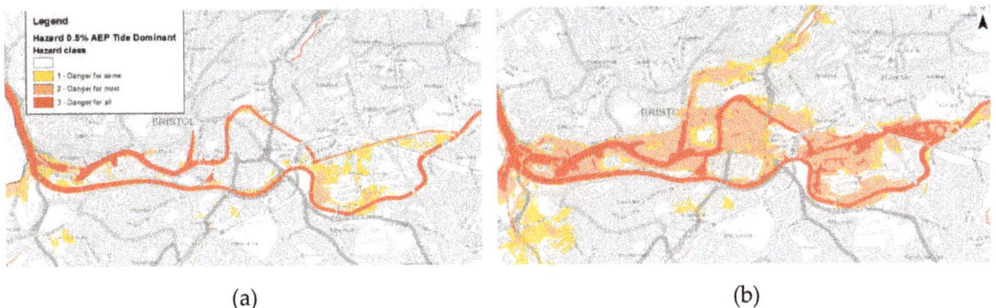

(a)   (b)

Figure 7. Hazard mapping for the 0.5% AEP tidal/fluvial event in (**a**) Present day (left) (**b**) 2115 (inclusive of climate change) (right).

3.1.2. Pluvial Flooding

More intense storms in the magnitude of 20–40% greater intensity are expected to occur more commonly by the turn of the next century [28]. This element is captured in the increased flood extents illustrated in Figure 8. In general, these show that under climate change and for very rare events, the depth and severity of flooding will increase more significantly than will the extents of flooding, due to the constraints imposed by the urban terrain.

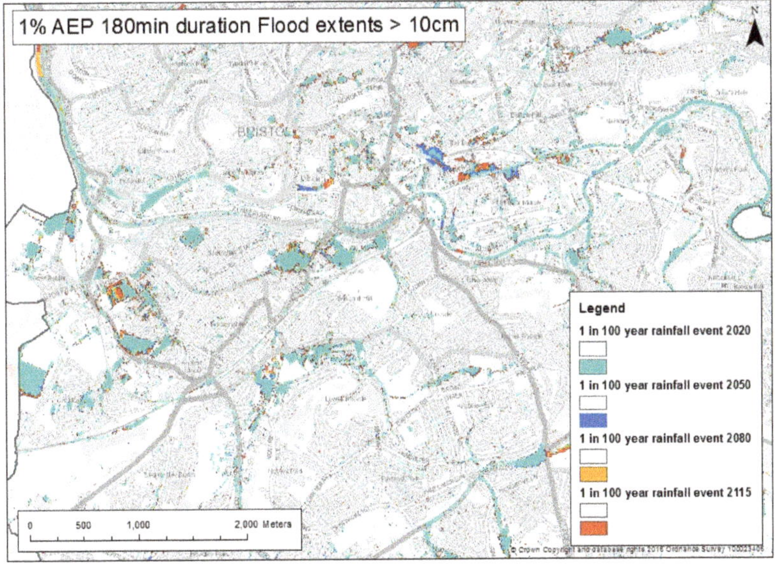

Figure 8. Pluvial flood extents increasing over time with the predicted impacts of climate change.

In the Ashton area, the likelihood of property flooding as a result of combining a high tide with a severe storm is estimated to be roughly four times more probable than at present by the 2050s and over ten times more probable by the 2080s under the climate change conditions assessed here.

3.2. Integrated Flood Models

Flood models are useful inputs for evaluating the potential impact on critical city services such as traffic and power supply. This then helps address the most vital elements and assists in targeting limited resources more effectively. It also helps promote the business case to encourage longer-term investment in strategic interventions or to help devise shorter-term remedies such as operational procedures.

3.2.1. Traffic

For pluvial flooding, the analysis in the Ashton area showed that for a 1 in 30-year rainfall event, (3.3% AEP) journey times went up considerably with the time of flood duration causing more prolonged traffic disruption. Delay times on the network increased by approximately 14, 24 and 25 min for the 20, 30 and 40 min duration rainfall events, respectively.

For fluvial flooding, the increase in reduced speed journeys and closed roads are shown in Figures 9 and 10 below.

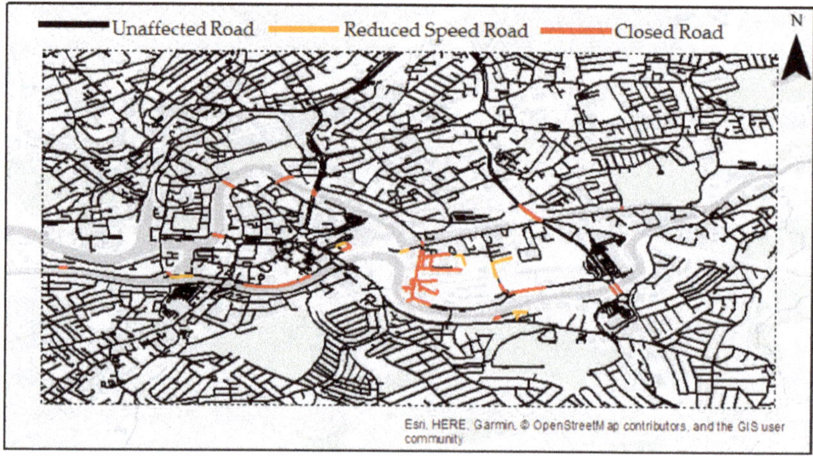

Figure 9. Flooding on the St Philip's Marsh Bristol Road Network for a 1 in 20 Year Fluvial Current Day event day.

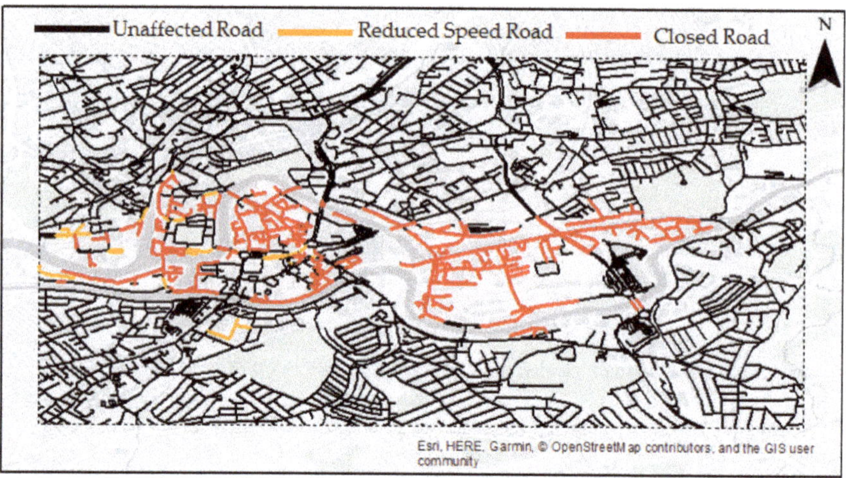

Figure 10. Flooding on the St Philip's Marsh Bristol Road Network for a 1 in 20 Year Fluvial Future Climate Change Event 2115.

3.2.2. Power

Utilising the location of critical infrastructure (such as electricity substations) and comparing against predicted flood outlines can help infer power outages associated with flood events. Substations located within the CAFRA and SWMP flood model outlines within central Bristol, St Philips and Ashton were analysed. Specific information relating to the criticality of individual major substations cannot be disclosed in greater detail due to security reasons, but the number of potential substations affected and the percentage over the total studied are given in Table 3 for the three Average Water Depth (AWD) categories established. In the most critical case (AWD > 1.60 m), the increase in severe flooding occurrences rises from 2 to 76 when increasing the return period from T20 to T1000, meaning a 17.2% increase over the total number of 11 kV substations. These results are also displayed in Figure 11.

Table 3. Number of substations affected resulting from the electrical sector analysis, and the percentage over the total of substations studied, according to different return periods, type of substations and average water depth (AWD) categories.

Water Depth	Type of Substation	Number of Occurrences			Percentage over Total		
		T20	*T200*	*T1000*	*T20*	*T20*	*T200*
	11 kV	80	49	41	18.6%	11.4%	9.5%
0.1 m < AWD ≤ 0.8 m	*33 kV*	0	1	1	0.0%	33.3%	33.3%
	132 kV	0	0	0	0.0%	0.0%	0.0%
0.8 m < AWD ≤ 1.6 m	*11 kV*	19	92	87	4.4%	21.4%	20.2%
	132 kV	0	1	1	0.0%	100.0%	100.0%
AWD > 1.6 m	*11kV*	2	56	76	0.5%	13.0%	17.7%

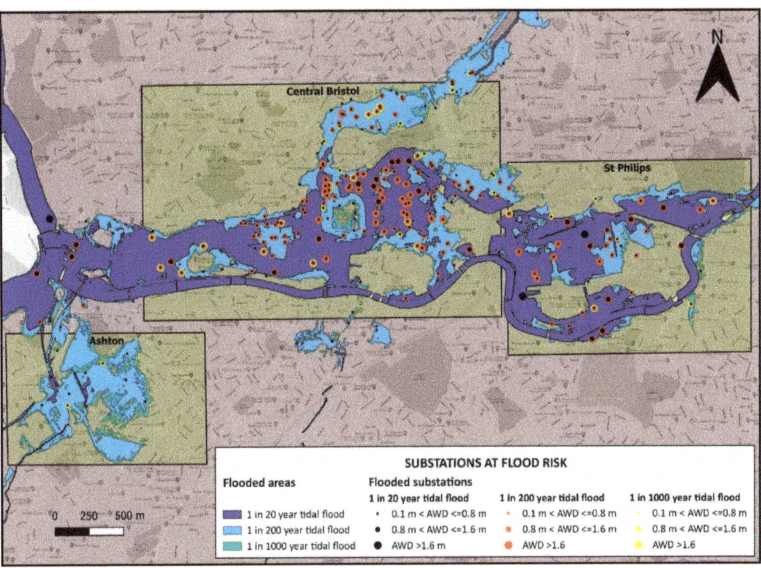

Figure 11. Map of substations potentially affected by flooding for various return periods in the 2115 future case scenarios. The points are sized according to the three categories of flood event magnitude established and coloured according to the year. This map shows the evolution of the flooding depending on the return period.

3.3. Focus Areas

3.3.1. St Philips Marsh

The benefits of protecting the riverside low spots subsequently by removing the over-spilling flood extents shown in Figure 12 in the Bristol central area were quantified through a damages assessment. Present-day damages are estimated through the MCM in the order of £40M whereas in 100 years with the rate of sea-level rise continuing, this will be around £400M. The cost-benefit ratio can be gauged from this when scoping out flood defences that may be suitable, feasible and economically affordable at this locality.

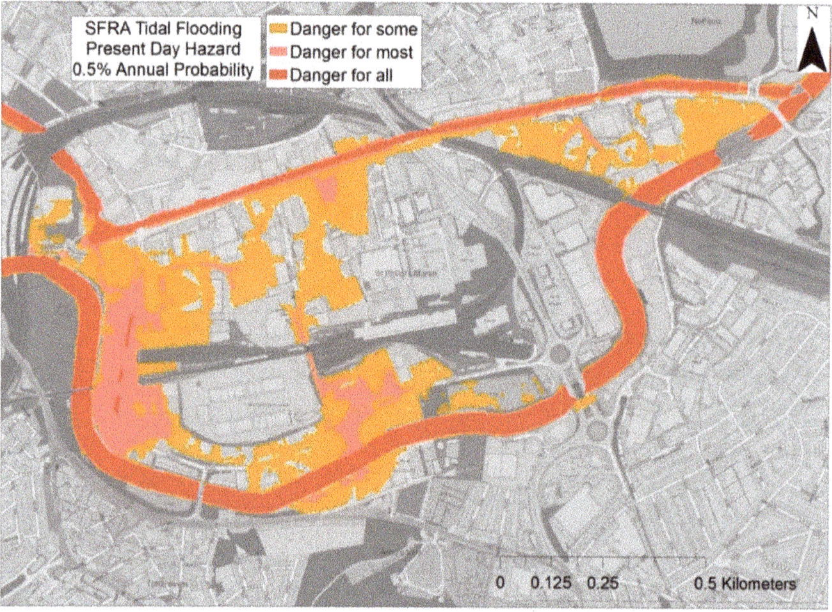

Figure 12. DeFRA/Environment Agency Flood Hazard Mapping showing the danger posed to people (please refer to Table 2) at St Philips Marsh for a 0.5%AEP tidal flood event.

3.3.2. Ashton

The increased pluvial flood depths and extents are very noticeable in the Ashton area in Figure 13 and the future tidal/fluvial flood hazard shown in Figure 14.

Critical parts of the Ashton area are at a lower elevation than the banks of the River Avon. The River Avon New Cut river channel banks are at an elevation of between 8.5 m and 14.0 m AOD. By contrast, the lowest ground elevation in Ashton is 6.3–6.8 m AOD (public open space/parkland), with roads and properties at 6.8–7.5 m AOD or higher. The river level frequently surpasses this level in the present-day during high Spring tides. In the extreme scenario, a (current) 1:200 year tidal/fluvial flood event could take the river level to about 9 m AOD, well over two metres higher than the lowest vulnerable ground level and would inundate the neighbourhood. With the effects of sea-level rise in 100 years, another metre may be added to the extreme tidal flood level. The average duration for which these critical levels are exceeded is reflected numerically in Table 4 and then visually in Figure 15 to outline the level of flooding experienced.

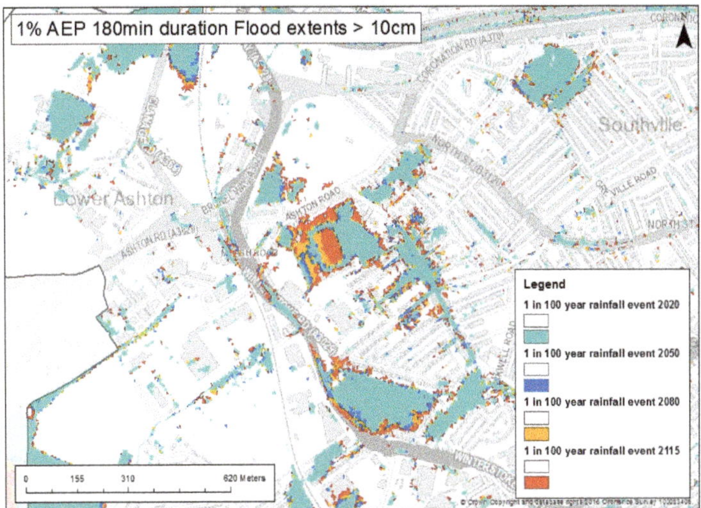

Figure 13. Flooding extents in the Ashton area with climate uplifts (Note: this model run excluded tidal effects in order to identify changes due to rainfall increase alone).

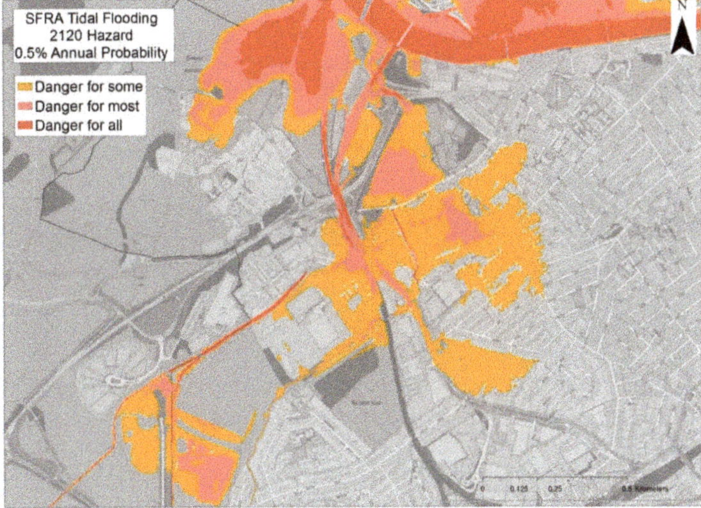

Figure 14. DeFRA/Environment Agency Flood Hazard Mapping showing the danger posed to people (please refer to Table 2) at Ashton for a 0.5% AEP tidal flood event in 2120 inclusive of climate change.

Table 4. Critical tide durations estimated in future epochs.

Decade	Critical Tide Level (7.5mAOD) Is Exceeded (%, Percentage of Time in a Year, on Average)	Extreme Tide Level (8.0mAOD) Is Exceeded (%, Percentage of Time in a Year, on Average)
2010	0.34	0.04
2050	0.63	0.13
2080	1.30	0.39
2110	2.04	0.79

Figure 15. Flooding in the Ashton area in the current day.

The complex interactions between drainage systems at Ashton include the influences of the river tide level on surface water and combined sewer overflow (CSO) outfalls causing "tidelocking" (the closure of non-return valves), resulting in backing up and surcharging of the system. Overflow from a culverted watercourse to the combined sewer system is another contributory factor as is the discharge from CSOs that will naturally increase with heightened rainfall. Inflow from natural watercourses to man-made drainage ditches, surface water sewers and culverts also occurs as does flooding out of watercourse channels to urban surfaces and flooding out of (combined and surface water) sewers to urban surfaces. The 2D modelling of these systems has improved the understanding of the complex interactions between surface flows and the drainage systems.

The Colliters Brook and surface water sewers discharge to the River Avon by gravity outfalls, protected by tide flaps. When high tide level exceeds the outfall level, flows back up within the Colliters Brook (which is in a culvert upstream of the outfall). Similarly, flows back up within the surface water systems discharging to the River Avon or the Colliters Brook. With increasing sea levels, surface water systems with tidal river outfalls will be compromised under high tide conditions, Figures 15 and 16 demonstrate this.

A major sewage pumping station (SPS) at Ashton Avenue takes combined sewage flow from the Ashton area which helps alleviate the existing local flood risk issues. During intense storms, if the pumping capacity is exceeded, flows are diverted to a gravity overflow system installed with tide flaps. The gravity overflow can only discharge when the energy head in the surcharged trunk sewer is higher than the river level. In extreme conditions (when all pumping capacity is beaten and very high tide conditions prevail), flow level in the sewer can back up-potentially to ground surface level. Thus, flooding of low-lying areas from the combined system could occur if the storm is of sufficient intensity/duration and is coincident with a high river level. This increases the risk of combined sewer flooding in severe storms which exceed the pumping capacity at Ashton Avenue SPS. Total pumping capacity is currently exceeded by a storm of roughly 1 in 5 years or greater under current rainfall conditions, with significant flooding predicted to occur roughly once in 30 years. Diversion of "clean" streamwater to the combined system also increases CSO spill frequency at the SPS under lesser storm events.

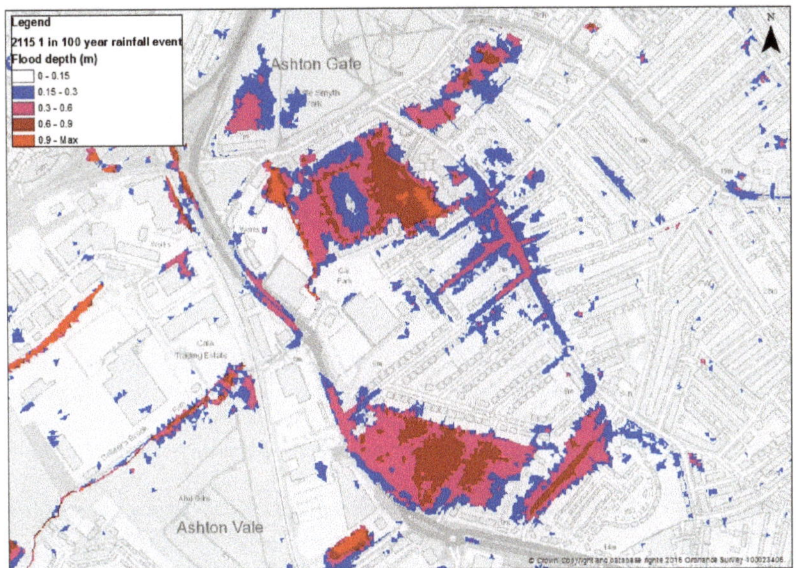

Figure 16. Flooding in the Ashton area for the 2115 scenario (inclusive of climate change).

Climate change poses two direct threats to flooding in Ashton:

- With projected sea-level rise, the duration of critical tidal levels (exceeding about 7.5 m AOD) will be longer, thus the gravity overflow at Ashton Avenue SPS will be able to operate for less time (as indicated in Table 4)
- Severe storms and wet-weather periods will be more frequent and intense, increasing the likelihood of
- Sewer flows exceeding the installed sewage pumping capacity
- Slightly higher flood flow levels in the river on top of the tidal effects

Within 100 years, the duration of a tide which is likely to cause flooding when the Ashton Avenue pumping station is beaten (i.e., 7.5 m AOD) shows a 6 x increase in the probability of occurrence. Furthermore, the duration of the tide which could cause serious flooding of properties (i.e., the 8.0 m AOD tide) would increase from 0.04% to 0.79% of the time–a 20× increase in the likelihood of occurrence compared to present day.

Sewer modelling has also indicated that under expected future rainfall conditions, the total installed pumping capacity at the Ashton Avenue pumping station could be exceeded roughly as follows:

- Current = exceeded once in 5 years
- 2050s = exceeded once in 2–3 years
- 2080s = exceeded once in 1–2 years
- 2110s = exceeded about once per year

4. Discussion

The main outputs of the SWMP model, that is, larger areas of pluvial flooding, have been verified by observations on-site during heavy rainfall events. The re-runs of the 2018 version of the SWMP model were also compared to the 2012 edition and the two correlated well. Through this analysis and interpretation of the results, an outlined package of adaptation measures and strategies based on these findings has been formulated. Examples of adaptations included in Ashton are:

- Provision of a surface water pumping facility to allow the watercourse/surface water system to discharge at all states of the tide (including sea-level rise)
- Re-grading of the Colliter's Brook open channel section to alter the gradient and widen the channel, providing more conveyance capacity
- Reinstatement of syphons and modifications to the structure which currently allows overflow from the watercourse/surface water system to the combined system
- Reducing the impermeable area in subcatchment upstream of Ashton Drive by 20% achievable through the introduction of sustainable drainage systems

Central area and St Philips Marsh:

- Construction of riverside flood defence walls to protect the low spots

Many of the outputs, methods and principles could be applied to any disruptive threats to the normal running of a city, thus allowing improved capacity to respond to shock events. Trying to reduce these impacts or enhance the recovery time by gaining a greater understanding of these systems and connections will offer improvement. Projections of key climate variables (rainfall, temperature) and sea-level rise for the epochs generated and how they will highlight the fragility and limitations on existing infrastructure and service functions were demonstrated. Knowledge of the problems faced assists in developing ways to sustain our key city functions and operations. Benefits of conducting the analysis include highlighting the criticality of points of the transport network for network management plans in the current day, such as the redirecting of traffic and issuing road closures that can be enacted ahead of a high tide warning. Another benefit is in identifying flooded electricity substations which, when resulting in failure, could indeed impact on another service that is reliant upon it as well within the wider network that it serves, and which may be well outside of the original flooded area. To make use of this, further investigation is required into other connected services. A key finding of the analysis is that there is a need for extra sewer and/or land-drainage pumping station pumping capacity to serve the area of Ashton in order to cope with future climate and tidal conditions. This is in addition to other means of reducing flood flows entering the area such as separation and use of SuDS.

The modelling begins to demonstrate the complexities of the city once different overlying functions are considered together as one. Understanding the 'domino effect' can then begin to quantify the cascading implications. Evaluating these connections can help build resilience in developing emergency response procedures and additionally inform strategic interventions. By integrating the models of urban management systems with flood models, an overall projection of city risk can be portrayed. Gaining a greater understanding of the resilience of city systems when we encounter disruptive events like flooding can, in principle, be applied in a similar process to other physical, social and economic challenges too. They will experience disturbances under such flood scenarios in the current day, which will worsen further still in the future with the predicted effects of climate change. From this, we can try and predict what some of the impacts will be if we were to experience extreme flood events, assess this and make plans to try and counteract it.

5. Conclusions

The key findings of this study include the need for an essential improvement of existing drainage infrastructure serving the area of Ashton in Bristol. The predicted effects of climate change and in particular, the impact of sea-level rise on tidal outfalls, will mean that a critical pumping station operating in the area will provide diminishing protection against extreme events as time progresses. The current pumping capacity will fail to deal effectively with the more intense storms and heightened river flows anticipated in future when outfalls become increasingly strained under rising sea-levels.

In the central St Philip's Marsh area, the "dry island" effect posed to this locality will have a larger knock-on impact to wider traffic flows in the adjoining road networks. In the future, under high spring tide or extreme tidal flood conditions, road closures will be far more prevalent and journey time delays escalated. Network management procedures will need to adapt to this and the requirement for longer-term strategic flood defences will become more pressing.

The number of electricity substations in central Bristol has shown to be at increased vulnerability to extreme tidal flooding when future flood extents and depths are considered highlighting how in the worst case (AWD > 160 cm) the number of substations affected could increase from 2 to 76 when increasing the return period from T20 to T1000. The substations identified as within the at-risk zones will need to ensure localized flood protection of the substation units up to the predicted future flood levels.

From this paper, it can be seen the importance and need for an integrated analysis for risk assessment related to city management during extreme events. Additionally, such analysis is of extreme relevance in order to detect the most critical zones and elements which may be impacted during such events; allowing to define and develop corrective strategies within the city with a holistic view.

Author Contributions: J.S. and R.H.; writing—original draft preparation, formal analysis, B.E., A.C., S.D., J.W., D.S.-M. and J.D.-G.; formal analysis, writing—review and editing. All authors have read and agreed to the published version of the manuscript.

Funding: This work was framed into the RESCCUE project, funded by the European Union's Horizon 2020 Research and Innovation Programme under Grant Agreement Number 700174.

Acknowledgments: The authors would like to acknowledge the RESCCUE project partner organisations and service providers operating in the city of Bristol, Western Power Distribution.

Conflicts of Interest: The authors declare no conflict of interest.

References

1. Djordjević, S.; Butler, D.; Gourbesville, P.; Mark, O.; Pasche, E. New policies to deal with climate change and other drivers impacting on resilience to flooding in urban areas: The CORFU approach. *Environ. Sci. Policy* **2011**, *14*, 864–873. [CrossRef]

2. Hallegatte, S.; Green, C.; Nicholls, R.J.; Corfee-Morlot, J. Future flood losses in major coastal cities. *Nat. Clim. Chang.* **2013**, *3*, 802–806. [CrossRef]

3. Committee on Climate Change. UK Climate Change Risk Assessment 2017 Synthesis Report. 2017. Available online: https://www.theccc.org.uk/wp-content/uploads/2016/07/UK-CCRA-2017-Synthesis-Report-Committee-on-Climate-Change.pdf (accessed on 6 March 2020).

4. United Nations Environment Programme (UNEP). Cities and Climate Change. Available online: https://www.unenvironment.org/explore-topics/resource-efficiency/what-we-do/cities/cities-and-climate-change (accessed on 6 March 2020).

5. Climate ADAPT. Climate Change Impacts on European Cities. Available online: https://climate-adapt.eea.europa.eu/knowledge/tools/urban-ast/step-0-2 (accessed on 6 March 2020).

6. World Bank. The Impact of Climate Change on Cities. Available online: https://siteresources.worldbank.org/INTUWM/Resources/340232-1205330656272/4768406-1291309208465/PartII.pdf (accessed on 6 March 2020).

7. Hunt, A.; Watkiss, P. Climate change impacts and adaptation in cities: A review of the literature. *Clim. Chang.* **2011**, *104*, 13–49. [CrossRef]

8. Jones, M.R.; Fowler, H.J.; Kilsby, C.G.; Blenkinsop, S. An assessment of changes in seasonal and annual extreme rainfall in the UK between 1961 and 2009. *Int. J. Climatol.* **2012**, *33*, 1178–1194. [CrossRef]

9. Westra, S.; Fowler, H.J.; Evans, J.P.; Alexander, L.V.; Berg, P.; Johnson, F.; Kendon, E.J.; Lenderink, G.; Roberts, N.M. Future changes to the intensity and frequency of short-duration extreme rainfall. *Rev. Geophys.* **2014**, *52*, 522–555. [CrossRef]

10. Bristol City Council. Mayor's Climate Emergency Action Plan. 2019. Available online: https://www.bristol.gov.uk/documents/20182/33379/Mayor%27s+Climate+Emergency+Action+Plan+2019+FINAL (accessed on 6 March 2020).

11. Bristol City Council. Bristol Resilience Strategy. Available online: https://www.bristol.gov.uk/documents/20182/1308373/Bristol+Resilience+Strategy/31a768fc-2e9e-4e6c-83ed-5602421bb3e3 (accessed on 6 March 2020).

12. Velasco, M.; Russo, B.; Martínez, M.; Malgrat, P.; Monjo, R.; Djordjević, S.; Fontanals, I.; Vela, S.; Cardoso, M.A.; Buskute, A. Resilience to Cope with Climate Change in Urban Areas—A Multisectorial Approach Focusing on Water—The RESCCUE Project. *Water* **2018**, *10*, 1356. [CrossRef]

13. Fewtrell, T.J.; Duncan, A.; Sampson, C.C.; Neal, J.C.; Bates, P.D. Benchmarking urban flood models of varying complexity and scale using high resolution terrestrial LiDAR data. *Phys. Chem. Earth* **2011**, *36*, 281–291. [CrossRef]

14. Pyatkova, K.; Chen, A.S.; Djordjević, S.; Butler, D.; Vojinović, Z.; Abebe, Y.A.; Hammond, M. *Flood Impacts on Road Transportation Using Microscopic Traffic Modelling Techniques*; Simulating Urban Traffic Scenarios; Springer: Cham, Switzerland, 2019; pp. 115–126.

15. Bristol City Council. *Bristol Local Flood Risk Management Strategy*; Bristol City Council: Bristol, UK, 2018.

16. Bristol City Council. Surface Water Management Plan. 2012. Available online: https://www.bristol.gov.uk/documents/20182/33916/2012.08.08%20SWMP_Final%20Phase%201%20Report-No%20Appendices_0.pdf/6d93f555-0558-4d0c-b94b-532620d3914c (accessed on 6 March 2020).

17. Bristol City Council. *Bristol Central Area Flood Risk Assessment*; Bristol City Council: Bristol, UK, 2015.

18. Bristol City Council. *South Gloucestershire Council Lower Severn Drainage Board: Avonmouth/Severnside SFRA Level 2 Summary Report*; Bristol City Council: Bristol, UK, 2011.

19. Gov.UK. Find out if You're at Risk of Flooding in England. Available online: https://www.gov.uk/check-flood-risk (accessed on 4 March 2020).

20. Innovyze. Infoworks ICM. Available online: https://www.innovyze.com/en-us/products/infoworks-icm (accessed on 6 March 2020).

21. Flood Modeller. Available online: https://www.floodmodeller.com/ (accessed on 6 March 2020).

22. TUFLOW. Available online: https://www.tuflow.com/ (accessed on 6 March 2020).

23. UK Department for Environment Food and Rural Affairs Standards. Available online: https://www.gov.uk/government/organisations/department-for-environment-food-rural-affairs (accessed on 30 January 2020).

24. Defra/Environment Agency. *Flood Risk Assessment Guidance for New Development Phase 2, Framework and Guidance for Assessing and Managing Flood Risk for New Development – Full Documentation and Tools*; R&D Technical Report FD2320/TR2.2005; Defra/Environment Agency: London, UK, 2005.

25. National Planning Policy Framework. Available online: https://www.gov.uk/government/publications/national-planning-policy-framework--2 (accessed on 30 January 2020).

26. UK Climate Projections 2009. Available online: https://assets.publishing.service.gov.uk/government/uploads/system/uploads/attachment_data/file/69257/pb13274-uk-climate-projections-090617.pdf (accessed on 30 January 2020).

27. Flood and Coastal Erosion Risk Management Handbook and Data for Economic Appraisal. 2019. Available online: https://www.mcm-online.co.uk/handbook/ (accessed on 30 January 2020).

28. Gov.UK. Flood Risk Assessments: Climate Change Allowances. Available online: https://www.gov.uk/guidance/flood-risk-assessments-climate-change-allowances (accessed on 30 January 2020).

29. SUMO. Simulation of Urban Mobility. Available online: http://sumo.sourceforge.net/ (accessed on 4 March 2020).

30. Lopez, P.A.; Behrisch, M.; Bieker-Walz, L.; Erdmann, J.; Flötteröd, Y.; Hilbrich, R.; Lücken, L.; Rummel, J.; Wagner, P.; WieBner, E. Microscopic Traffic Simulation using SUMO. In Proceedings of the 2018 21st International Conference on Intelligent Transportation Systems (ITSC), Maui, HI, USA, 4–7 November 2018; pp. 2575–2582. [CrossRef]

31. Find Open Data. Available online: https://data.gov.uk/ (accessed on 21 December 2019).

32. OpenStreetMap. Available online: https://www.openstreetmap.org/ (accessed on 21 September 2019).

33. TransCAD Transportation Planning Software. Available online: https://www.caliper.com/tcovu.htm (accessed on 20 December 2019).

34. Pyatkova, K.H. Flood Impacts on Road Transportation. Ph.D. Thesis, University of Exeter, Exeter, UK, 2019.

35. House of Commons Transport Committee. Keeping the UK Moving: The Impact on Transport of the Winter Weather in December 2010. Available online: http://www.publications.parliament.uk/pa/cm201012/cmselect/cmtran/794/794.pdf (accessed on 15 April 2020).

36. QGIS. Available online: https://www.qgis.org/en/site/ (accessed on 6 March 2020).

37. Western Power Distribution. Available online: https://www.westernpower.co.uk/ (accessed on 6 March 2020).

38. Sánchez-Muñoz, D.; Domínguez-García, J.L.; Martinez-Gomariz, E.; Russo, B.; Stevens, J.; Pardo, M. Electrical grid risk assessment against flooding in Barcelona and Bristol cities. *Sustainability* **2020**, *12*, 1527. [CrossRef]
39. Energy Research Partnership. Future Resilience of the UK Electricity System. 2018. Available online: https://erpuk.org/wp-content/uploads/2018/11/4285_resilience_report_final.pdf (accessed on 15 April 2020).
40. Martinez-Gomariz, E.; Gómez, M.; Russo, B. Experimental study of the stability of pedestrians exposed to urban pluvial flooding work. *Nat. Hazards* **2016**, *82*, 1259–1278. [CrossRef]
41. Chanson, H.; Brown, R. New criterion for the stability of a human body in floodwaters. *J. Hydraul. Res.* **2015**, *53*, 540–541. [CrossRef]
42. Russo, B.; Gómez, M.; Macchione, F. Pedestrian hazard criteria for flooded urban areas. *Nat. Hazards* **2013**, *69*, 251–265. [CrossRef]
43. Arrighi, C.; Oumeraci, H.; Castelli, F. Hydrodynamics of pedestrians' instability in floodwaters. *Hydrol. Earth Syst. Sci.* **2017**, *21*, 515–531. [CrossRef]

Article

Flood Risk Assessment in an Underground Railway System under the Impact of Climate Change—A Case Study of the Barcelona Metro

Edwar Forero-Ortiz [1,2,*], Eduardo Martínez-Gomariz [1,2], Manuel Cañas Porcuna [3], Luca Locatelli [4] and Beniamino Russo [4,5]

[1] Cetaqua, Water Technology Centre, Carretera d'Esplugues, 75, 08940 Cornellà de Llobregat, Barcelona, Spain; eduardo.martinez@cetaqua.com
[2] Flumen Research Institute, Universitat Politècnica de Catalunya, Calle del Gran Capità, 6, 08034 Barcelona, Spain
[3] TMB (Transports Metropolitans de Barcelona), Carrer 60, núm. 21-23, sector A, Pol. Ind. de la Zona Franca, 08040 Barcelona, Spain; mcanas@tmb.cat
[4] AQUATEC (SUEZ Advanced Solutions), Paseo de la Zona Franca, 46-48, 08038 Barcelona, Spain; luca.locatelli@aquatec.es (L.L.); brusso@aquatec.es (B.R.)
[5] Grupo de Ingeniería Hidráulica y Ambiental (GIHA), Escuela Politécnica de La Almunia (EUPLA), Universidad de Zaragoza, Calle Mayor, 5, 50100 La Almunia de Doña Godina, Zaragoza, Spain
*   Correspondence: eaforero@cetaqua.com or edwar.forero@gmail.com; Tel.: +33-783-53-72-22

Received: 19 May 2020; Accepted: 28 June 2020; Published: 30 June 2020

Abstract: Flooding events can produce significant disturbances in underground transport systems within urban areas and lead to economic and technical consequences, which can be worsened by variations in the occurrence of climate extremes. Within the framework of the European project RESCCUE (RESilience to cope with Climate Change in Urban arEas—a multi-sectorial approach focusing on water), climate projections for the city of Barcelona manifest meaningful increases in maximum rainfall intensities for the 2100 horizon. A better comprehension of these impacts and their conditions is consequently needed. A hydrodynamic modelling process was carried out on Barcelona Metro Line 3, as it was identified as vulnerable to pluvial flooding events. The Metro line and all its components are simulated in the urban drainage models as a system of computational link and nodes reproducing the main physical characteristics like slopes and cross-sections when embedded in the current 1D/2D hydrodynamic model of Barcelona used in the project RESCCUE. This study presents a risk analysis focused on ensuring transport service continuity in flood events. The results reveal that two of the 26 stations on Metro Line 3 are exposed to a high risk of flooding in current rainfall conditions, and 11 of the 26 stations on Metro Line 3 are exposed to a high risk of flooding in future rainfall conditions for a 20-year return period event, which affects Metro service in terms of increased risk. This research gives insights for stakeholders and policymakers to enhance urban flood risk management, as a reasonable approach to tackle this issue for Metro systems worldwide. This study provides a baseline for assessing potential flood outcomes in Metro systems and can be used to evaluate adaptation measures' effectiveness.

Keywords: flood risk assessment; climate change; 1D/2D hydrodynamic model; Metro system; subway; urban mobility

---

1. Introduction

Current trends in the analysis of climate-driven events on urban societies, infrastructures, and services have guided study towards the direct and indirect impacts of these events, resulting in disruptions within the interdependent infrastructure systems. These studies draw our awareness to the importance of analyzing the impacts generated by extreme events, such as heavy rainfall, flooding, storm surge, and other extreme natural phenomena which severely compromise city services. This introduction presents the importance of assessing Metro systems' flooding impacts, pointing to the research gap on this topic and supporting the motivations leading this study.

Climate change represents a severe threat to cities and their resilience, which are complex systems characterized by constant flow and are the result of a lot of dynamic variables that change in space and time [1,2]. As discussed by Wan et al. [3], transportation resilience can be described as the capacity of a transportation system to "absorb disturbances," sustaining its fundamental structure and purpose, plus recovering to an expected service level following occurring disruptions. As an initial action for a risk/resilience appraisal [4], a potential hazard for urban resilience is the occurrence of flooding events which might impact the continuity of Metro services. Metro systems, as one strategic transport service on urban areas, represent one key variable for assessing resilience in the context of physical infrastructure used by citizens. Decreasing the risks that may affect the continuity of the Metro service, therefore, will increase resilience at the municipal level [5]. Although there are a variety of words for underground transport railway systems (e.g., subway, underground, or tube), this paper will use the term "Metro systems" when referring to them.

Metro systems' importance worldwide for cities' operability is evident. According to 2018 figures [6], 178 cities in 56 countries have a Metro transportation system, with an average ridership of 168 million per day, and increasing annual use of 19.5%. Figure 1 presents the currently in-service number of Metro systems in Europe. Between China, India, and Iran, Asia is building 16 new systems, shaping the decarbonizing urban transport effort as a response to climate change [6].

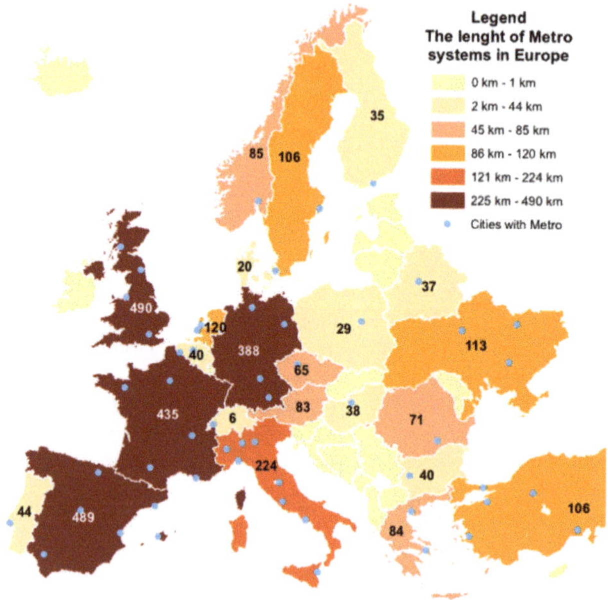

Figure 1. Metro systems in operation in Europe by 2019 [6].

Following the accelerated growth of cities and their Metro systems, and the increasing growth and complexity of Metro networks, the ability to study and improve the vulnerability of systems are more complex, according to the heterogeneity of the different components of a large scale system, such as a transport network [7]. The probability of disruptions should consider many factors, not only the number of affected Metro stations, but also the distance between them [8], along with several other factors.

Considering hazards that Metro systems face worldwide, there is a notable lack of research on water-related hazards [5], including those related to flooding events and their impact on Metro systems' resilience [9]. In a context of climate change, flooding events can produce increasing significant socioeconomic risks in urban areas. Worse still, urbanization heightens the likelihood of water disasters such as floods because development decreases the amounts of permeable surfaces where water can soak into the soil, thereby originating runoff that contributes to flooding. This situation poses a critical risk to Metro systems, which can suffer from diverse types of flooding, such as riverine (fluvial) and rain-related (pluvial) flooding, and tidal surges.

Climate change endangers the Barcelona Metro system when subjected to flooding events. Flood events have caused property damage along with service disruptions, by the inundation of underground infrastructures (e.g., Metro tunnels and facilities) [10,11]. Therefore, it is essential to evaluate flood risks in Metro systems to plan for flood disasters and set mitigation strategies efficiently. Figure 2 illustrates how Metro flooding events affect both Metro systems' infrastructure and users, exposing them to collateral effects such as hard falls due to the loss of stability of passengers crossing through water flow [12,13], material damage, and reduction of service provision due to damage caused to both the users and the companies that administer the systems. These events have gradually increased in recent years due to cities' growth and climate change [14–16].

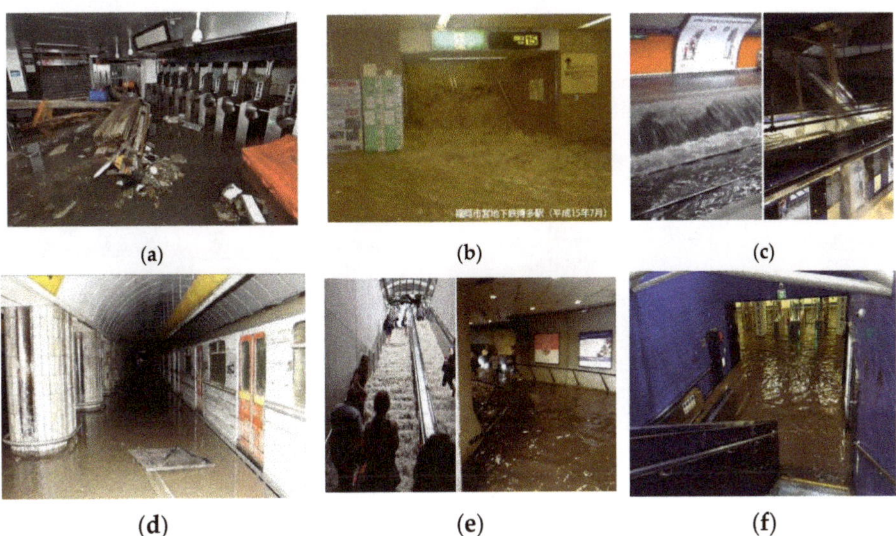

Figure 2. Flood events in major global Metro systems: (a) New York [17]; (b) Tokyo [18]; (c) Madrid [19]; (d) Prague [20]; (e) Washington [21]; (f) Brussels [22].

The main objective of this research is to analyze Metro stations' vulnerability facing floods under extreme rainfall events, then, exposing the system to climate change flooding influenced events. Other research approaches focus on ensuring transport service continuity using other public transport modes, responding to Metro system flooding events [23–26]. In order to address the cited research gaps, this study proposes and analyzes interactions connecting pluvial flood events and the water level thresholds inside Metro station tunnels that might result in service disruption, through a hydrodynamic model including the Metro system. Barcelona Metro Line 3 is the line most vulnerable to flood events due to extreme rainfall, according to TMB (Transports Metropolitans de Barcelona, Barcelona Metropolitan Transport—the leading public transport operator in Barcelona) records. Hence, in this study, the Metro

Line 3 is integrated into the 1D/2D hydrodynamic model developed for the entire drainage system of the city of Barcelona within the European project RESCCUE (RESilience to cope with Climate Change in Urban arEas—a multi-sectorial approach focusing on water) [27].

This study proposes a new methodology for a flood risk assessment in an underground Metro system, based on a surface pluvial flooding 1D/2D hydrodynamic modelling and the Metro infrastructure real data. The new flood model, derived from the integration of the infrastructure of Line 3 of the Barcelona Metro with the city's current drainage model, was calibrated based on flood observed from photos and visual inspection by TMB operators. The Metro elements through which floodwater enters into the tunnels (i.e., staircases, ventilation grates, hallways) were simulated based on their simplification, depicting the Metro entrances as pipelines addressing the Metro tunnels and stations, as an approach that can be applied to other case studies.

As a vital component of the flood risk assessment, future scenarios using rainfall projections under the climate change impact were applied. The model calibration aimed at reproducing observed water levels in the studied Metro stations, water levels in various real events were adequately reflected by the model. Also, it was possible to identify the likely effects of climate change on underground flooding by analyzing the effects of future rainfall conditions. Finally, a risk assessment of the Metro station was performed, adaptation strategies were proposed to reduce floods' impact and to prevent disruptions in Metro service, increasing the resilience of city transportation services.

2. Materials and Methods

This study incorporates Barcelona Metro Line 3 real data into the hydrodynamic 1D/2D model of Barcelona considering the connections among possible water inflows and Metro tunnels and stations and considering them as pipelines. To summarize the overall structure of the model, Figure 3 lists the necessary steps performed for the proposed flood risk assessment.

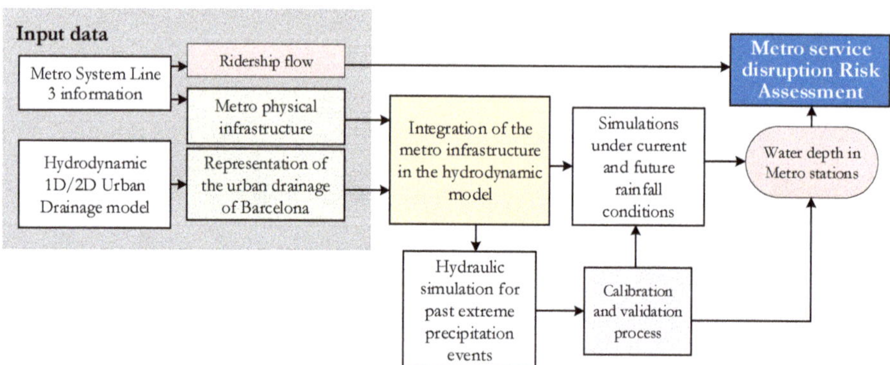

Figure 3. The overall model structure for flood risk assessment in Metro systems.

This chapter presents a list of the precipitation events in Barcelona that have caused disruptions in the Metro service due to the entry of water into the system, considering both the sources and methods of study, with the hydrodynamic model implementation, the Metro Line 3 data acquisition, and components' introduction processes to the model. Furthermore, this chapter describes the model calibration and validation process, along with the flood risk assessment under the impact of climate change.

2.1. Historical Data of Metro Systems' Flood Events in Barcelona

Table 1 summarizes a two-decade historical data of Metro system flood events. Only those events that caused service disruption have been considered. The table was obtained linking Barcelona Metro system records from system administrator TMB, with internet media. Service disruption information is available for request to TMB, due to its confidential nature.

Table 1. Barcelona Metro service disruptions due to flooding events in the last 20 years.

Date	Average duration – Cumulative Rainfall	Disruptions in Metro Stations – Line Affected (L, Line Number)	Source
2002-10-10	1 h – 170 mm	Via Júlia (L4) – Trinitat Nova (L4) – Rambla Just Oliveras (L1) – Maria Cristina (L3)	[28]
2004-09-14	1 h – 65 mm	Cornellà Centre (L4)	[29]
2009-10-22	1 h – 70 mm	Sant Antoni (L2) – Paral·lel (L3) – Rambla Just Oliveras (L1) – Trinitat Vella (L1) – Verdaguer (L5)	[30]
2011-07-19	1.5 h – 80 mm	Verdaguer (L4) – Bogatell (L4)	[31]
2018-07-16	1 h – 40 mm	Canyelles (L3) – Valldaura (L3)	[32]
2018-09-16	2 h – 75 mm	Poble Sec (L3) – Paral·lel (L3) – Espanya (L3) – Drassanes (L3) – Liceu (L3)	[33]
2018-10-09	1 h – 36 mm	Espanya (L1) – Santa Eulàlia (L1) – Torrasa (L1) - Paral·lel (L3) – Badal (L5)	[34]
2018-11-15	2 h – 95 mm	Poble Sec (L3) – Paral·lel (L3) – Espanya (L3) – Santa Eulàlia (L1) – Liceu (L3) - Vallcarca (L3)	[35]
2019-07-27	0.5 h – 43 mm	Sagrada Familia (L2) – Espanya (L3) – Verdaguer (L5)	[36,37]
2019-08-27	1 h – 50 mm	Espanya (L3) – Collblanc (L5)	[38]
2019-11-14	1 h – 53 mm	Arc de Triomf (L1) – Paral·lel (L3)	[39]
2019-12-04	5 h – 102 mm	Tetuan (L2) – Lesseps (L3) – Joanic (L4) – Urgell (L1) – Espanya (L3, L1) – Arc de Triomf (L1) – Sants (L5) – Sant Roc (L2)	[40]

Table 1 suggests Barcelona Metro flooding events frequency has increased over the last two years, indicating that these events could maintain or increase their frequency or intensity, due to climate change influence on rainfall variations [16]. Barcelona rainfall events frequency and intensity increases, doubling frequency downpours and peak rainfall rates up to 20% higher [41], rainfall intensities rise due to climate change impacts [42].

2.2. Overview of the Hydrodynamic Model

A 1D/2D urban drainage model of the whole municipality of Barcelona was developed, calibrated, and validated using the hydraulic modelling software Innovyze InfoWorks ICM®and local observation data from different rain events. This model is an update of the model used for the Drainage Master Plan of Barcelona of 2006.

A rainfall-runoff model was used to compute the runoff from buildings that were assumed to directly drain into the 1D hydraulic model representing the drainage network. The 2D model was used to reproduce overland flow on streets, parks, and further areas located at terrain elevation that can be flooded. The 1D and the 2D models continuously interact with each other through model nodes that physically represent surface inlets (like gullies or maintenance holes) to the drainage system. At these nodes, the exchange of water between the 1D and the 2D model was computed using different modelling options available in InfoWorks (nodes can be defined as Inlets 2D, 2D, Gullies 2D) that compute water flows as a function of local water levels and inlet geometries.

A peculiarity of this 1D/2D model is that part of the rainfall is directly applied to the streets and parks represented by the 2D overland flow model, and part to the rainfall-runoff model used for buildings, roofs, and elevated areas. Conventional urban drainage modelling approach generally applies rainfall directly to rainfall-runoff models that compute runoff diverting it into the 1D hydraulic model, and flooding can only occur through a maintenance hole surcharge. In Barcelona, it is believed that urban floods are partly caused by deficient surface drainage capacity due to, for instance, a reduced number of street gullies.

This model includes approximately 2164 km of pipes, 67,967 pipes, 66,158 nodes, 18 detention tanks with a total volume of approximately 461,600 m^3, 489 weirs, 22 pumps, 47 sluice gates, and 120 outfalls. The full 1D Saint-Venant equations are used to solve the sewer flow. Rainfall-runoff processes were simulated using a single non-linear reservoir model with routing coefficients that are a function of surface roughness, surface area, ground slope, and catchment width. Initial hydrological losses were simulated to be approximately 1 mm (a little higher in pervious areas and lower in impervious ones).

Continuous losses were only applied to pervious areas and were simulated using the Horton model. The 2D model has an unstructured mesh with 662,071 cells created from a Digital Terrain Model (DTM) with a resolution of 2x2 m^2 obtained by a special combination of a 3-D scanning and laser scanning (LIDAR) provided by the Cartographic and Geological Institute of Catalonia (with an accuracy of 0.15 m for altitudes). The 2D cells have areas in the range of 5–125 m^2 in the urban area and 100–10,000 m^2 in the upstream rural areas.

The 1D/2D model was quantitatively and qualitatively calibrated and validated using data from four different historical rain events. In Barcelona, approximately 25 rainfall gauges and more than 100 water level sensors and flood event videos and photos are available. Calibration and validation details can be found in deliverable 2.2 of the EU RESCCUE project [27].

2.3. Description of how the Metro Service is Affected by Flooding

Depending on the type of tunnel, existing floor, sleeper type, and other variables unique to each Metro system in the world, the characteristics of the infrastructure that perform primary flood control in Barcelona Metro service tunnels are summarized as follows.

The dimensions presented in Figure 4 are estimated, since they may vary depending on the sort of tunnel, among other factors, and should be taken as a theoretical reference. Therefore, depending on the existing drainage in each case, it will be more or less rapid for the water to reach the base of the rail.

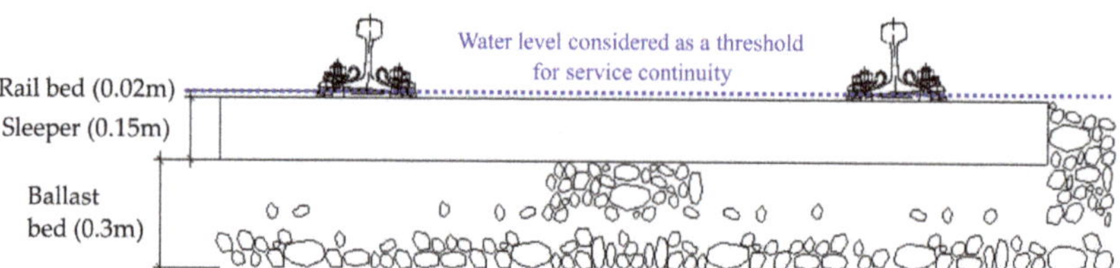

Figure 4. Approximate scheme of the Barcelona Metro's railway infrastructure.

One of the essential parts of the train movement is railway signaling, which always allows safe movement by regulating the speed and location of trains. The basis of the railway signaling is based on the track circuits, whose theoretical schematic configuration is shown in Figure 5 (it does not precisely represent the reality for the whole Line 3 of the Barcelona Metro), where the track is electrically isolated in sections of a certain length. In the case of Metro Line 3, the track circuits are about 25 and 35 meters long.

The railway signaling equipment is qualified to work under wet conditions; therefore, the circuit could work in case of flooding. Depending on the amount of water in the surrounding area and its conductivity, it may not complete the circuit, then diverting the electric current between the power supply and the receptor, producing the receptor is not over-excited by the electric current and the block is assumed occupied. It is not conceivable to set only a water level which produces the false occupation phenomenon. However, it could be established as an inaccurate reference level when the water level reaches half of the railway rail (0.08 m from the rail bed), we could have an occasional false occupation.

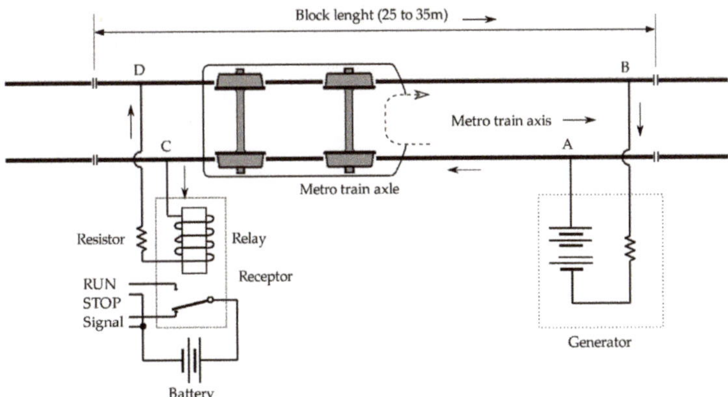

Figure 5. Occupied block—track circuit scheme. Modified from [43].

In the case of having a higher water level, in many cases, there would be false occupation, and therefore no Metro service would be provided. Due to these circumstances, this study takes a level of 0.15 m as an approximate reference level in which, in any road configuration, it is inevitable that because of a flooding event, false occupation is produced. It must be considered that even if we could find a situation where the water reaches the height of the head of the rail, there would be no false occupation, due to there being no short circuit of the electrical circuit. At this point, it would be necessary to analyze whether it is mechanically viable for trains to transit, given the possibility of slippage in the wheel-rail contact and the impossibility of visualizing the rail with the danger that this entails. Furthermore, concerning the ATP (automatic train protection) system, the closer the water gets to the head of the rail, the easier it is for the train's ATP antennas not to be able to obtain correct track circuit information, causing the train to brake for lack of obtaining the ATP code. When the water covers the whole head of the rail, accurate ATP interpretation is not possible.

2.4. Integration of the Metro Infrastructure in the Hydrodynamic Model

Water presence in the tunnel and the stations of a Metro system affects the continuity of the service. Thus, this study represents Metro network components into the hydrodynamic model through the creation of hydraulic geometry elements, comparable to real infrastructure components, evaluating a conceivable water entry event into the system, through hydrodynamic modelling. Principal floodwater access to the Metro system during extreme precipitation events are the ventilation grates and stations' entrances, according to TMB and visual inspections carried out by TMB personnel during extreme rainfall events.

The data required to calculate the flow intercepted at the ventilation grates depends on the modelled inlet type. InfoWorks ICM®offers various methods to estimate captured flow by inlets, two of them based on the Federal Highway Administration (FHWA) Urban Drainage Design Manual, Hydraulic Engineering Circular No. 22 (HEC-22) [44]. This study applies another method, an alternative to the HEC-22 continuous grate inlet based on the equations from work carried out by Gomez et al. [45] at the Technical University of Catalonia (UPC), implemented in the InfoWorks®ICM software package.

Three-dimensional numerical modelling experimental campaigns [46,47] executed at UPC characterizes Barcelona drainage grates' hydrodynamic behavior [45,48,49], including physical obstruction conditions determining hydrodynamic efficiency decrease on partially clogged grated inlets [50]. The ventilation grates as a water inlet use this previous research outcome since the model integrates hydrodynamic equations representing urban drainage grates, comparable to ventilation grates.

Metro tunnels' representation in the hydrodynamic model introduces pipes with identical dimensions and geometry to the real infrastructure, considering tunnels' cross-section and elevation data as Figure 6 shows. Metro network elements' depiction in the hydrodynamic model process disregards entrance to service tunnel variability, which simplifies the flood impact analysis on the stations' platform and tunnels.

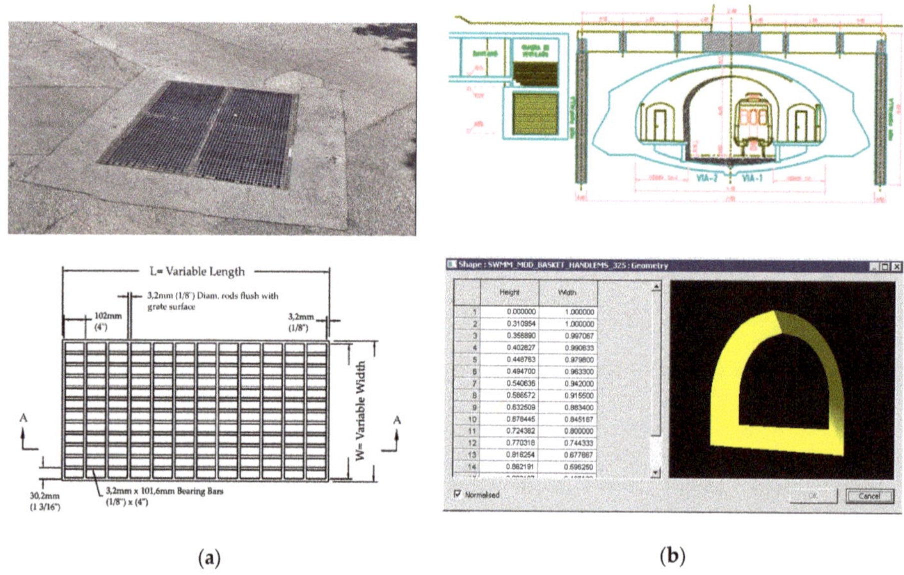

(a)   (b)

Figure 6. (**a**) Image (up) and essential features (down) of a typical surface grate for ventilation purposes in the Barcelona Metro system; (**b**) Catalunya Metro station cross-section drawing (up); tunnel depiction shape in ICM®software (down).

Reaching a proper representation of Metro entrances requires gathering information such as longitude, altitude, stairs slope, and the elevation difference with the tunnel service, to set the pipe connecting them. An evaluation and calibration process for different inlet types concludes, in this case, that the continuous curb-opening inlet typology provides a better representation of Metro station entrances' hydrodynamic behavior.

2.5. Calibration and Validation Process

The calibration process consists of adapting model parameters to achieve correspondence among simulated flood behavior and recorded or observed flooding. The calibration parameters are listed in Table 2. This process trains the model concerning selected hydraulic conditions that are compared with the observed data. Specifically, the computed water levels in both the surroundings of the Metro stations' access and the tunnels were validated in this study.

TMB selects the pluvial flooding event dated 2018-09-06 as a flood event suitable for calibration purposes in two stations of Line 3 (Paral·lel and Drassanes). Throughout this precipitation, the water depth in the Metro tunnels of these two stations induced a service disruption, the presence of water already generates the false occupation phenomenon. Metro operators can remark on how risky it is to have a station open to possible water flows. However, it is the false occupation that disrupts the service. With a variable water depth in the tunnel service, although the train can pass through the station it cannot stop to pick up or drop off passengers, as it is technically incapable of stopping.

Table 2. Selected calibration parameters of the hydrodynamic model, including calibrated values.

Model Element	Calibration Parameter	Range (min–max)	Calibrated Value
Pipes—mains are connecting from water entries to tunnels, tunnels representation.	Pipe roughness—Metro accesses and ventilation grates	0.01–0.2 (Manning's roughness coefficient $s/m^{1/3}$)	0.01
	Pipe roughness—Metro tunnels		0.02
Mesh for the 1D/2D coupled model	Maximum triangle area	100–10,000 m^2	10,000 m^2
	Minimum element area	5–125 m^2	125 m^2

A calibration method employing a large data sample is not feasible in this model, due to the water level measurements shortage inside tunnels, hence this study tests to check its performance in real-world applications after calibration and before practice utilization, as the validation phase. Validation process applies another recorded surface flood event to ensure that the model acceptably reproduces recorded results, in this case, the pluvial flooding event dated 2018-08-17. This surface flooding event affected potential water entry points to the Metro system, but regular Metro service continued. The water level was not enough for the TMB operators to set the service interruption, nor was the water accumulated in the tunnels ample for the Metro system's self-protection system to produce a disruption in service.

2.6. Flood Risk Assessment under the Impact of Climate Change

One of the most used approaches in assessing the vulnerability of underground mass transport systems is the identification of the effects of service disruptions [9], giving a primary role to the ridership flow at Metro stations. Some authors have mainly been interested in questions concerning risk assessment for Metro stations [51,52] in a holistic manner. Metro system flooding events can lead to risk conditions involving both ridership and infrastructure, for instance, its impact on passengers' stability and safety [12,53]. Other authors focus their research on the optimization of evacuation processes during flood events in underground infrastructure [54–58]. However, this study defines risk according to the interruption of the Metro service, as a result of rainwater flooding on the urban transport system.

The concept of risk used in this research encompasses the combination of hazard and vulnerability. Hazard is characterized by flooding water depths (as the intensity of the hazard) along the Metro tunnel related to design storms of different return periods (probability of flooding events). Vulnerability criterion focuses on the ridership number affected by service interruptions. Therefore, the flood risk assessment process for Metro stations involves a hazard assessment, following the vulnerability assessment, and the procedure to obtain the risk data and maps for Barcelona Metro Line 3, as Figure 7 illustrates. For the hazard assessment, one, five, 10, and 20 years return period design storms are employed as inputs for Barcelona hydrodynamic 1D/2D coupled model for this study.

These design storms were selected following the recommendations of the municipal operator of the city drainage system (Barcelona Cicle de l'Aigua, SA—BCASA). The principal criteria for establishing this hazard estimation is to ensure continuity of Metro services, as the primary concern of the Metro administrator. One of the operational thresholds which defines Metro service operation is the water depth within the Metro tunnels. Considering a flood event, after the water depth reaches 0.15 m inside Metro station tunnels, Depending on the state of the railway road track in front of the train, if there is false occupation due to water, the train would not reach that circuit because it would stop before arriving, impacting the entire line service. Through operational experience and based on technical assumptions with support from TMB, the hazard ranges are defined as follows: water depths between 0 and 0.15 m are defined as low, between 0.15 and 0.30 m as medium, and depths greater than 0.30 m are high. These values are variable in a real operation scheme and depend on the track circuit, so they are only taken as a reference.

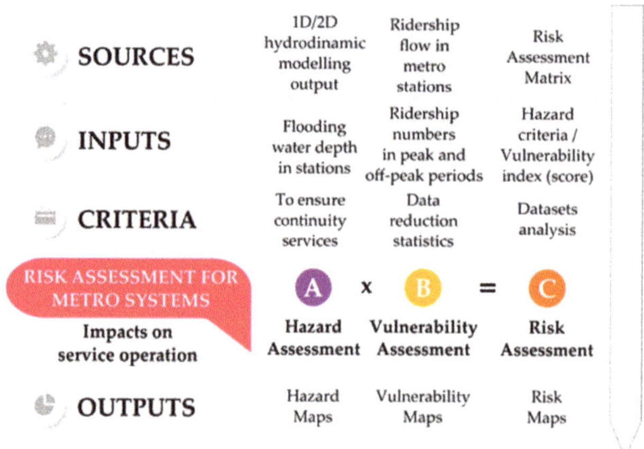

Figure 7. Risk concept definition and datasets required in each stage.

For the vulnerability assessment, registered Line 3 station user entrance data, for every hour in the months November and December 2019, reflect the ridership flow, defining peak hours (18:00–19:00) and off-peak hours (12:00–13:00). Vulnerability assessment development uses ridership flows as inputs. It applies the Jenks natural breaks algorithm as a standard method for dividing a dataset into three homogenous classes; low, medium, or high. The hazard and vulnerability limits proposal and matrices are summarized in Table 3. Methods for risk assessment can be qualitative or quantitative, both having limitations [59]. To defining risk as to the probability or threat to a hazard occurring in a vulnerable area, flood risk was assessed through a risk map related to a determined scenario and return period by combining hazard and vulnerability maps. Metro stations and ridership are affected by floods in Barcelona, and their assessment defines hazard, vulnerability, and risk levels by significance levels such as "high," "medium," and "low," evaluating the resultant risk level against qualitative criteria.

Table 3. Hazard and vulnerability criteria for Metro stations service exposed to flooding.

Index / Score	Hazard, Depending On Flood Return Periods (Water Depth)	Vulnerability for Off-Peak Hours (Ridership Flow)	Vulnerability for Peak Hours (Ridership Flow)
1 (low)	0–0.15 m	116–578 users/h	96–591 users/h
2 (medium)	0.15–0.30 m	578–1075 users/h	591–1244 users/h
3 (high)	> 0.30 m	1075–1516 users/h	1244–2701 users/h

Hazard range assignment relates to Metro trains' capacity to operate under tunnel flooding conditions, considering the maximum water levels for the inundation events and the consequent service disruption. For the low hazard level defined by floodwater levels between 0 m and 0.15 m, service continuity disruption starts, and the Metro operation is subject to substantial supervision until the water level reaches 0.1 m when the service might be disrupted. At the medium hazard level, service disruption is likely because trains can no longer stop, affecting service availability on affected stations. High hazard level corresponds to floodwater level exceeding 0.3 m, under this water depth condition, the rail is likely covered by water, thus services are disrupted and trains cannot ride.

The creation of risk and vulnerability maps implies multiplying the vulnerability index (1, 2, or 3, corresponding to the low, medium, and high vulnerability) by the risk index (1, 2, or 3, corresponding to the low, medium and high). The total risk ranges from 1 to 9, where higher levels indicate a higher risk. This approach is compiled in the risk matrix shown in Table 4.

Table 4. The risk matrix for Metro stations and ridership.

	Hazard		
Vulnerability	**Low**	**Medium**	**High**
Low	Low	Low	Medium
Medium	Low	Medium	High
High	Medium	High	High

The entry of water into the subway system due to surface flooding is since the flood levels are higher than the elevation of the Metro infrastructure, which allows access to water, e.g., ground ventilation grates and user entrances to the system. This study identifies these water entry accesses according to different rainfall conditions. It suggests some non-structural and structural measures [5], combining both as a successful strategy facing the increased risk of flooding, according to Barcelona Metro system-specific factors.

3. A Case Study for the Barcelona Metro System

The purpose of this chapter is to present the physical characteristics of the study area, including a comprehensive summary of the Metro system, the hydrological conditions over the study area, a comprehensive summary of the rainfall data, and the boundary conditions applied in the integrated hydrodynamic 1D/2D model.

3.1. Characteristics of the Study Area

Barcelona has a Mediterranean climate, with gentle winters and mild summers. Barcelona soil cover experienced massive urbanization through the last century. Nowadays, the city land-use is about 80% of impervious areas and 20% of pervious areas. This morphology causes a brief time of concentration in catchment areas. As a result, medium and heavy storm events produce quick stormwater storage in flatland areas, provoking urban flash floods.

3.2. Underground Infrastructure Network: Metro System

Based on TMB confidentiality policies, this research does not offer Metro system details such as ventilation grates, accesses, tunnel dimensions, incoming water volumes, and ridership flows. The system runs by one operator (TMB) whose fare integrates into the unified six-zone system created by the Metropolitan Transport Authority. This research has TMB support, manager of the Metro system as a benchmark public mobility company in Europe and the world. With 13 lines, consisting of ten Metro lines, a cable car, a tram and a funicular, with a length of 121.4 km for the entire network, and 160 stations, the Metro network is the second largest conventional Metro network in Spain after Madrid, and since December 2009, the first with automated lines.

In 2019, 411.95 million passengers used the Barcelona Metro network managed by TMB. Barcelona Metro Line 3 links Zona Universitària and Trinitat Nova stations along a 17.8-kilometre double-track stretch with 26 stations underground. Line 3 at the peak hour averages 26 trains with 95 million passengers being second in terms of ridership volume, reaches a commercial speed of 26.5 km/h, and uses a rigid catenary system for the power supply and standard track width [60]. Figure 8 reveals flooding events' marked concentration on one specific area of Metro stations, which coincides with the most frequent surface pluvial floods area; during rainfall events, stormwater flow exceeds sewer pipe capacity.

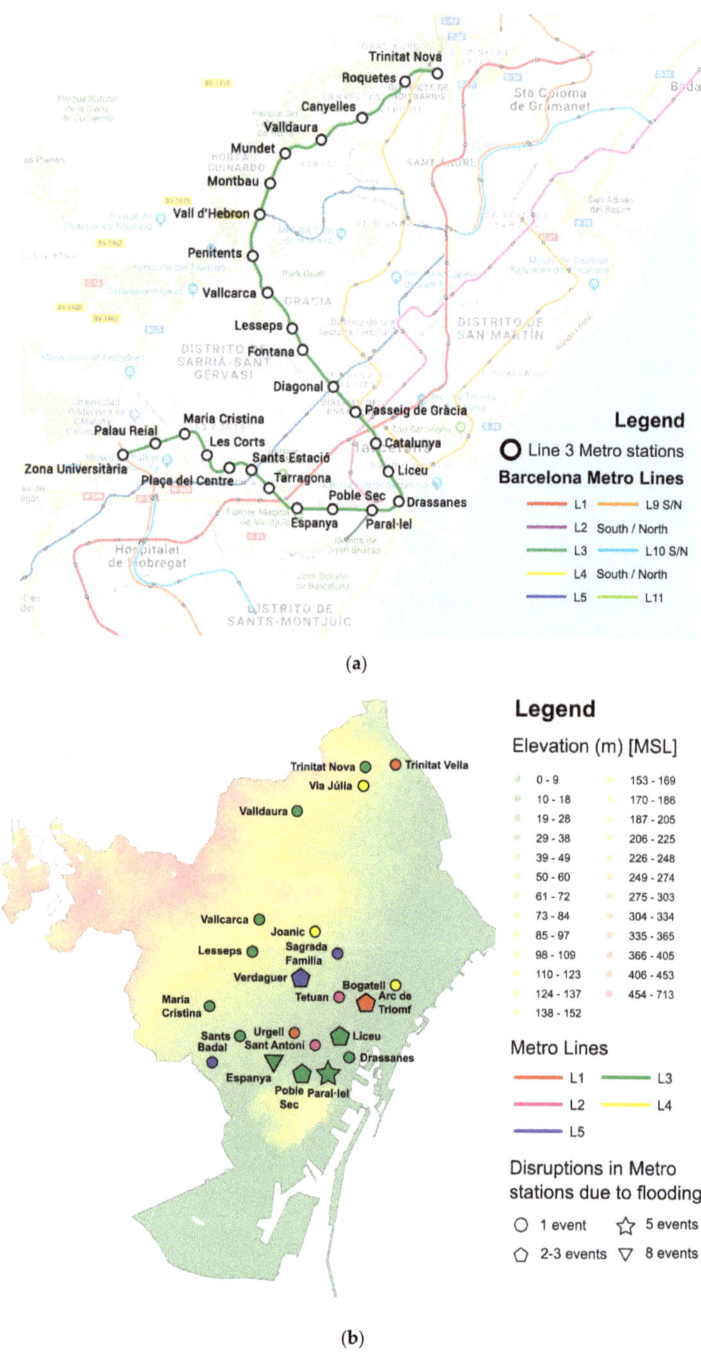

Figure 8. (**a**) Location of the stations on Line 3 of the Barcelona Metro system; (**b**) Barcelona Metro stations affected by flooding in the last 20 years, the lines to which they belong, and the number of service interruptions due to flooding at stations. The size of the nodes for every affected Metro station is proportional to the number of disruptions and related to the geometric element.

3.3. Data Acquisition

As part of the research, a precise record of Barcelona Metro Line 3's possible water accesses was obtained based on TMB provided data. Historical flood events confirm surface runoff reaches the Metro system by specific accesses and ventilation grates, set by visual reports authored by newspapers, also registered and validated by TMB staff. Metro Line 3 elements' representation as part of the coupled 1D/2D Barcelona drainage network model considers water flow through Metro station entrances, lifts, and ventilation grates, as Figure 8 shows. This inclusion allows improving model outcomes, as flow dynamics description between surface flow and urban drainage improves [61]. Figure 9 indicates the spatial location of the possible water entry points to Line 3 of the Metro system.

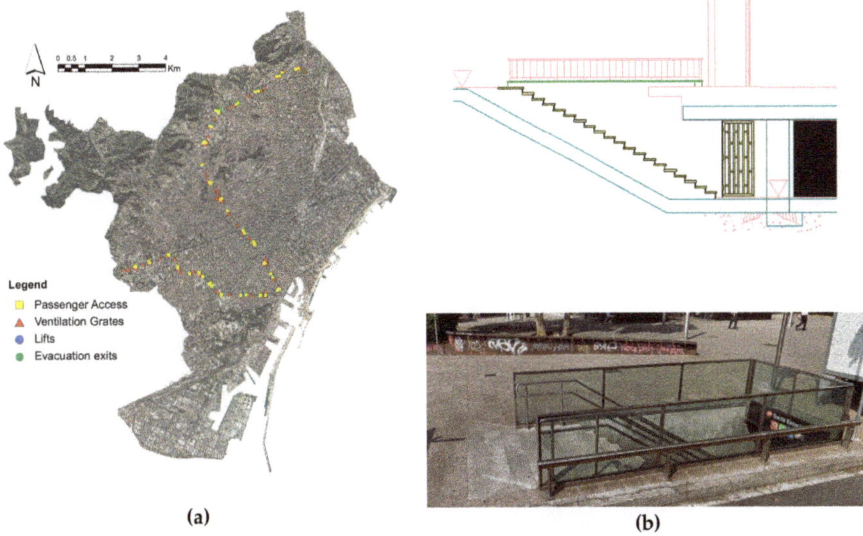

Figure 9. (**a**) Integrated Metro elements into the hydrodynamic model, such as possible water entry points for Barcelona Metro Line 3; (**b**) schematic representation of the Metro station accesses analyzed in the study.

3.4. Rainfall Data and Boundary Conditions

The 1D/2D hydrodynamic model boundary conditions consisted of the result of the rainfall-runoff model (storm water management model based on the non-linear reservoir was chosen among the options provided by InfoWorks ICM®) applying the rain gauge data and the observed water depth in the Metro tunnels downstream. The upstream boundary conditions applied in the calibration and validation phase are two historical rainfall events with surface flooding over Barcelona.

3.4.1. General Rainfall Conditions in Barcelona

The annual average rainfall is 460 mm. The Mediterranean rainfall pattern shows short-duration high-intensity events and spatial high-variability; 50% of the annual precipitation happens throughout variable rainfall events [62]. These events, in combination with city morphological characteristics and impervious areas, produce high flows in the sewer system. All these factors increase urban flood risk in city flatland areas. Barcelona has rainfall data since 1927 from Fabra Observatory; this long-time data series allowed the creation of the intensity duration frequency (IDF) curves for the city. Figure 10 shows new IDF curves based on 81 years data series (1927–1992 and 1995–2009) for some return periods [42]. Intensity values from these IDF curves are currently employed in local sewer network studies.

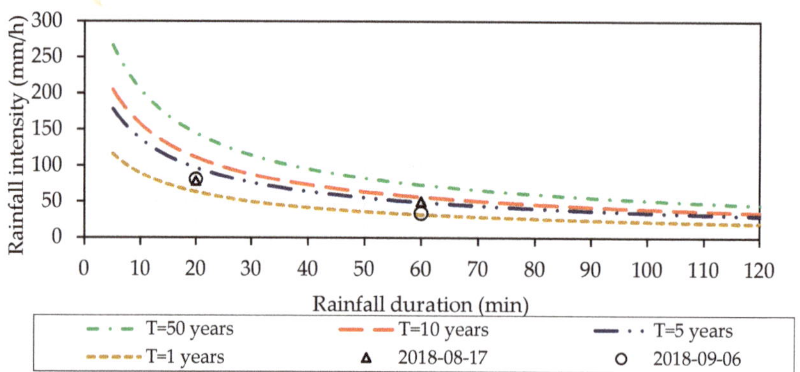

Figure 10. Intensity duration frequency (IDF) curves relating return periods (in years) with durations (in minutes) for Barcelona [63]. Points with information of two real precipitation events (2018-09-06, 2018-08-17), to compare them with the IDF curves.

3.4.2. Calibration and Validation Data

Two recent rainfall events induced surface flooding in Barcelona. One of these events (2018-09-06) caused Metro service disruptions in five stations. Ergo, this study uses this flooding event to calibrate the hydrodynamic model considering a water level of 0.15 m, for which service availability is likely to be affected by flooding in Metro service tunnels. Figure 11 shows calibration, and validation rainfall events' hydrological features, including rain gauges' covered area, Thiessen polygon distribution, and cumulative rainfall depth.

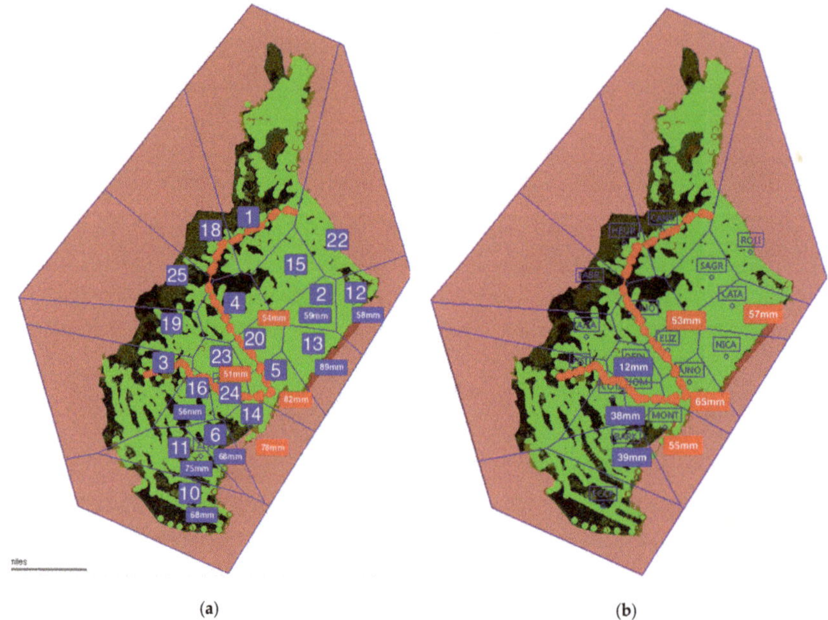

(a)   (b)

Figure 11. Study area map indicating rain gauges used for rainfall data acquisition process, and the cumulative rainfall (mm) for every flooding event. Red dots reveal possible water entry points to Metro Line 3. Blue lines show the Thiessen polygon division for each gauge region (25). (a) Flooding event for calibration 2018-09-06; (b) flooding event for validation 2018-09-06.

The other flooding event (2018-08-17) produced surface flooding; nevertheless, this event did not affect Line 3 Metro stations. On this basis, this research considers this event valid for validation purposes, understanding the process to validate the surface 2D overland flow in the vicinity of the Metro station. For this event, the model registers surface flooding, but it does not indicate floods inside the station.

3.4.3. Design Storms

Climatic variables' projections unto 2100 are forecasting within project RESCCUE. This study embraces these rainfall projections to create design storms (using alternating blocks method, according to recommendations produced near the study area [64]). After running the hydrodynamic model for the future expected conditions, including climate change influence, hence obtaining the pluvial floods to assess expected future impacts.

Extreme events analysis is considering return periods of one, five, 10, and 20 years as possible adaptation frameworks. The same rain intensity is distributed over the whole area for these synthetic events, with the rain gauges used for rainfall data acquisition process, and the cumulative rainfall (mm) for an event. These synthetic rains are conceived with a five-minute temporal resolution to reach a high-level accuracy, duration up to 2 h 30 min, and a maximum rainfall intensity as Figure 12 illustrates for each one of the return periods.

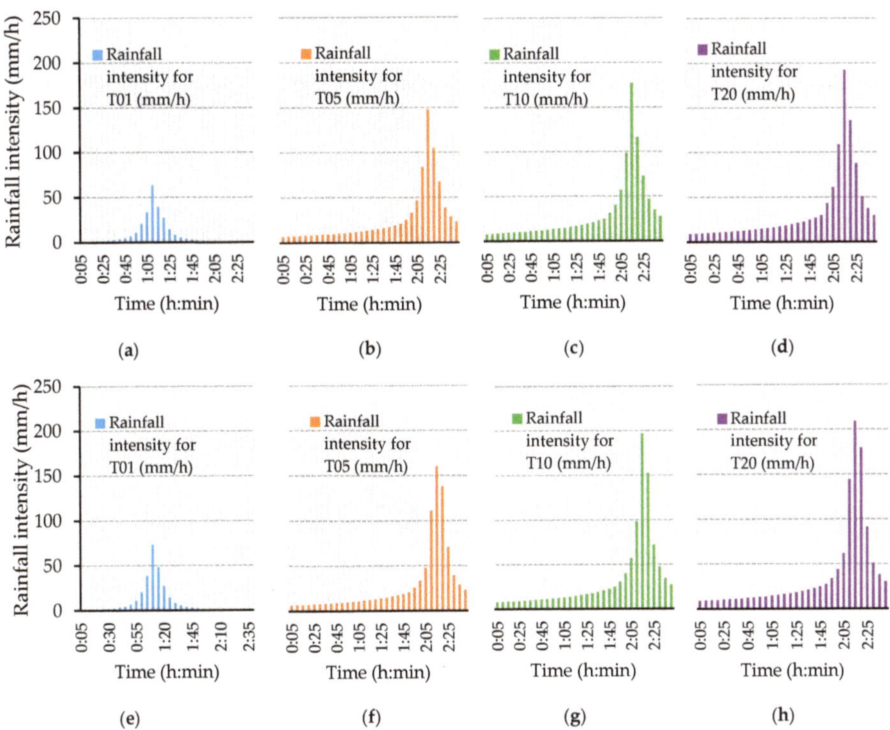

Figure 12. Synthetic rainfall events characteristics. Five-minute rainfall intensity hyetographs for events under current rainfall conditions: (**a**) A 1-in-1 year event; (**b**) a 1-in-5 year event; (**c**) a 1-in-10 year event; (**d**) a 1-in-20 year event. Five-minute rainfall intensity hyetographs for events under future rainfall conditions: (**e**) A 1-in-1 year event; (**f**) a 1-in-5 year event; (**g**) a 1-in-10 year event; (**h**) a 1-in-20 year event.

4. Results

Section four highlights the research findings, focusing on the three key themes: The calibration and validation results obtained through hydrodynamic modelling of two historical flood events, from which there are real data to carry out the required corroboration; and the outcomes of the simulations under current and future rainfall conditions. This section outlines the modelling process results achieved via the application of the proposed methodology, their interpretation as well as the conclusions that can be drawn for three points: The calibration and validation phase. Besides, it shows the hydraulic modelling outputs from the application of the synthetic rainfall conditions driven by the impact of climate change into the drainage model.

With the results of the projected water depth for the conditions of the historical event used for the calibration of the model and the projected water depth for future flooding events, one of the two fundamental criteria is obtained. Along with the ridership flow to obtain the hazard and vulnerability assessment, the flood risk assessment results for Line 3 Metro stations are obtained, following the methodology introduced in Section 3.

4.1. Calibration and Validation Processes

The first set of analyses examined the real events' pluvial flooding impact; one of them (2018-09-06) caused both surface flooding, along with two stations flooding on Metro Line 3, as indicated in Figure 13. The other flood event (2018-08-07) was merely superficial, comparable to the first event, but it did not generate flooding on Metro stations.

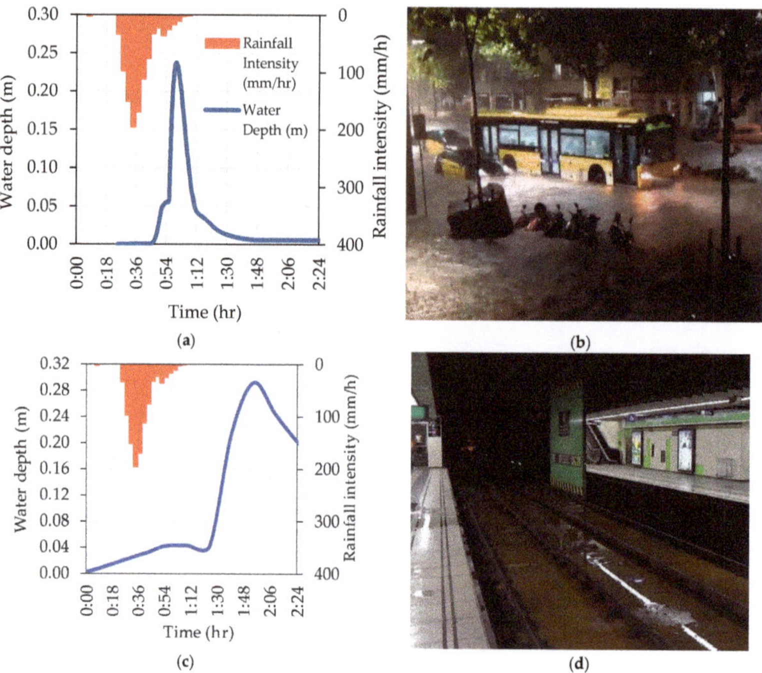

Figure 13. Calibration process summary output, for flooding event 2018-09-06: (**a**) Water depth modelling outcome in access 323_1 that serves the Paral·lel Metro station; (**b**) photograph of the flood event in the Metro access surroundings, corroborating the water depth obtained in the modelling; (**c**) water depth modelling outcome in Paral·lel Metro station tunnel; (**d**) photograph of the flood event in the Paral·lel Metro tunnel, corroborating the water depth obtained in the modelling.

This real-life hydraulic behavior replication in the hydrodynamic modelling output is successful in obtaining similar water depths at the surface level, through the calibration process applying this model for the Barcelona master drainage plan and the executed process for this study. Moreover, for the first event, the model identified water entries to Line 3 surrounding Paral·lel and Drassanes Metro stations areas, including water in their tunnels, as can be seen from the table (below). Then, water depth results in the 2D hydraulic element surrounding the ventilation grates and Metro stations' accesses are summarized, in addition to the water level in the cited Metro stations, are consistent with the flood event records.

4.2. Simulations under Current and Future Rainfall Conditions

Table 5 presents the hydrodynamic modelling process results for various return periods, which includes the Line 3 infrastructure points number where water accesses are identified, the maximum water depth in the 2D element surrounding water access points, and the maximum water depth projected in Metro stations' tunnels, for scenarios with climate change influence or without it. In the same way, it details the variation among every return period and the scenarios involving climate change impact, identifying such impact directly. Three things are evident from the changes among the results with the climate change influence and those not considering such influence. Firstly, rainfall intensities and water flow increment due to the climate change impact will impact Metro infrastructure, suggesting that the Metro system should develop adaptation measures against climate change effects. Secondly, the magnitude of the maximum water depth estimation varies between the differing return periods, with the climate change projections yielding a more significant water depth than estimated via non-influenced projections. Finally, the maximum water depth in Metro station tunnels increases between the different approaches; this is particularly evident when comparing T20 and T20 CC, which show a 54% increase estimation for the projected water depths.

Table 5. Outcome summary for hydrodynamic simulations performed under current and future rainfall conditions.

	2018-09-06	T01	T01 CC	Δ	T05	T05 CC	Δ	T10	T10 CC	Δ	T20	T20 CC	Δ
Water entry points	17	10	11	+1	12	13	+1	14	21	+7	17	25	47
Maximum water depth in the surroundings of the entry points (m)	0.29	0.03	0.04	+0.01	0.26	0.30	+0.04	0.31	0.32	+0.01	0.34	0.38	12
Maximum water depth in Metro station tunnels (m)	0.28	0	0	0	0.30	0.33	+0.03	0.45	0.46	+0.01	0.78	1.20	54

The results suggest that the magnitude of change is dependent upon the rainfall intensity for every outcome category. For example, the water depth changes appear most substantial for the T20 CC generated scenario, with the water depth estimates being shifted further than with the other approaches, without the climate change influence. However, the varying water depth associated with each return period is also highlighted, with the results differing between each return period.

4.3. Risk Assessment

Applying the outlined methodologies, Figures 14 and 15 describe the comparison of the risk assessment outcomes among the real event (2018-09-06) and the 1-in-20 years return period flooding event with climate change impact, as the most critical scenario analyzed. Differences between the projections provided by each flooding scenario and applying a peak ridership flow pattern are highlighted by the risk assessment presented in Figure 16 as a summary, detailing the risk obtained for Barcelona Metro Line 3 stations facing flooding events. The findings presented here highlight the considerable influence of the water depth in the water entry points' surrounding areas when assessing the climate change impact upon flood risk. Although all the flood events used here do not result in quite different pictures of future flood risk on the Metro accesses or ventilation grates, there are some significant differences in terms of the risk magnitude.

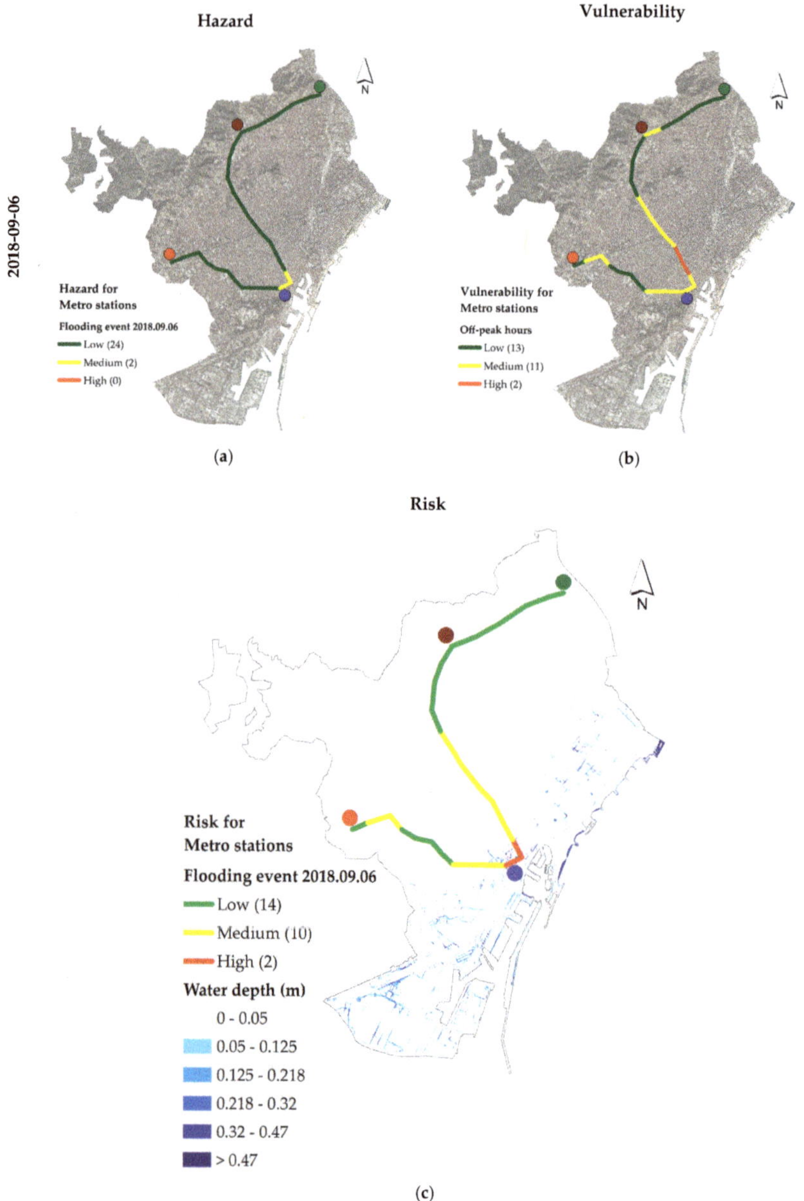

Figure 14. Metro stations risk assessment map for real event pluvial flooding scenario (2018-09-06): (**a**) Hazard map considering real flooding event; (**b**) vulnerability map applying off-peak ridership flow; (**c**) risk map with flooding water depth.

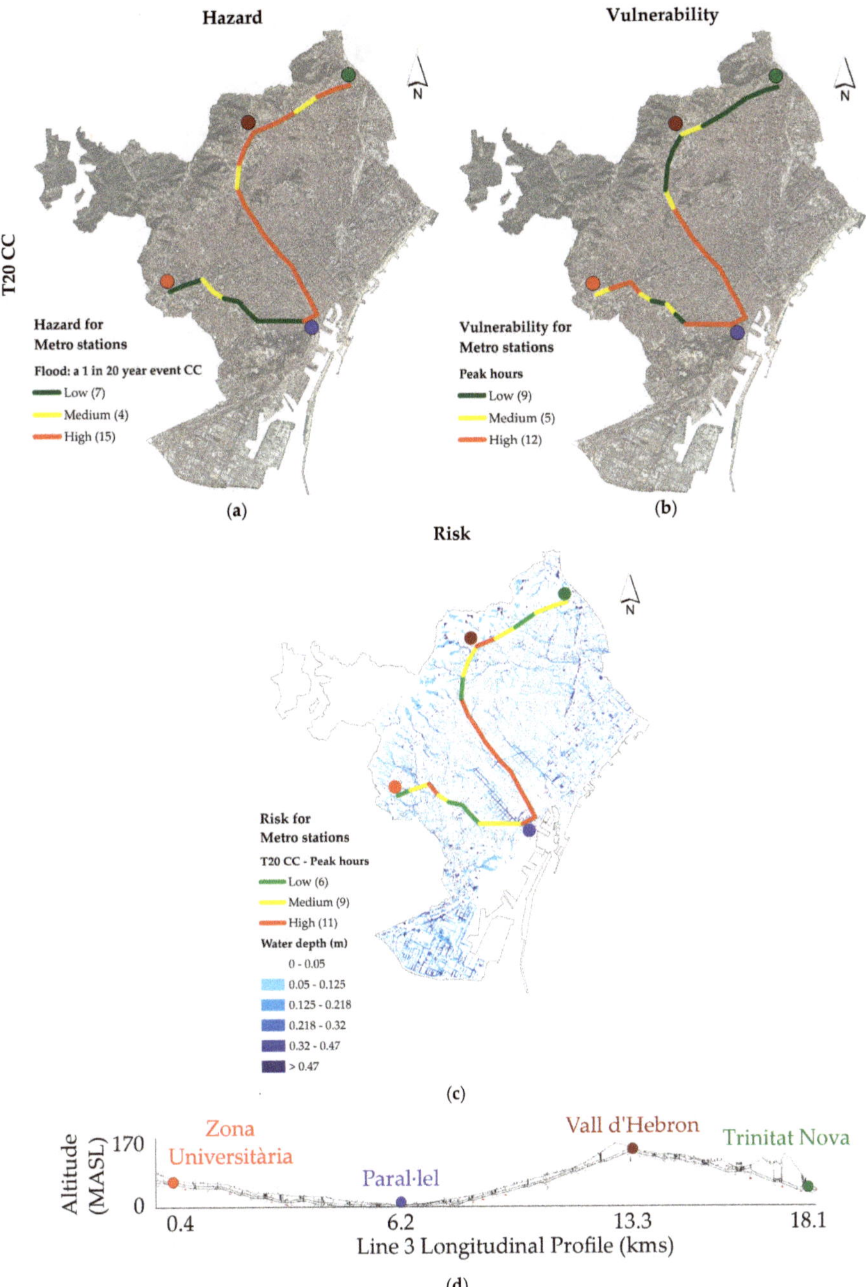

Figure 15. Metro stations risk assessment map for synthetic event scenario with climate change influence (a 20 year return period event—"T20 CC"): (**a**) Hazard map considering future flooding conditions with climate change influence; (**b**) vulnerability map applying peak ridership flow; (**c**) risk map with flooding water depth; (**d**) Barcelona Metro Line 3 longitudinal profile scheme, with the location of the Metro stations located at the highest and lowest points of the network.

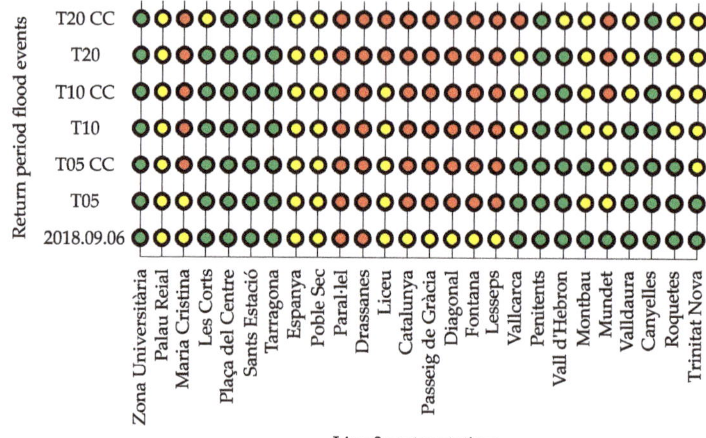

Figure 16. Risk assessment outcome for flood events on Barcelona Metro Line 3 stations, arranged by return period (T, in years) and climate change (CC) influence on the projected flood event. The green color means a low risk, the yellow color symbolizes medium risk, and the red color denotes a high risk.

The small difference between the maximum water depth affecting the water entry points, comparing T05 CC and T20 CC scenarios (0.08 m), produces significant differences for the risk assessment results. As Figure 15 illustrates, with a significant risk for 14 stations in T05 CC scenario and increasing to 20 stations having a significant risk in T20 CC scenario. It therefore seems necessary to incorporate a range of approaches in adaptation measures that are focused on coping with this increment in water depth. The majority of Metro Line 3 stations are highly susceptible to floods. Considering all risk scenarios, the maximum water depth in the surroundings of the water entry points, such as ventilation grates and Metro stations access, is not larger than 0.38 m.

5. Discussion

This chapter draws together and discusses the key findings, analyzing the results obtained in the study, and indicates how adaptation measures can be applied, according to local conditions. To the best of our knowledge, this is the first study to have quantified the flooding risk to underground Metro stations for water intrusion hazard due to rainfall events, obtaining water depths estimation for projected extreme rainfall events. The approach used in this study differs substantially from other approaches studied previously [5]. In particular, although it uses geographic information systems (GIS) as the basis for collecting and organizing information from the Barcelona Metro system, the use of GIS is a secondary component which does not form part of the analysis in this study, contrasting other studies in the field [25].

This study considered only the water depth in the Metro station tunnels, obtained from hydrodynamic modelling, as a variable to characterize the hazard. Although the hazard due to flow velocity can be taken into account since there is sufficient information to carry out such an analysis from the hydrodynamic model, the objective of the analysis carried out in this research is the interruption of the Metro service. Therefore, the flow velocity is not a relevant parameter for this analysis, as it would be if this study were focused on passenger safety.

A hydrodynamic modelling software, combined with innovative analysis on how Metro systems can be represented into a drainage system, and assumptions about the fragility of transport infrastructure derived from pluvial flooding events have been used to assess flood risk in a Barcelona Metro line. The study demonstrates the potential for conducting transport infrastructure risk analysis through hydrodynamic modelling and a depiction of the real physical conditions for representing Metro systems.

One of the main conditions of this study was to define a simplification of Metro system water accesses, such as ventilation grates, depicted as drainage grates adapted to each grate condition; the stations' access, neglecting the complexity inherent to each station, such as hallways, stairs, and additional architectural components before entering the station tunnel. All these elements were considered as pipes until arriving at the main tunnel; this enabled the Metro network to be integrated into the complex hydrodynamic model.

We have found that improving some ventilation grates and Metro access designs, by increasing their minimum height to 0.35 m, could yield a potential decrease in flooding events for Metro Line 3 tunnels and stations. Of course, care should be taken with the interpretation of these results, as local conditions for each of the ventilation grates and accesses must be considered in detail. This simplification could affect water transport times from water entry points to tunnels.

Service disruptions due to flooding events are not only by water depth increase in Metro tunnels, but also by impacts on ridership stability inside stations, corridors, and hallways. Flooding events can also affect the electrical equipment and increase maintenance costs, as Figure 2 illustrates. Service disruption does not occur mainly because of the water level itself but because of the misleading interaction between water and the Metro blocks-based electrical system. The higher the water depth, the more likely a block can be misleadingly occupied, and thus the service disrupted.

The outcomes of this study display that by accounting for increased possible water entry sources, water depth range forecasts in Metro tunnels widen. It is reasonable to assert that the principal source of flooding is a local level phenomenon of water entry. However, differences in topographic altitude also influenced flooding cases in the Metro tunnels for our case study. The locations of water entry points to the Metro identified in the real flood event (2018-09-06) are near the two most affected stations, Paral·lel and Drassanes. Nevertheless, flood events with projected high-rainfall intensities also affect high sub-basin areas in our case. These outcomes suggest an association between water entry from the upper sub-basin and the increased water entry risk for upper and lower Metro stations. It is therefore likely that such connections exist between a water transport phenomenon into the system tunnels and the increased water depth in Paral·lel and Trinitat Nova Metro stations.

Flooding events also impact the system's vulnerability, as the presence of water in the Metro tunnels affects the continuity of the service, also impacting the passengers' flow. By preventing passengers' access to the stations due to an extreme precipitation event and the consequent flooding that this causes, the vulnerability decreases because the flow of passengers entering the system also decreases. By preventing water from entering the Metro system, both the flooding hazard and the vulnerability expressed in the flow of ridership decrease. Nevertheless, such vulnerability is never zero, and in fact, it is high, since the disruption of the service also impacts passengers who are already on the trains, inhibiting their adequate transportation.

Our findings draw attention to the importance of considering that the low level of some ventilation grates and Metro stations' accesses may increase probabilities for water entering in flood events, as Figure 17 illustrates.

Moreover, the significant water depths in upper-located stations such as Vall d'Hebron associated with high-return periods under observed conditions may further question rainfall events' role in impacting water intrusion to the Metro system. Addressing the spatial and temporal full range uncertainty related to future rainfall events poses a significant challenge, therefore, we believe that having a homogeneous rainfall distribution over the city is adequate to cope with most possible climate change situations to come.

Concerning the research methods, some limitations need to be acknowledged. The interaction at the hydraulic level between the 1D and 2D model is not fully understood, for example, during the transformation of the surface flood water level to the entry points of the Metro system. Due to the lack of modelling elements that competently render physical structures that are not common in hydraulic systems. A possible source of bias for the study is the influence that unexamined physical details of stairs, access corridors, and vents, as water access points to the Metro, could have physical details beyond the roughness, sizes, and lengths that were considered.

(a)

(b)

Figure 17. The visual contrast among a Paral·lel station entrance facing a flood event, and its regular service; (**a**) flooding event 2009-10-22; (**b**) current situation.

This study emphasizes that this risk assessment combines water level projections with present ridership flows and traveler density. Outcomes should hence not be used as projected impacts. Instead, they ponder the current risk, which we define as a threat indicator for Metro system sustainability in the future. Future research should consider and integrate all the major possible flood hazard modes for Metro systems, such as the influence of groundwater infiltrations. A full climate risk assessment is required for the Metro system. It must have a high priority within a comprehensive, Barcelona-wide climate risk assessment and adaptation effort, involving all governmental agencies.

6. Conclusions

This study presents a novel integrated model to simulate flooding in the underground Metro system of Barcelona. This model assumes that the Metro flooding only occurs from surface flooding that can drain into the underground Metro infrastructure through Metro staircases, ventilation grids, and elevators. The new model integrates a conventional 1D/2D urban flood model with a Metro model that is simulated as a system of pipes and links just as if it was a drainage network. The integrated model can simulate the spatial and temporal variability of Metro flood water depth and velocity. Simulated maximum flood depth is used as input for the flood risk assessment presented.

This research analyzes the current flooding of the Barcelona Metro Line 3 infrastructure, considering the climate change influence on future rainfall intensities, for flood risk assessment. This research proposed a framework for flood simulation of underground Metro systems, showing that Metro infrastructure can be simulated as a "drainage system" for risk assessment purposes of Metro systems. Notwithstanding the relatively limited data sample analyzing one of 12 Metro lines, this work offers valuable insights into developing targeted interventions aimed at reducing Metro service disruptions frequency and impact. This piece of research includes strategies to enhance flood risk management as a reasonable approach to tackle this issue, not only for Barcelona Metro but for Metro systems worldwide. This study provides a baseline for assessing potential flood outcomes in Metro systems, notwithstanding, any proposed adaptations are not studied in detail. However, the model can be used to evaluate adaptation measures' effectiveness. To estimate the adaptation costs under various climate change and socioeconomic development situations, compatible adaptation measures with the urban planning recommended by the Barcelona City Council are necessary.

One might thus conclude that decreasing Metro service flood hazard by heightening this infrastructure is further up the list of priorities, after ensuring pedestrian mobility at the ground level and easy access to Metro stations. Still, there are significant benefits from improving the flood hazard of flooding-exposed Metro infrastructure performing some adaptation measures, either increasing accesses' heights or achieving other structural measures such as hydraulic barriers by demand.

Author Contributions: Conceptualization, E.F.-O. and E.M.-G.; methodology, E.F.-O., E.M.-G. and B.R.; software, E.F.-O. and L.L.; validation, E.M.-G., L.L. and B.R.; formal analysis, E.F.-O. and E.M.-G.; investigation, E.F.-O., M.C.P. and L.L.; resources, E.M.-G., and M.C.P.; data curation, E.F.-O. and M.C.P.; writing—original draft preparation, E.F.-O.; writing—review and editing, E.F.-O., E.M.-G. and L.L.; visualization, E.F.-O. and L.L.; supervision, E.M.-G.; project administration, E.F.-O. and E.M.-G.; funding acquisition, E.M.-G. and B.R. All authors have read and agreed to the published version of the manuscript.

Funding: This research was funded by the RESCCUE project, which is sponsored by the European Union's Horizon 2020 research and innovation program under grant agreement No. 700174, whose support is gratefully recognized.

Acknowledgments: The contents of this research are a part of the findings of the project RESCCUE, which has obtained funding from the EU H2020 (grant agreement n. 700174). Part of the content is also presented in this article with the explicit consent of Transports Metropolitans de Barcelona (TMB). Re-use of the knowledge enclosed in this paper for commercial and/or non-commercial purposes is allowed and free of charge, on the requirements of compliance by the re-user of the research, not distortion of the original meaning or information of this research and the non-liability of the RESCCUE project partners and/or TMB for any consequence stemming from the re-use. The RESCCUE project partners and TMB do not accept any liability for the errors, consequences, or omissions herein contained.

Conflicts of Interest: The authors declare no conflict of interest. The funders had no role in the design of the study; in the collection, analyses, or interpretation of data; in the writing of the manuscript, or in the decision to publish the results.

References

1. Michener, W.K.; Blood, E.R.; Bildstein, K.L.; Brinson, M.M.; Gardner, L.R. Climate change, hurricanes and tropical storms, and rising sea level in coastal wetlands. *Ecol. Appl.* **2016**, *7*, 770–801. [CrossRef]
2. Chapin, F.S.; Kofinas, G.P.; Folke, C.; Carpenter, S.R.; Olsson, P.; Abel, N.; Biggs, R.; Naylor, R.L.; Pinkerton, E.; Stafford, D.M.; et al. Resilience-based stewardship: Strategies for navigating sustainable pathways in a changing world. In *Principles of Ecosystem Stewardship: Resilience-Based Natural Resource Management in a Changing World*; Springer New York: New York, NY, USA, 2009; pp. 319–337. ISBN 9780387730325.
3. Wan, C.; Yang, Z.; Zhang, D.; Yan, X.; Fan, S. Resilience in transportation systems: A systematic review and future directions. *Transp. Rev.* **2018**, *38*, 479–498. [CrossRef]
4. Markolf, S.A.; Hoehne, C.; Fraser, A.; Chester, M.V.; Underwood, B.S. Transportation resilience to climate change and extreme weather events – Beyond risk and robustness. *Transp. Policy* **2019**, *74*, 174–186. [CrossRef]
5. Forero-Ortiz, E.; Martínez-Gomariz, E.; Cañas Porcuna, M. A review of flood impact assessment approaches for underground infrastructures in urban areas: A focus on transport systems. *Hydrol. Sci. J.* **2020**. [CrossRef]
6. UITP World Metro Figures. 2018. Available online: https://www.uitp.org/sites/default/files/cck-focus-papers-files/StatisticsBrief-Worldmetrofigures2018V4_WEB.pdf (accessed on 15 May 2019).
7. Vespignani, A. The fragility of interdependency. *Nature* **2010**, *464*, 984–985. [CrossRef] [PubMed]
8. Sun, D.J.; Guan, S. Measuring vulnerability of urban metro network from line operation perspective. *Transp. Res. Part A Policy Pract.* **2016**, *94*, 348–359. [CrossRef]
9. Forero-Ortiz, E.; Martínez-Gomariz, E. Hazards threatening underground transport systems. *Nat. Hazards* **2020**, *100*, 1243–1261. [CrossRef]
10. Transport for London. Providing Transport Services Resilient to Extreme Weather and Climate Change. 2011. Available online: http://www.energyforlondon.org/wp-content/uploads/2016/07/TfL-adaptation-report-.pdf (accessed on 29 January 2020).
11. Secretary Transport UK. *Transport Resilience Review—A Review of the Resilience of the Transport Network to Extreme Weather Events*; Government of the United Kingdom: London, UK, 2014; ISBN 9781474106610.
12. Martínez-Gomariz, E.; Gómez, M.; Russo, B. Experimental study of the stability of pedestrians exposed to urban pluvial flooding. *Nat. Hazards* **2016**, *82*, 1259–1278. [CrossRef]
13. Chen, Q.; Xia, J.; Falconer, R.A.; Guo, P. Further improvement in a criterion for human stability in floodwaters. *J. Flood Risk Manag.* **2018**, *12*, e12486. [CrossRef]
14. Madsen, H.; Lawrence, D.; Lang, M.; Martinkova, M.; Kjeldsen, T.R. Review of trend analysis and climate change projections of extreme precipitation and floods in Europe. *J. Hydrol.* **2014**, *519*, 3634–3650. [CrossRef]

15. Willems, P.; Olsson, J.; Arnbjerg-Nielsen, K.; Beecham, S.; Pathirana, A.; Gregersen, I.B.; Madsen, H.; Nguyen, V.-T.-V. *Impacts of Climate Change on Rainfall Extremes and Urban Drainage Systems*; PB-IWA Publishing: London, UK, 2012; Volume 68, ISBN1 978-1-78040-126-3. ISBN2 978-1-78040-125-6.

16. Nissen, K.M.; Ulbrich, U. Increasing frequencies and changing characteristics of heavy precipitation events threatening infrastructure in Europe under climate change. *Nat. Hazards Earth Syst. Sci.* **2017**, *17*, 1177–1190. [CrossRef]

17. Joe Weisenthal Hurricane Sandy: Photos of New York Subway Flooded. Available online: https://www.businessinsider.com/hurricane-sandy-photos-of-new-york-subway-flooded-2012-10?IR=T (accessed on 4 November 2019).

18. Japanese Water Management and Land Conservation Measures against Inundation of Underground Space-Past Flooding Cases. Available online: http://www.mlit.go.jp/river/bousai/main/saigai/jouhou/jieisuibou/bousai-gensai-suibou01-kako.html (accessed on 12 November 2019).

19. Tremending Metro de Madrid: Flooding after This Week's Summer Storms. Available online: https://www.publico.es/tremending/2017/07/07/facebook-metro-de-madrid-nada-las-inundaciones-tras-las-tormentas-de-verano-de-esta-semana/ (accessed on 7 November 2019).

20. Vaníček, I.; Jirásko, D.; Vaníček, M. Interaction of transport infrastructure with natural hazards (landslides, rock falls, floods). *Ce/Papers* **2018**, *2*, 135–164. [CrossRef]

21. NBC. Metro Station Flooding: Nearby Parking Lot Expansion Could Be Part of Cause. Available online: https://www.nbcwashington.com/traffic/transit/Metro-Station-Flooding-Nearby-Parking-Lot-Expansion-Could-Be-Part-of-Cause-384015451.html (accessed on 4 December 2019).

22. The Bulletin. Bulletin Brussels Metros Flood with Water after Tuesday Storms. Available online: https://www.thebulletin.be/brussels-metros-flood-water-after-tuesday-storms (accessed on 14 November 2019).

23. Wu, J.; Fang, W.; Hu, Z.; Hong, B. Application of bayesian approach to dynamic assessment of flood in urban underground spaces. *Water* **2018**, *10*, 1112. [CrossRef]

24. Lyu, H.M.; Wang, G.F.Q.G.; Shen, J.S.; Lu, L.H.; Wang, G.F.Q.G. Analysis and GIS mapping of flooding hazards on 10 May 2016, Guangzhou, China. *Water* **2016**, *8*, 447. [CrossRef]

25. Lyu, H.M.; Sun, W.J.; Shen, S.L.; Arulrajah, A. Flood risk assessment in metro systems of mega-cities using a GIS-based modeling approach. *Sci. Total Environ.* **2018**, *626*, 1012–1025. [CrossRef] [PubMed]

26. Lyu, H.-M.; Shen, S.-L.; Yang, J.; Yin, Z.-Y. Scenario-based inundation analysis of metro systems: A case study in Shanghai. *Hydrol. Earth Syst. Sci. Discuss.* **2019**, 1–30. [CrossRef]

27. Velasco, M.; Russo, B.; Martínez, M.; Malgrat, P.; Monjo, R.; Djordjevic, S.; Fontanals, I.; Vela, S.; Cardoso, M.A.; Buskute, A. Resilience to cope with climate change in urban areas-A multisectorial approach focusing on water-The RESCCUE project. *Water* **2018**, *10*, 1356. [CrossRef]

28. EFE. Power Cuts and Flooding on Roads, Railways and Metro due to Heavy Rains in Catalonia. Available online: https://www.elmundo.es/elmundo/2002/10/09/sociedad/1034147378.html (accessed on 16 November 2019).

29. Vanguardia. Heavy Rains Cause Problems in the Accesses to Barcelona Due to Floods and Traffic Light Breakdowns. Available online: https://www.lavanguardia.com/vida/20040914/51262798057/las-fuertes-lluvias-provocan-problemas-en-los-accesos-a-barcelona-por-inundaciones-y-averias-de-sema.html (accessed on 3 November 2019).

30. Vanguardia. An Autumn Storm Floods the Ronda Litoral and Barcelona's Metro for Three Hours. Available online: https://www.lavanguardia.com/sucesos/20091022/53809334313/una-tormenta-otonal-inunda-la-ronda-litoral-y-el-metro-barcelones-por-tres-horas.html (accessed on 23 November 2019).

31. Vanguardia. The Strong Storm Causes Effects on Trains and Subway in the Area of Barcelona. Available online: https://www.lavanguardia.com/sucesos/20110719/54187726290/el-fuerte-temporal-causa-afectaciones-en-trenes-y-metro-del-area-de-barcelona.html (accessed on 14 November 2019).

32. Vanguardia. Firefighters Make about a Hundred Exits Due to Flooding in the Area of Barcelona. Available online: https://www.lavanguardia.com/vida/20180716/45935565609/bomberos-efectuan-unas-cien-salidas-por-inundaciones-en-el-area-de-barcelona.html (accessed on 26 November 2019).

33. Vanguardia. Storm in Barcelona: The Strong Storm that Has Hit the Capital City. Available online: https://www.lavanguardia.com/local/barcelona/20180906/451668443082/tormenta-lluvia-rayos.html (accessed on 19 November 2019).

34. Vanguardia. Heavy Rainfall in Barcelona. Available online: https://www.lavanguardia.com/local/barcelona/20181009/452266860153/lluvia-torrencial-barcelona-temporal-aguaceros-inundaciones-video-seo-lv.html (accessed on 25 November 2019).

35. Vanguardia. The Emergency Number 112 Records 157 Rain Calls up to 8 Hours. Available online: https://www.lavanguardia.com/local/barcelona/20181115/452943966963/el-112-registra-157-llamadas-por-la-lluvia-hasta-las-8-horas.html (accessed on 18 November 2019).

36. EFE. The Rain is Pouring into the Metro and Flooding Roads and Shops in Barcelona. Available online: https://elpais.com/ccaa/2019/07/27/catalunya/1564240513_745071.html (accessed on 7 November 2019).

37. Metrópoli Abierta Waterfalls in the Verdaguer and Sagrada Família Metro Stations. Available online: https://www.metropoliabierta.com/el-pulso-de-la-ciudad/metro-verdaguer-sagrada-familia-inudada-lluvia_18588_102.html (accessed on 12 November 2019).

38. ABC. Heavy Rains from the Cold Drop Cause Flooding in the Barcelona Metro. Available online: https://www.abc.es/espana/catalunya/barcelona/abci-fuertes-lluvias-gota-fria-causan-inundaciones-metro-barcelona-201908271615_noticia.html (accessed on 1 December 2019).

39. Abierta, M. Rain Affects Metro, Bus and Renfe Services. Available online: https://www.metropoliabierta.com/movilidad/lluvia-afecta-servicios-metro-bus-renfe-barcelona_21490_102.html (accessed on 9 December 2019).

40. Vanguardia. Flooded Streets and Closed Metro Stations: The Incidents Due to the Storm in Barcelona. Available online: https://www.lavanguardia.com/local/barcelona/20191204/472058751703/incidencias-temporal-lluvias-barcelona-dana.html (accessed on 21 December 2019).

41. Ajuntament de Barcelona. Barcelona, a City Committed to Combatting Climate Change. 2013. Available online: https://carbonn.org/uploads/tx_carbonndata/Barcelona%20committed%20to%20combat%20climate%20change-Mitigation&Adaptation%20actions_05.pdf (accessed on 12 December 2019).

42. Rodríguez, R.; Navarro, X.; Casas, M.C.; Ribalaygua, J.; Russo, B.; Pouget, L.; Redaño, A. Influence of climate change on IDF curves for the metropolitan area of Barcelona (Spain). *Int. J. Climatol.* **2014**, *34*, 643–654. [CrossRef]

43. Scalise, J. How track circuits detect and protect trains. *Railw. Walk Rail Talk* **2014**, *1*, 1–7.

44. FHWA. *Urban Drainage Design Manual, Hydraulic Engineering Circular No. 22 (HEC-22)*; FHWA: Washington, DC, USA, 2009.

45. Manuel, G.; Beniamino, R. Hydraulic efficiency of continuous transverse grates for paved areas. *J. Irrig. Drain. Eng.* **2009**, *135*, 225–230.

46. Gómez, M.; Recasens, J.; Russo, B.; Martínez-Gomariz, E. Assessment of inlet efficiency through a 3D simulation: Numerical and experimental comparison. *Water Sci. Technol.* **2016**, *74*, 1926–1935. [CrossRef] [PubMed]

47. Tellez, J.; Gómez, M.; Russo, B.; Redondo, J.M. Performance assessment of numerical modelling for hydraulic efficiency of a grated inlet. In Proceedings of the 12th Workshop on Synthetic Turbulence Models: Synthetic Flows for Heat and Mass Transfer, Caen, France, 3–4 July 2017; p. 1.

48. Russo, B.; Gómez, M.; Tellez, J. Methodology to estimate the hydraulic efficiency of nontested continuous transverse grates. *J. Irrig. Drain. Eng.* **2013**, *139*, 864–871. [CrossRef]

49. Gómez, M.; Macchione, F.; Russo, B. Methodologies to study the surface hydraulic behaviour of urban catchments during storm events. *Water Sci. Technol.* **2011**, *63*, 2666–2673. [CrossRef] [PubMed]

50. Gómez, M.; Rabasseda, G.H.; Russo, B. Experimental campaign to determine grated inlet clogging factors in an urban catchment of Barcelona. *Urban Water J.* **2013**, *10*, 50–61. [CrossRef]

51. Avci, O.; Ozbulut, O. Threat and vulnerability risk assessment for existing subway stations: A simplified approach. *Case Stud. Transp. Policy* **2018**, *6*, 663–673. [CrossRef]

52. Compton, K.L.; Faber, R.; Ermolieva, T.Y.; Linnerooth-bayer, J.; Nachtnebel, H. *Uncertainty and Disaster Risk Management: Modeling the Flash Flood Risk to Vienna and Its Subway System*; International Institute for Applied Systems Analysis (IIASA): Laxenburg, Austria, 2009; ISBN 9783704501486.

53. Chanson, H.; Brown, R. Stability of individuals during urban inundations: What should we learn from field observations? *Geosciences* **2018**, *8*, 341. [CrossRef]

54. Ishigaki, T.; Onishi, Y.; Asai, Y.; Toda, K.; Shimada, H. Evacuation criteria during urban flooding in underground space. In Proceedings of the 11th International Conference on Urban Drainage Modelling, Palermo, Italy, 23–26 September 2008; p. 7.

55. Ishigaki, T.; Asai, Y.; Nakahata, Y.; Shimada, H.; Baba, Y.; Toda, K. Assessment of safety on evacuation route during underground flooding. In Proceedings of the 16th APD-IAHR, Nanjing, China, 20–23 October 2008; pp. 141–146.
56. Ishigaki, T.; Baba, Y.; Toda, K.; Inoue, K. Experimental study on evacuation from underground space in urban flood. In Proceedings of the 31st IAHR Congress, Seoul, Korea, 1 September 2005.
57. Ishigaki, T.; Asai, Y.; Nakahata, Y.; Shimada, H.; Baba, Y.; Toda, K. Evacuation of aged persons from inundated underground space. *Water Sci. Technol.* **2010**, *62*, 1807–1812. [CrossRef]
58. Mizuguchi, M.; Kasuya, T.; Omori, T.; Sawada, M.; Saito, T. Study of potential flooding of underground pedestrian space in the area around Tokyo station. In Proceedings of the Advances in Underground Space Development, Singapore, 7–9 November 2012; pp. 258–266.
59. Demirel, H.; Kompil, M.; Nemry, F. A framework to analyse the vulnerability of European road networks due to Sea-Level Rise (SLR) and sea storm surges. *Transp. Res. Part A Policy Pract.* **2015**, *81*, 62–76. [CrossRef]
60. TMB. *Dades Bàsiques/Metro System Basic Data 2019*; Transports Metropolitans de Barcelona: Barcelona, Spain, 2019; p. 1.
61. Chang, T.J.; Wang, C.H.; Chen, A.S.; Djordjević, S. The effect of inclusion of inlets in dual drainage modelling. *J. Hydrol.* **2018**, *559*, 541–555. [CrossRef]
62. Dünkeloh, A.; Jacobeit, J. Circulation dynamics of Mediterranean precipitation variability 1948–1998. *Int. J. Climatol.* **2003**, *23*, 1843–1866. [CrossRef]
63. Casas Castillo, M.; Rodríguez Solá, R.; Navarro, X.; Redaño Xipell, Á. Influencia del Cambio Climático en las Curvas IDF y en la Lluvia de Diseño del Área Metropolitana de Barcelona. Segundo Informe Para el Proyecto SW0801. 2010. Available online: https://upcommons.upc.edu/handle/2117/15055?show=full (accessed on 17 January 2020).
64. Balbastre-Soldevila, R.; García-Bartual, R.; Andrés-Doménech, I. A comparison of design storms for urban drainage system applications. *Water* **2019**, *11*, 757. [CrossRef]

Article

Urban Resilience to Flooding: Triangulation of Methods for Hazard Identification in Urban Areas

Maria do Céu Almeida [1,*], Maria João Telhado [2], Marco Morais [2], João Barreiro [3] and Ruth Lopes [4]

[1] Urban Water Unit, National Civil Engineering Laboratory, LNEC, Av. Brasil 101, 1700-066 Lisbon, Portugal
[2] Lisbon City Council, Câmara Municipal de Lisboa, CML, Praça José Queirós, n.º1 – 3º piso – Fração 5, 1800-237 Lisboa, Portugal; joao.telhado@cm-lisboa.pt (M.J.T.); marco.morais@cm-lisboa.pt (M.M.)
[3] CEris, Instituto Superior Técnico, University of Lisbon, Av. Rovisco Pais, 1049-001 Lisboa, Portugal; joao.barreiro@tecnico.ulisboa.pt
[4] HIDRA, Av. Defensores de Chaves, 31 – 1º Esq., 1000-111 Lisboa, Portugal; r.lopes@hidra.pt
* Correspondence: mcalmeida@lnec.pt

Received: 31 January 2020; Accepted: 10 March 2020; Published: 12 March 2020

Abstract: The effects of climate dynamics on urban areas involve the aggravation of existing conditions and the potential for emergence of new hazards or risk factors. Floods are recognized as a leading source of consequences to society, including disruption of critical functions in urban areas, and to the environment. Consideration of the interplay between services providers ensuring urban functions is essential to deal with climate dynamics and associated risks. Assessment of resilience to multiple hazards requires integrated and multi-sectoral approaches embracing each strategic urban sector and interactions between them. A common limitation resides in the limited data and tools available for undertaking these complex assessments. The paper proposes a methodology to undertake the spatial characterization of the flood related hazards and exposure of both essential functions and services providers in urban areas, in the context of limitations in data and in ready-to-use tools. Results support the resilience assessment of these hazards, taking into account interdependencies and cascading effects. The approach is applied to Lisbon city as the study case. Results are promising in demonstrating the potential of combining data and knowledge from different sources with dual modelling approaches, allowing us to obtain trends on the magnitude of effects of climate scenarios and to assess potential benefits of adaptation strategies. Quantification of the effects is reached, but results need to be assessed together with the underlying levels of uncertainty. The methodology can facilitate dialogue among stakeholders and between different decision levels.

Keywords: climate change; flooding; hazard mapping; risk identification; sustainability; urban resilience

---

1. Introduction

Comprehensive understanding of the vision for urban areas encompasses a number of widely used concepts. The concepts of sustainability, resilience, adaptability, safety, transformation and liveability are more or less implicit as main guidelines for action [1,2], even if the understanding of these concepts is not generalized and is often vague or narrow [3]. Translation of these concepts into development strategies for the complex and dynamic urban systems can be an unreachable aim. However, as emphasized in Reference [3], "cities have proven to be remarkably resilient complex systems: many cities have existed for thousands of years and have persevered in the face of natural and human-induced disasters to become stronger and in some cases more resilient". Despite these debates, there is a need for a shift to a more sustainable and resilient path [1].

Aligned with this vision, and aware of the challenges that climate change imposes on urban areas, the option of dealing with specific associated hazards, while ensuring the involvement of multiple urban services and interested parties, allows for reducing the dimension of the problem.

Water related risks are amongst those that are significantly dependent on climate related events. The intrinsic dynamic nature of climate in each region already challenges the managers of urban systems. Climate change effects on the urban areas potentially aggravate existing conditions and lead to the emergence of new hazards or risk factors. Floods are a leading source of adverse consequences for urban areas [4,5].

City resilience, understood as the ability to absorb, adapt and recover from disruptive events in a path towards increasing sustainability [6], provides a broad conceptual structure to assess and support planning in urban areas. Any integrated and sustainable approach to increase resilience needs to be supported by sound knowledge. However, the complexity and dynamics of urban areas imply the acceptance of several limitations on data, tools and resources. Planning for increasing resilience can benefit from the use of diverse information and methods to add reliability and validity to the results. Furthermore, explicit consideration of uncertainties of climate phenomena and meteorological events [7] is critical, even if quantification is not always feasible.

Focusing on the water sector, events such as intense rainfall, storm surge, sea level rise and temperature increase are of concern in urban areas. Aggravation of these climate conditions has the potential to increase the likelihood and consequences of severe events, as well as to reduce the performance of urban water systems during less extreme conditions [7]. Cities and services providers must be prepared to cope with these challenges. The interplay between services providers, ensuring urban functions, is essential to face climate related events and to assess the resilience to multiple hazards. This requires an integrated and multi-sectoral approach taking into account strategic urban sector and their interactions [8–10], but significant gaps have been found in risk-based approaches [11]. As emphasized by Reference [9], conceptual approaches based solely on vulnerability and precaution are limited and the adoption of the concept of resilience as a new paradigm allows for the implementation of more integrated risk management in a systemic manner (p.237).

Cities and towns rely on water systems, some incorporating components built over 200 years ago, built gradually following population growth. These systems are functional and represent a high asset value. Rehabilitation rates are already lower than needs today and expectations are for continued deterioration and low investments [12], especially in sewerage. Service failures affect society, for instance, whenever volumes exceed transport, treatment or storage capacities, and excess water often results in flooding. Consequences of flood events are multidimensional, including adverse effects on the water services, health and safety of populations, socio-economy and environment. Mobility, wastes and electricity supply are some of the sectors often affected by flooding and potentially propagating their effects [8,11,13–16].

Assessment of resilience and selection of options to improvement depends on two key steps [17,18]: characterization of flood events, which determines exposed assets and population; and estimation of vulnerability to allow estimating magnitude of damages for specific types of events. Limitations on risk identification determine the robustness of subsequent analysis, from exposure to impacts estimation.

The work presented herein is part of a broader approach (RESCCUE: *Resilience to cope with climate change in urban areas – a multisectorial approach focusing on water* project) to enable city resilience assessment, planning and management by incorporating new and existing knowledge of the urban systems performance under climate change conditions in a water-centred multi-risk assessment of strategic urban services performance using a comprehensive resilience platform [10]. The assessment of urban resilience from a multisector approach is carried out for current and future climate scenarios, and includes multiple risks.

Sound assessment of the resilience to flooding requires the systematic identification of risks and corresponding hazards, risk factors and risk sources for dealing with current and expected future levels of risk in order to increase the resilience of cities, using the available information. This paper

Sustainability **2020**, 12, 2227

presents the developments in terms of risk identification related with climate change effects and services interdependencies, specifically for flood related hazards. The approach aims at setting a practicable methodology in a background of limitations in data and in ready-to-use tools using the city of Lisbon as a study case, including the spatial characterization of these hazards. The results are essential to supporting the assessment of the resilience to these hazards of essential urban functions such as mobility, wastes management and electricity supply, taking into account interdependencies and cascading effects. The paper details developments relevant for the mobility and wastes management sectors. The electrical sector, deemed essential for the city, has a Quality Service Zone Type A for the design and planning of the network, implying the existence of incremental layers of resilience and the robustness of the grid, while minimizing the impact in case of disruption. A Type A Zone has high-level redundancy in the electricity supply service [19].

2. Lisbon City Overview

Lisbon, the capital of Portugal, is one of 18 the municipalities of the biggest Portuguese metropolitan area and has the second largest European port on the Atlantic Ocean. Lisbon is a city shaped by influences of a large number of cultures over time and by the extensive riverfront on the river Tagus estuary. Characteristic figures for Lisbon city are listed in Table 1. The city has a temperate climate, classified as Mediterranean climate (Köppen climate classification: Csa), characterized by dry and hot summers and wet and fresh winter periods.

Climate change trends for Lisbon include the increase in average air temperature, decrease in annual and non-wet season rainfall, increase in wet-season rainfall and in intense rainfall events frequency, average sea level rise and increase of coastal floods frequency. The combined action of intense rainfall, wind, sea level rise with tides and storm surges is especially relevant for Lisbon's context and geographical position.

Table 1. Lisbon city characteristics [20–22].

Area (Km^2)		85		Economic indicators (2013)	
Population (2011) (inhab.)	Residents	547,733		GDP (millions of euros)	63,902
	Commuters balance	+378,226		Gross value added GVA (millions of euros)	56,154
	Disabled (%)	17.1		GDP per capita (thousands of euros)	22.7
Tourism (2011)	tourists/year	2,949,579		Apparent labour productivity (per person employed) (GVA/Employment, 2011)	41.7
	tourist nights/year	6,789,166			
Age distribution	<15 years old	12.9%		**Employment indicators, 2011**	
	>65 years old	23.9%		Employment (thousands of persons)	1,385.8
Land slope	Average: 5.7°	Maximum: 81°		Employment (% country)	29%
Altitude (m)	Minimum: 0	Maximum: 217		Water distribution service connections	≈80,000
Land use values	Consolidated urban	90%		**Wastewater infrastructures**	
	Buildings (n.)	52,496		Combined sewer network served area (%)	73
	Vehicles/day (2012)	648,615		Treatment plants	3

The Municipality is involved and proactively committed to increase the resilience of the city to achieve the 17 Sustainable Development Goals by actively working in relevant areas such as participating in international initiatives such as C40 and 100 Resilient Cities, largely investing from strategic to practical actions to increase city sustainability [23,24].

3. Methodology and Data

3.1. Methodology Main Steps

The overall methodology proposed to undertake the spatial characterization of the flood related hazards has the following main steps:

(i)   Identification of flood related hazards, risk factors and risks using the selected affected sectors as case studies, namely, electricity supply, urban mobility and wastes collection.

(ii)   Selection of metrics for hazards characterization and mapping.

(iii) Selection of representative scenarios to characterize current and future situations.

(iv)   Mapping of hazards and calculation of metrics to support further work on resilience assessment using GIS.

The methodology adopted for hazards identification takes the data obtainable for the investigation and available tools into consideration to diagnose and evaluate the effect of climate change scenarios in terms of flood related hazards as a route to ascertain resilience of urban services to these events. Both data and models have inherent limitations and uncertainties and research design is grounded on the use of complementary methods, in a triangulation-based approach [25], using multiple methods to study the research problem [26]. Methodological triangulation allows adding reliability and validity to the results and cross checking of results [27], taking advantage of overlapping and intersecting layers of geographic information. For the study case of Lisbon, the set of methods, available for the triangulation to support flood risk identification, are identified. The structure adopted in this study is given in Figure 1.

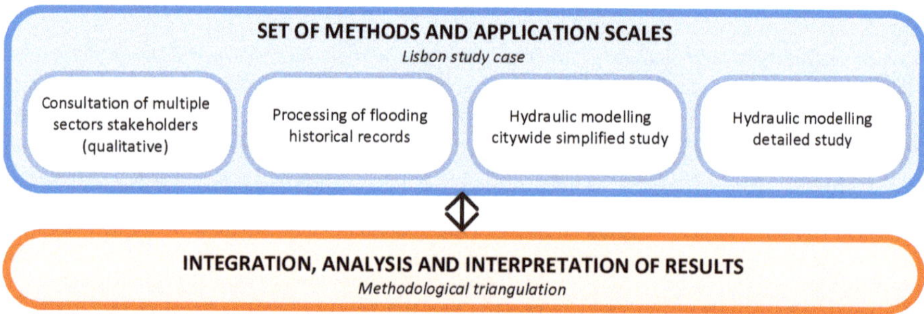

Figure 1. Methods triangulation for characterization and mapping of flooding related hazards.

Assumptions in this study include the focus on the water cycle and flood related hazards, while the risk sources analysed are rainfall and coastal overtopping [10]. The emphasis here is on the mobility and waste management sectors, but the methodology is applicable to other urban sectors.

3.2. Tools and Data to Support Risk Identification

The first step is to identify the data and tools available for the study case to support the methodology refinement; the methodology takes into account current and future situations while considering climate change.

In terms of tools, two types of hydraulic mathematical models are available for the city of Lisbon [28,29]: (1) the City wide 1D GIS model; (2) the Downtown catchments JL using 1D/2D combined model (SWMM and Basement) [30,31]. These models have a number of limitations in the data for model building and confirmation, but the two models do represent a balance between spatial scope, level of detail and data availability. The former (1) covers the city as a whole but adopts a simplified hydraulic model and sewer network; the later (2) uses a more robust hydraulic formulation and includes network as well as overland flow simulation.

For the study case of Lisbon, the set of methods available to support description and mapping of flood related hazards are indicated in Figure 2, together with main outputs and hazard characteristics.

The approach adopted in method 1, a qualitative method, allows for collaborative crosscheck, where groups or individuals with different points of view investigate common issues involving interrelated systems and services, increasing the validation and consolidation of the aspects evaluated [32]. With this approach, identification of flood related hazards and risk factors for strategic urban services was developed in a collaborative process involving representatives of the Lisbon city services included in the study. Stakeholders with direct involvement are listed in Table 2. Following the consultation sessions with stakeholders, the analysis of results for the study case resulted in an interdependencies matrix. The structure and aims of the study and focus on issues related to flooding in this specific city has determined the level of involvement of the stakeholders.

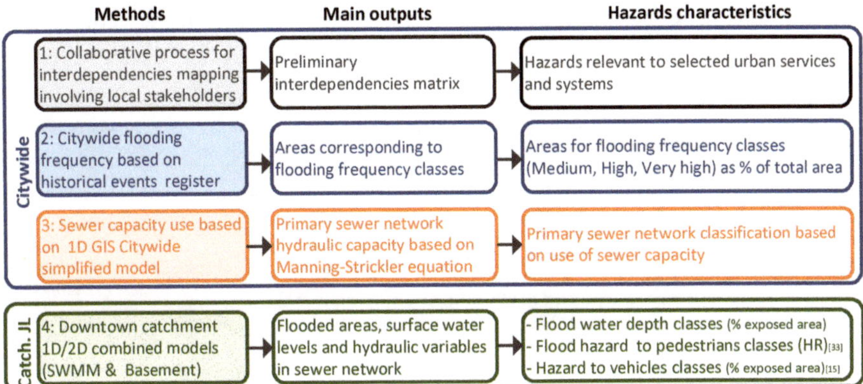

Figure 2. Methods for characterization and mapping of flood related hazards and outputs.

Table 2. Main stakeholders involved in the Lisbon city study case.

Service	Stakeholders	Level of Involvement
Municipality *	Lisbon Municipality (CML)	High
Energy supply	Distribution System Operator (EDP D)	High
Rain and wastewater systems	CML and ADTA	High
Water supply	EPAL	Low
Public transport	CML, CARRIS, METRO	Medium
Communications	MEO Altice, Vodafone, NOS	Low

* Includes a range of urban services, e.g., civil protection, wastes, public lighting, urban planning, mobility and environment.

For method 2, data from a historical events register, currently updated by the civil protection services, was used to assess citywide flood frequency.

In method 3, the citywide model approach (1) the hydraulic modelling simplified study of the drainage system (1D GIS model), was based on a conceptual model due to the complexity of Lisbon's drainage system and data limitations. This simulation tool uses GIS routines and was implemented on ArcMap ™ software. This model was built with the primary sewer network, to enable modelling of major physical and hydrodynamic properties of the system. This model includes 421 sub-catchments, 797 junctions and 753 sewers, which make up a total of 173 km, around 12% of the whole sewer network length. From the 797 junctions, 218 are head junctions and 48 are final junctions, which discharge to a main trunk system (primary sewer conveying wastewater to the treatment plant), the Tagus River (receiving water body) or to neighbouring councils' sewers. Secondary sewers (cross-sections smaller than 800 mm) are not included.

In method 4, for the Lisbon Downtown catchments J-L, the 1D/2D SWMM and BASEMENT combined model was set to allow for estimating the flooded areas and its water levels at the surface. These catchments were selected because they are two of the most flood prone catchments in Lisbon, encompassing historical and touristic downtown areas with relevant infrastructure and services. The model for this area includes 32 sub-catchments, 331 sewers and 318 nodes, from which six corresponded to outfalls [28,29].

3.3. Selection of Metrics for Hazards Characterization and Mapping

Depending on the available information and model used, different criteria are applicable, resulting in a more comprehensive understanding of the hazards. The urban functions selected to illustrate the application of this methodology are those related to urban mobility and wastes collection.

In Table 3, a summary of metrics selected for flood related hazards for different approaches and scenarios is given.

Table 3. Definition of flood related hazards for different approaches and scenarios.

Data/model	Criteria: Metric, Scale Classes	Scenarios
Lisbon flooding historical records	**Flooded areas**: frequency, 3 classes: medium, high, very high	Current situation
Citywide 1D GIS based	**Use of sewer transport capacity**: $C = Q_{wet}/Q_{full}$ (%), 4 classes, low $C \leq 0.5$, moderate $0.5 < C \leq 1.0$, high $1.0 < C \leq 1.5$, very high $C > 1.5$	Current situation and climate change
Downtown catchments J and L 1D/2D CMSB	**Water level**: water depth, d(m), at critical time, 5 classes: very low $d \leq 0.2$, low $0.2 < d \leq 0.4$, moderate $0.4 < d \leq 0.6$, high $0.6 < d \leq 0.8$, very high $0.8 < d \leq 1.0$	
	Hazard to pedestrians: Flood hazard rating $HR=d \times(v+0.5)+DF$ (d - water depth (m), v - overland flow velocity (m/s), DF - debris factor [33], 4 classes: low $HR \leq 0.75$, moderate $0.75 < HR \leq 1.25$, high $1.25 < HR \leq 2$, very high $HR > 2$	
	Hazard to vehicles: $F(flow\ depth\ D, flow\ velocity|\vec{v}|)$ [15], 3 classes: low $D \leq 0.28$ and $D \times |\vec{v}| \leq 0.40$, moderate $D \leq 0.28$ and $0.40 < D \times |\vec{v}| \leq 0.55$, high $D > 0.28$ or $D \times |\vec{v}| > 0.55$	
Estuary water level	**Area as a function of simulated water level** available modelling results for the scenarios of estuary water level are from a study promoted by CML [34].	

For the current situation, the historical data on flood events allowed obtaining hazard maps with areas as a function of flooding frequency, used to cross-validate the results from the simulations. A systematic recording of water levels was not undertaken and information on the water levels reached in each flood event is not available. For surface flows, the water level metric allowed for evaluation of consequences in properties.

3.4. Selection of Representative Scenarios

The selection of representative scenarios took two aspects into account: the infrastructure and climate. For the former, two situations were analysed: existing infrastructure and adaptation strategies (CAS, or climate adaptation strategy). For the latter, two situations were studied: the current situation and a future situation where climate change is accounted for. From the results of available studies on climate change to Lisbon, to characterize current situation and climate change [34,35], representative scenarios were selected, for both the current situation and future situation, for rainfall and for Tagus river estuary levels. These scenarios are aligned with those used by the Municipality for climate adaptation planning purposes.

For both climate situations, three return periods were selected (10 years, 20 years and 100 years) to take into account the variations in precipitation intensity. The actual values for existing infrastructure with climate change were defined as relative changes to current situation values. To limit the number of hydraulic simulations, an average estuary water level was adopted for each climate situation. The reference period for the future situation taking climate change into account is 2071-2100 (worst-case scenario).

Three scenarios were analysed and compared: CS, used as the baseline, with the current system and climate characteristics; BAU, business as usual (for the system) assuming a future situation with climate change; and CAS, a future situation including the implementation of selected strategies for climate adaptation and assuming climate change scenarios.

3.5. Mapping of Hazards and Data for Calculation of Global Metrics for Resilience Assessment

The mapping of flood related hazards and calculation of metrics to be used as input to resilience assessment was carried out using the municipality GIS and data from several city databases. The global metrics adopted depended on the information available, but essentially consisted of aggregation by length and area, number or aggregated values for the variables used.

4. Results and Discussion

4.1. Preliminary Interdependencies Matrix

The broad identification of interdependencies in the city of Lisbon considers seven main urban services or infrastructure for which a matrix of exposure and interdependencies was developed: electricity supply; telecommunications; water cycle; wastes; mobility; green and blue infrastructure; urban equipment; public lighting; and heritage areas. A simplified version of the matrix is presented in Table 4. Systematic analysis took several urban services into consideration, including plausible cascading events. Globally, runoff and flooding can directly or indirectly affect most services and infrastructure.

4.2. Main Results for Processing of Flooding Historical Records and Comparison with CS Simulation Results

The results of the processing of historical records for the three return periods compared with the results of the model (1) simulations for current situation (CS as baseline) (Figure 3, for T=100 years) allow cross validating the results. Overall, the simulation results match with historical records for flood prone areas, especially for those near the river Tagus or areas where floods are more severe. As is evident from Figure 3, simulation results using a simplified network with only primary sewers do not provide full spatial coverage, with some areas with historical records likely to be associated with secondary sewers or rainwater inlets insufficiency. Additional model limitations such as simplified hydraulics, and the eventual occurrence of sedimentation and blockages not represented in the model, explain the apparent spare flow capacity in areas where flooding occurs regularly.

Historical records provide valuable information since the areas for each class of flood frequencies incorporate several records associated with each event and an extensive number of observations, and represent a proxy of the area for each flood frequency.

Table 4. Summary interdependencies matrix for Lisbon urban services and infrastructures (R: rain; W: wind; SLR: sea level rise; WWTP: Wastewater treatment plants).

SERVICE|SUBSYSTEM|Component	SERVICE OR INFRASTRUCTURE FAILURE	EXPOSURE TO	POTENTIAL DERIVED RISKS AND CASCADING EFFECTS
Energy | Electricity transport and distribution | Substations, overhead lines, underground cables	Damages, collapse, interruption of energy supply	**R**: flood **W**: storms	**Water supply**: failures of electromechanical and control systems; **Urban drainage**: failures of pumping and control systems; **WWTP**: failures of electromechanical elements and control systems; **Street lights**: failures regarding function and control systems; **Communications**: cellular towers, central offices, other critical communications for monitoring and controlling electricity delivery
Communications | Network and nodes (operational centres)	Damage, collapse, interruption of communications	**R**: runoff, flood **W**: storms	Effects on several urban services depending on communications
Urban water cycle | Wastewater and rainwater systems | Sewer systems	Limited conveyance capacity, high street runoff (level and velocity), CSOs	**SLR**; **R: Rain**: high inflows, runoff	**Mobility** (road, rail): disturbance and interruptions, flooding of underground infrastructures (metro, train, parking, tunnels); **Wastes**: overturn, dragging and damage on wastes; **Electrical energy**: damage to equipment and lines **Other**: pedestrian ways, parking lots, playgrounds, CSOs
Urban water cycle | Wastewater and rainwater systems | Pump stations	Electrical or mechanical failures due to flooding (pumping capacity and CSO), salinity degrading components, excessive inflows	**SLR**; **R**: high inflows, runoff, flood	**Mobility**: traffic disturbances and interruptions; flooding of underground infrastructures (metro, train, parking, tunnels) **Receiving water pollution; Recreational uses affected**
Urban water cycle | Wastewater and rainwater systems | WWTP	Lower treatment efficiency and CSO due to excessive flows and dilution; lower treatment efficiency and corrosion of infrastructures by salt water intrusion	**SLR**; **R**: high inflows, flood	Receiving water pollution Recreational uses affected
Wastes collection | Cleaning, containers	Container damage, displacement and overturn	**R**: runoff, flood **W**: storms	**Urban drainage**: obstruction of components and surface flows **Mobility** (road, rail): traffic disturbances and interruptions
Mobility | Roadways | Main roads, secondary roads, tunnels	Runoff, flooding and windstorm: disruption, interruption of mobility functions	**SLR**; **R**: runoff, flood; **W**: storms	Several urban services can be affected by cascading effects if maintenance or repair tasks are required during failures
Mobility | Roadways |Traffic signs	Wind can generate failures of traffic control systems	**R**: runoff, flood; **W**: storms	

Table 4. *Cont.*

SERVICE|SUBSYSTEM|Component	SERVICE OR INFRASTRUCTURE FAILURE	EXPOSURE TO	POTENTIAL DERIVED RISKS AND CASCADING EFFECTS
Mobility | Railways |Surface and underground	Flood and storms can cause interruption of public and private transportation	**SLR**; **R**: runoff, flood; **W**: storms	
Mobility | Railways | Traffic signs	Wind can generate failures of traffic control systems	**R**: runoff, flood **W**: storms	
Green and blue infrastructure and urban equipment | Several components (trees, street lighting)	Collapse of trees Damage and collapse	**R**: runoff, flood **W**: storms	**Urban drainage**: obstruction of components (e.g., inlets, sewers); **Electrical energy**: damage to equipment and lines; **Mobility** (road, rail): traffic disturbances and interruptions; **Communication**: damage to equipment and lines

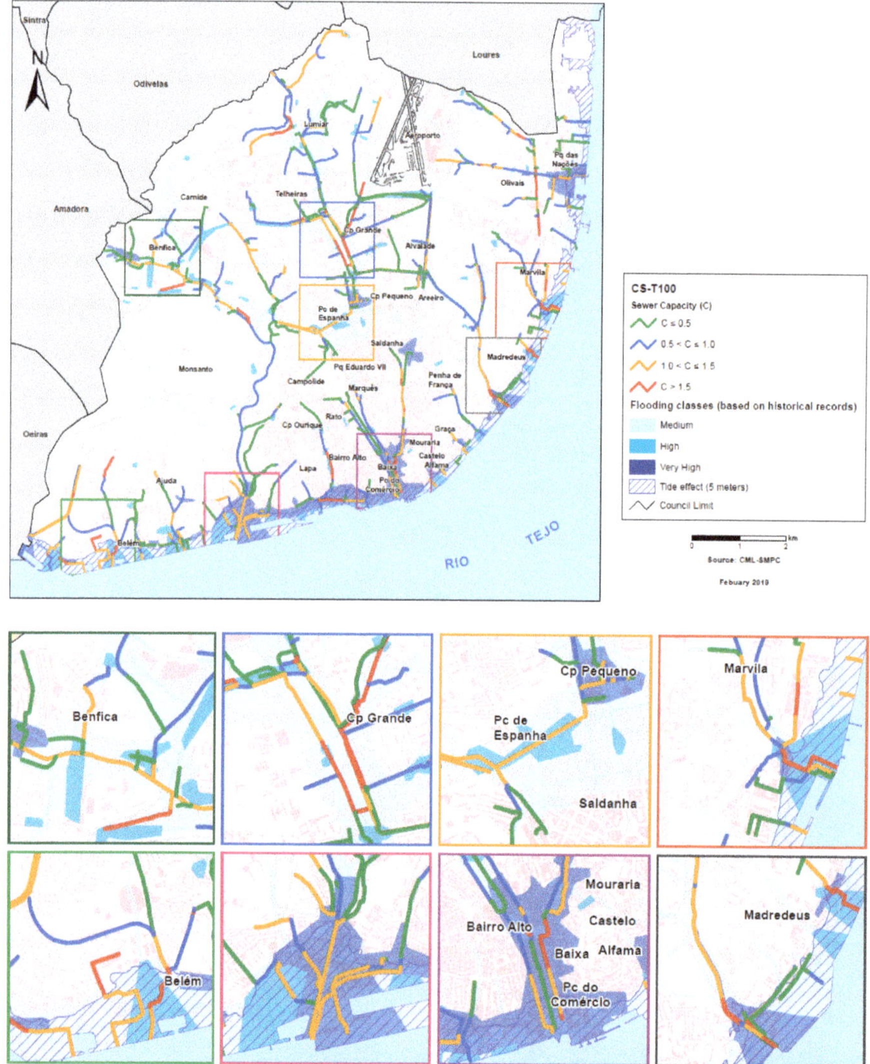

Figure 3. Use of sewer capacity at primary sewers (current situation (CS)-T100) and flood hazards based on historic observations.

4.3. Main Results for Current Situation and Business as Usual with Climate Change

Comparing the results of model 1 (1D GIS based), for the current situation (CS as baseline) and climate change scenarios (BAU, business as usual), i.e., BAU-CS, for each return period, allows estimating the effects of climate change scenarios simulated for the whole city, despite the limitations previously mentioned. The results show an aggravation in the metric C (use of sewer capacity) as response to increased flows generated, for the three return periods (Figure 4 and Table 5). The urban drainage overall performance decreases 8.6%, 7.8% and 10.0%, respectively for the return periods 10, 20 and 100 years. These percentages correspond to the sum of the values of classes "high" and "very high" from Table 5 and represent the relative increase in sewer length where capacity exceeds full pipe capacity and is a proxy of the relative effect of the simulated climate change scenarios.

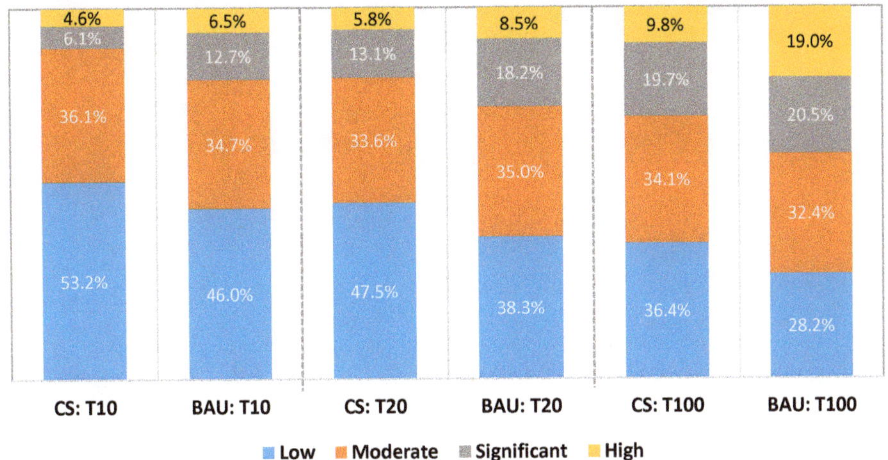

Figure 4. Citywide results for use of sewer capacity (model 1D GIS): CS and business as usual (BAU) situations.

Table 5. Citywide results for use of sewer capacity (model 1D GIS): comparison between CS and BAU.

C Range	Use of Sewer Capacity	Return Period (%(BAU-CS))		
		T010	T020	T100
$C \leq 0.5$	Low	−7.2	−9.2	−8.3
$0.5 < C \leq 1.0$	Moderate	−1.4	+1.4	−1.7
$1.0 < C \leq 1.5$	High	+6.7	+5.1	+0.8
$C > 1.5$	Very high	+1.9	+2.7	+9.2

For Lisbon downtown catchments J and L, results of the assessment of flood water level hazard using model 2 (Table 6) show a predominance of very low water levels and a slight increase in flooded areas associated with climate change scenarios (BAU). The apparent small magnitude in the effect of precipitation and sea level increase is explained by the aggregation in hazard classes, a factor not affected by variations in level within the classes of hazard, which is observed in detailed simulation results such as the example in Figure 5, the map showing the results at a critical time step, for T100. The overall magnitude of this hazard is identical in BAU and CS simulation results.

Table 6. Catchments J and L results for flood water level hazard (model 1D/2D): CS and BAU situations summary.

Flood Water Level (m)	Hazard Class	T010 (%)*		T020 (%)*		T100 (%)*		Δ (BAU-CS) (%)		
		CS	BAU	CS	BAU	CS	BAU	T010	T020	T100
$d \leq 0.2$	Very low	96.5	96.0	95.8	95.4	93.1	92.4	−0.57	−0.47	−0.67
$0.2 < d \leq 0.4$	Low	2.9	3.4	3.5	3.9	5.9	6.4	+0.53	+0.40	+0.58
$0.4 < d \leq 0.6$	Moderate	0.5	0.5	0.5	0.6	0.8	0.8	+0.07	+0.05	+0.06
$0.6 < d \leq 0.8$	High	0.1	0.1	0.1	0.1	0.2	0.3	−0.02	+0.02	+0.03
$0.8 < d \leq 1.0$	Very high	0.0	0.0	0.0	0.0	0.0	0.0	−0.01	+0.00	+0.00

(*) Percentage in each class.

The results for hazard to pedestrians are shown in Table 7. The magnitude of this hazard increases for all return periods, reaching around 4.0, 3.6 and 2.9 for the T010, T020 and T100 return periods, showing a slight decrease with the return period. Results for the assessment of hazard to vehicles (Table 8), follow a similar trend.

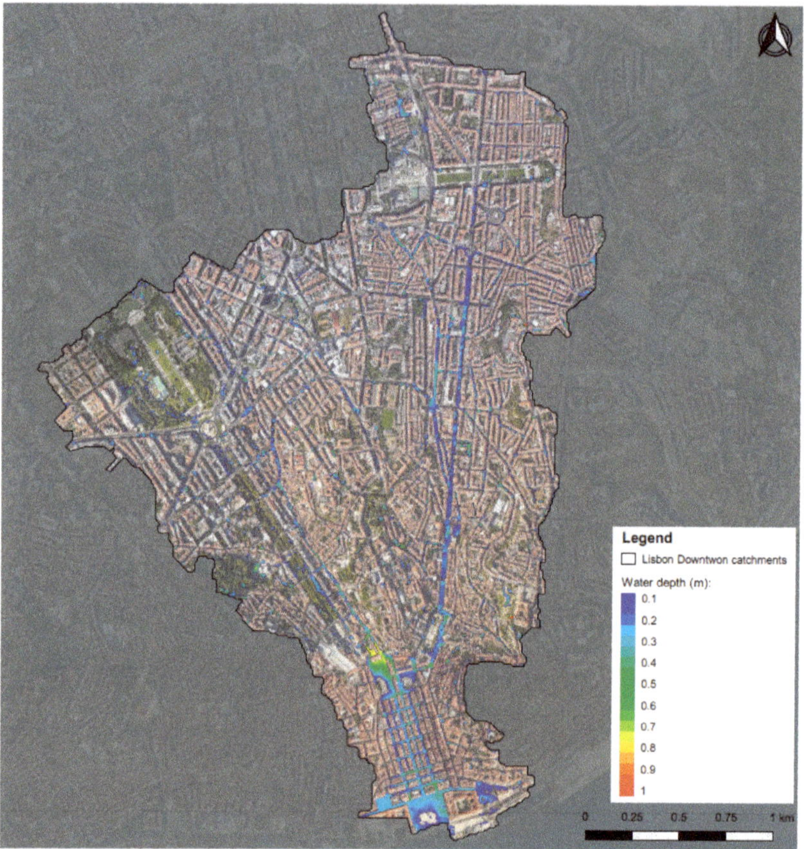

Figure 5. Results from model 1D/2D, catchments J and L. for the BAU situation (T100): water level at a critical time.

Table 7. Catchments J and L results for hazard to pedestrians (model 1D/2D): CS and BAU situations summary.

HR Range	Hazard Class	T010 (%)*		T020 (%)*		T100 (%)*		Δ (BAU-CS) (%)		
		CS	BAU	CS	BAU	CS	BAU	T010	T020	T100
$HR \leq 0.75$	Low	79.3	75.2	76.0	72.5	67.3	64.4	−4.0	−3.6	−2.9
$0.75 < HR \leq 1.25$	Moderate	18.5	21.7	20.9	23.5	26.9	28.9	+3.2	+2.6	+2.0
$1.25 < HR \leq 2$	Significant	2.2	3.0	3.0	3.9	5.7	6.6	+0.8	+0.9	+0.8
$HR > 2$	Extreme	0.1	0.1	0.1	0.1	0.2	0.2	0.0	0.0	0.0

(*) Percentage in each class.

Considering the results from the methods applied, it can be concluded that current situation is already unfavourable in terms of flooding frequency in various locations in Lisbon, but the magnitude of the hazards is generally low. Generally, these results are in alignment with available historical observations. The overlapping of the maps for the use of sewer capacity ((model 1D GIS) with results of model 1D/2D, for catchments J and L, confirms the most overloaded parts of the sewer network.

The climate change scenarios simulated do not impose significant increase in the magnitude of flood related hazards to properties and infrastructures, pedestrians and vehicles. Consequently, implications for the mobility and to waste sectors, compared to current situation, are expected to be low.

Table 8. Catchments J and L results for hazard to vehicles (model 1D/2D): CS and BAU situations summary.

Class Range	Hazard Class	T010 (%)*		T020 (%)*		T100 (%)*		Δ (BAU-CS) (%)		
		CS	BAU	CS	BAU	CS	BAU	T010	T020	T100
$D \leq 0.28$ and $D \times |\vec{v}| \leq 0.40$	Low	80.3	76.9	77.8	74.9	70.5	68.3	−3.4	−2.9	−2.3
$D \leq 0.28$ and $0.40 < D \times |\vec{v}| \leq 0.55$	Moderate	13.2	15.2	14.5	16.0	18.1	18.9	+1.9	+1.5	+0.8
$D > 0.28$ or $D \times |\vec{v}| > 0.55$	High	6.5	7.9	7.7	9.0	11.3	12.8	+1.4	+1.4	+1.5

(*) Percentage in each class.

4.4. Main Results for Selected Adaptation Strategies and Climate Change Scenarios

The three adaptation strategies selected for testing the proposed methodology, are some of those that can be simulated by hydraulic modeling and considered to be relevant by the stakeholders. These strategies are proposed in the Lisbon Master Plan 2016-2030 [36]. Even during risk identification, as presented in this paper, it is valuable to assess the impact of adaptation strategies in the flood related hazards.

The first strategy, CAS1—Adaptation of green infrastructure, corresponds to a significant increase of the total green area in the city (Figure 6). The second strategy, CAS2—Peak flow attenuation through the construction of two retention basins, includes the construction of two small retention basins, one of which has the main purpose of retaining solids (Figure 7).

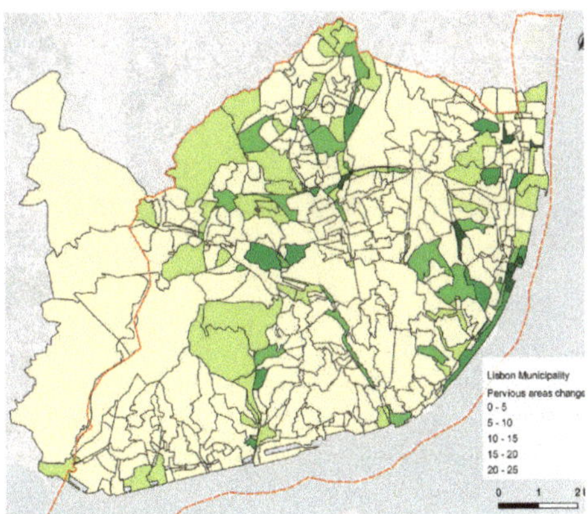

Figure 6. Green areas relative change: % increase from BAU to CAS1.

The third strategy modelled, CAS3—Construction of new components in drainage system, proposes the construction of a large interception tunnel and improvement in the inlets to the sewer network (Figure 8).

The three strategies were simulated with the citywide simplified model 1 but for the detailed model 2 of catchments J and L, only CAS3 is relevant.

Results of model 1 (1D GIS based), for BAU, CAS1, CAS2 and CAS3, for each return period, for the metric C, use of sewer capacity, are given in Figure 9, in terms of the relative variations to CS. The results show an aggravation in the metric C for BAU situation as presented previously as response to increased flows generated in the scenarios of climate change, for the three return periods. CAS1 has only some effect in the areas downstream of the catchments but even in those areas the reduction is limited, since the area upstream to the basins is small and the basin volumes are also small. CAS2 has no substantial influence and results are similar for all return periods. This is attributed to the small influence of the green areas on the hydrological processes for intense rainfall events. CAS3 is the only strategy contributing to decrease the length of sewers in the most severe class. However, since the effect is mainly expected in the areas downstream of the tunnels, the effect is not evident when evaluated for the whole city. In Table 9, the results for the areas downstream of the tunnels are presented and the effect of the tunnels is clearly effective in the reduction of flooding.

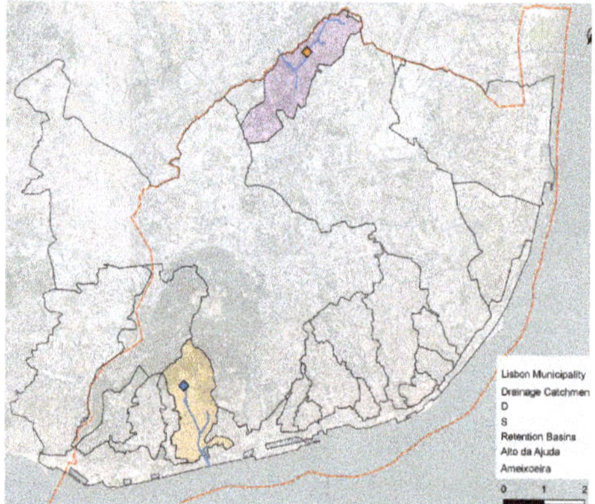

Figure 7. Location of the retention basins planned in the CAS2.

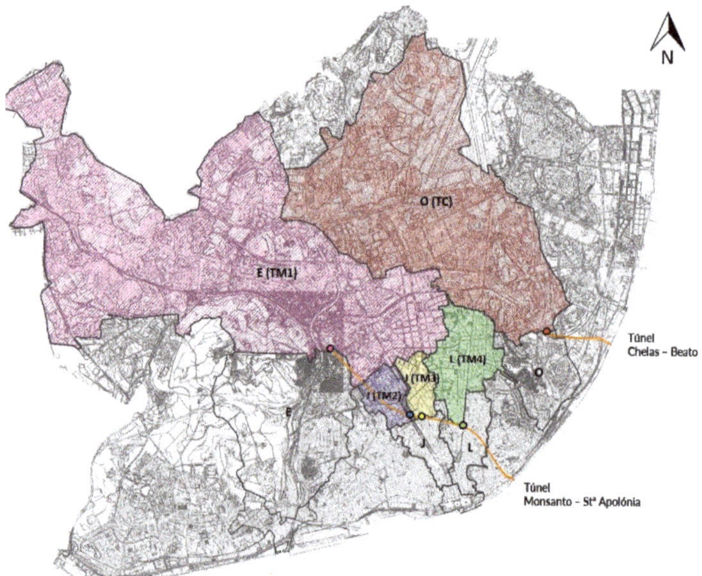

Figure 8. Tunnels: associated drainage catchments and intersection locations [37].

Globally, the results indicate that from the three strategies analysed, only CAS3 has a significant effect on flood related hazards and is limited to the areas downstream of the tunnels.

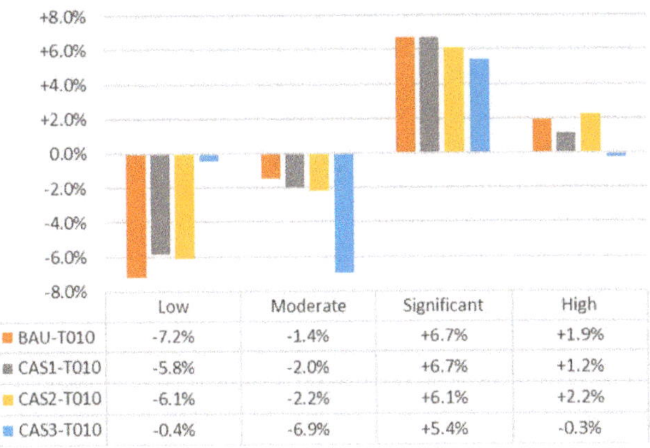

	Low	Moderate	Significant	High
BAU-T010	-7.2%	-1.4%	+6.7%	+1.9%
CAS1-T010	-5.8%	-2.0%	+6.7%	+1.2%
CAS2-T010	-6.1%	-2.2%	+6.1%	+2.2%
CAS3-T010	-0.4%	-6.9%	+5.4%	-0.3%

(**a**) Use of sewer capacity: results for T010

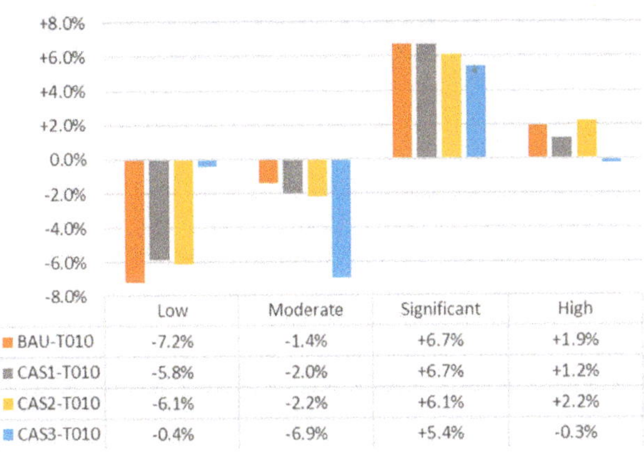

	Low	Moderate	Significant	High
BAU-T010	-7.2%	-1.4%	+6.7%	+1.9%
CAS1-T010	-5.8%	-2.0%	+6.7%	+1.2%
CAS2-T010	-6.1%	-2.2%	+6.1%	+2.2%
CAS3-T010	-0.4%	-6.9%	+5.4%	-0.3%

(**b**) Use of sewer capacity: results for T020

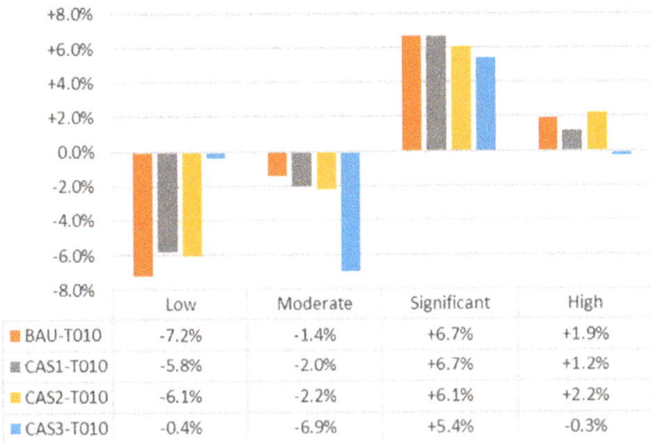

	Low	Moderate	Significant	High
BAU-T010	-7.2%	-1.4%	+6.7%	+1.9%
CAS1-T010	-5.8%	-2.0%	+6.7%	+1.2%
CAS2-T010	-6.1%	-2.2%	+6.1%	+2.2%
CAS3-T010	-0.4%	-6.9%	+5.4%	-0.3%

(**c**) Use of sewer capacity: results for T100

Figure 9. Citywide results for use of sewer capacity (model 1D GIS): results for BAU and CAS situations compared with CS.

Table 9. Results for use of sewer capacity (model 1D GIS) in catchments downstream of the tunnels: comparison between CS and CAS3.

C Range	Use of Sewer Capacity	Return Period (%(CAS3-CS))		
		T010	T020	T100
$C \leq 0.5$	Low	+36.0	+41.3	+38.3
$0.5 < C \leq 1.0$	Moderate	−36.6	−35.1	−30.6
$1.0 < C \leq 1.5$	High	+1.4	−4.4	−8.3
$C > 1.5$	Very high	−0.8	−1.8	+0.6

The effect of CAS3 on the catchments J and L is also beneficial but is not improving significantly under the current situation. This can be explained by the existence of duckbill tidal valves that require a certain pressure upstream to allow flows downstream to receiving waters. The CAS-3 results for downtown catchments detailed modelling (model 2) show the increase of the carrying capacity at the downstream sewers as the main effects of construction of the diversion tunnels, as obtained for the 1D GIS Model simulations. Nevertheless, in some cases, considerable water depths still occur.

The results for flood hazard to pedestrians show a slight decrease for all return periods. The results for the hazard to vehicles follows a similar trend.

5. Final Remarks

Risk identification is a key phase for risk-based approaches as well as to assessing resilience. Limitations in data and in ready-to-use tools often limits the development of these assessments. The methodology applied allows us to proceed with the spatial characterization of the flood related hazards and exposure of essential functions and services providers in urban areas, as a first step to support the assessment of the resilience to specific hazards, taking into account interdependencies and cascading effects. The combination of methods and existing data can add benefits often not obtained with more sophisticated methods and data. The combination of data and knowledge from different sources with dual modelling approaches can provide helpful results about the magnitude hazards, using metrics that relate with the urban functions and corresponding services, for different situations and scenarios.

An encouraging outcome of the methodology is the understanding achieved by stakeholders that are not familiar with water issues, about the relevance of flood related hazards for their modus operandi and the value of information they could obtain from other stakeholders to increase the resilience and reliability of their services.

The results obtained for the situations analysed modified the beliefs of involved people about the effect of some strategies to reduce flooding, while improving their understanding of hydrological and hydraulic processes and their relevance for managers and operators in Lisbon.

Areas for improvement include procedures to record flooding events and corresponding meteorological information, information on sewer network operational condition promotion of closer collaboration between stakeholders. Tools such as hydraulic models need to be recognized as effective in supporting current systems management.

The characterization of the hazards is instrumental to the adaptation of different sectors and, in this application, allowed us to realize that despite climate change impacts, the current situation is already affecting a number of city functions and services.

Author Contributions: M.d.C.A. supervised this study, developed the methodology and its application, analysed the results, drafted the manuscript and finalized it. M.J.T., M.M., J.B. and R.L. contributed to application of the methodology and analysis of the results, and provided suggestions on the draft manuscript. M.J.T. and M.M were responsible by undertaking data collection and overall mapping in ArcMap ™ software. J.B. and R.L. were responsible by undertaking the simulations of the drainage systems. All authors have read and agreed to the published version of the manuscript.

Funding: This research was funded by EUROPEAN UNION'S HORIZON 2020 RESEARCH AND INNOVATION PROGRAM, call H2020-DRS-2015 under the Grant Agreement number 700174 and acronym RESCCUE.

Acknowledgments: The work presented was developed within the EU H2020 RESCCUE project – Resilience to Cope with Climate change in Urban areas. Acknowledgment is due to Portuguese RESCCUE partners, as well as to all external contributors.

Conflicts of Interest: The authors declare no conflict of interest.

References

1. UN. Transforming Our World: The 2030 Agenda for Sustainable Development. A/RES/70/1 United Nations. 2015. Available online: https://sustainabledevelopment.un.org/content/documents/2125203020Agenda20for20Sustainable20Development20web.pdf (accessed on 10 January 2020).

2. Rogers, C.D.F. Engineering future liveable, resilient, sustainable cities using foresight. *Proc. Inst. Civ. Eng. Civ. Eng.* **2018**, *171*, 3–9. [CrossRef]

3. Elmqvist, T.; Andersson, E.; Frantzeskaki, N.; McPhearson, T.; Olsson, P.; Gaffney, O.; Folke, C. Sustainability and resilience for transformation in the urban century. *Nat. Sustain.* **2019**, *2*, 267–273. [CrossRef]

4. IPCC. *Global Warming of 1.5 °C. Summary for Policymakers*; Intergovernmental Panel on Climate Change: Geneva, Switzerland, 2018.

5. WEF. *The Global Risks Report 2019*, 14th ed.; World Economic Forum: Geneva, Switzerland, 2019.

6. UN-Habitat City Resilience Profiling Programme. Guide to the City Resilience Profiling Tool. United Nations Human Settlements Programme (UN-Habitat). 2018. Available online: http://urbanresiliencehub.org/wp-content/uploads/2018/07/CRPT-Guide-18.07-Pages.pdf (accessed on 24 September 2018).

7. IPCC. *5th Assessment Report*; Intergovernmental Panel on Climate Change: Geneva, Switzerland, 2014.

8. Kachali, H.; Storsjö, I.; Haavisto, I.; Kovács, G. Inter-sectoral preparedness and mitigation for networked risks and cascading effects. *Int. J. Disaster Risk Reduct.* **2018**, *30*, 281–291. [CrossRef]

9. Serre, D.; Heinzlef, C. Assessing and mapping urban resilience to floods with respect to cascading effects through critical infrastructure networks. *Int. J. Disaster Risk Reduct.* **2018**, *30*, 235–243. [CrossRef]

10. Velasco, M.; Russo, B.; Martínez, M.; Malgrat, P.; Monjo, R.; Djordjevic, S.; Fontanals, I.; Vela, S.; Cardoso, M.A.; Buskute, A. The RESCCUE Project: RESilience to cope with Climate Change in Urban arEas–a multisectorial approach focusing on water–The RESCCUE Project. *Water* **2018**, *10*, 1356. [CrossRef]

11. Pearson, J.; Punzo, G.; Mayfield, M.; Brighty, G.; Parsons, A.; Collins, P.; Jeavons, S.; Tagg, A. Flood resilience: consolidating knowledge between and within critical infrastructure sectors. *Environ. Syst. Decis.* **2018**, *38*, 318–329. [CrossRef]

12. Burns, P.; Hope, D.; Roorda, J. Managing infrastructure for the next generation. *Autom. Constr.* **1999**, *8*, 689–703. [CrossRef]

13. Pregnolato, M.; Ford, A.; Glenis, V.; Wilkinson, S.; Dawson, R. Impact of Climate Change on Disruption to Urban Transport Networks from Pluvial Flooding. *J. Infrastruct. Syst.* **2017**, *23*, 1–13. [CrossRef]

14. Otto, A.; Kellerman, P.; Thieken, A.H.; Costa, M.M.; Carmona, M.; Bubeck, P. Risk reduction partnerships in railway transport infrastructure in an alpine environment. *Int. J. Disaster Risk Reduct.* **2019**, *33*, 385–397. [CrossRef]

15. Martínez, E.; Gómez, M.; Russo, B.; Djordjevic, S. A new experiments-based methodology to define the stability threshold for any vehicle exposed to flooding. *Urban Water J.* **2017**, *14*, 930–939. [CrossRef]

16. Martínez, E.; Russo, B.; Gómez, M.; Plumed, A. An approach to the modeling of stability of waste containers during urban flooding. *J. Flood Risk Manag.* **2020**, *13*, 1–18. [CrossRef]

17. Hammond, M.J.; Chen, A.S.; Djordjević, S.; Butler DMark, O. Urban flood impact assessment: A state-of-the-art review. *Urban Water J.* **2015**, *12*, 14–29. [CrossRef]

18. Rosenzweig, B.R.; McPhillips, L.; Chang, H.; Cheng, C.; Welty, C.; Matsler, M.; Iwaniec, D.; Davidson, C.I. Pluvial flood risk and opportunities for resilience. *Wires Water* **2018**, *5*, 1–18. [CrossRef]

19. Evans, B.; Chen, A.; Djordjevic, S.; Webber, J.; Almeida, M.C.; Morais, M.; Telhado, M.J.; Silva, I.; Duarte, N.; Martínez-Gomariz, E.; et al. *Impact Assessments of Multiple Hazards in Case Study Areas with Adaptation Strategies*; RESCCUE Project: Barcelona, Spain, 2020.

20. INE. Censos 2011 Resultados Definitivos–Portugal. In *Census 2011 Definitive Results*; National Statistics Institute: Lisbon, Portugal, 2012.

21. CML. The economy of Lisbon in numbers 2014 (A economia de Lisboa em números 2014). Available online: http://observatorio-lisboa.eapn.pt/ficheiro/Lisboa_em_numeros_2014_final_01.pdf (accessed on 20 April 2017).

22. Telhado, M.J.; Baltazar, S.; Fernandes, F.; Cardoso, M.A.; Almeida, M.C.; Vieira, P.V. *Lisbon Municipality Contribution to the Demonstration of the WCSP, RIDB, RRDB, GIS Applications for Risk Assessment in Lisbon*; PREPARED project: Nieuwegein, The Netherlands, 2014.

23. The Rockfeller Foundation. Cities Taking Action. How the 100RC Network is Building Urban Resilience. 2017. Available online: http://100resilientcities.org/wp-content/uploads/2017/07/WEB_170720_Summit-report_100rc-1.pdf (accessed on 1 February 2020).

24. Silva, M.M.; Costa, J.P. Urban Flood Adaptation through Public Space Retrofits: The Case of Lisbon (Portugal). *Sustainability* **2017**, *9*, 816. [CrossRef]

25. Turner, S.F.; Cardinal, L.B.; Burton, R.M. Research Design for Mixed Methods: A Triangulation-based Framework and Roadmap. *Organ. Res. Methods* **2017**, *20*, 243–267. [CrossRef]

26. Jick, T.D. Mixing qualitative and quantitative methods: triangulation in action. *Adm. Sci. Quarterly* **1979**, *24*, 602–611. [CrossRef]

27. Duffy, M.E. Methodological triangulation: A vehicle for merging quantitative and qualitative research methods. *J. Nurs. Scholarsh.* **1987**, *19*, 130–133. [CrossRef] [PubMed]

28. Russo, B. (Ed.) *Multi-Hazards Assessment Related to Water Cycle Extreme Events for Current Scenario*; RESCCUE Project: Barcelona, Spain, 2018.

29. Russo, B. (Ed.) *Multi-hazards Assessment Related to Water Cycle Extreme Events for Future Scenarios (Business as Usual)*; RESCCUE Project: Barcelona, Spain, 2019.

30. Huber, W. *Storm Water Management Model (SWMM) Bibliography. Athens, Ga: Environmental Research Laboratory, Office of Research and Development*; U.S. Environmental Protection Agency: Washington, DC, USA, 1985.

31. VAW. *BASEMENT–Basic Simulation Environment for Computation of Environmental Flow and Natural Hazard Simulation. Version 2.8*; ETH Zurich: Zurich, Switzerland, 2018.

32. Hosseini, S.; Barker, K.; Ramirez-Marquez, J.E. A review of definitions and measures of system resilience. *Reliab. Eng. Syst. Saf.* **2016**, *145*, 47–61. [CrossRef]

33. Defra, E.A. *Flood Risk Assessment Guidance for New Development. Phase 2 Framework and Guidance for Assessing and Managing Flood Risk for New Development–Full Documentation and Tools*; Defra and Environment Agency: London, UK, 2005.

34. Antunes, C.; Catita, C.; Rocha, C. *Estudo de Avaliação da Sobrelevação da Maré–Determinação da Cartografia de Inundação e Vulnerabilidade da Área Ribeirinha de Lisboa Afetada Pela Sobrelevação da Maré Como Consequência da Futura Subida do Nível Médio do Mar. Relatório Técnico EMAAC*; Câmara Municipal de Lisboa: Lisbon, Portugal, 2017; 46p.

35. Paradinas, C.; Monjo, R.; Gaitán, E.; Carravilla, C.; Torres, L. *Projection of Climate Extremes in the City of Lisbon: A Comparative Study*; RESCCUE Project: Barcelona, Spain, 2019.

36. Hidra, Engidro, Bluefocus. *Plano Geral de Drenagem de Lisboa 2016–2030*; Câmara Municipal de Lisboa: Lisbon, Portugal, 2015.

37. Hidra. *Tender Documents for the Construction Works of the Drainage Tunnels and Associated Interventions in Lisbon*; Câmara Municipal de Lisboa: Lisbon, Portugal, 2018.

sustainability

Article

Socio-Economic Assessment of Green Infrastructure for Climate Change Adaptation in the Context of Urban Drainage Planning

Luca Locatelli [1,*]**, Maria Guerrero** [2]**, Beniamino Russo** [1]**, Eduardo Martínez-Gomariz** [2,3]**, David Sunyer** [1] **and Montse Martínez** [1]

[1] AQUATEC—Suez Advanced Solutions, Ps. Zona Franca 46-48, 08038 Barcelona, Spain; brusso@aquatec.es (B.R.); dsunyer@aquatec.es (D.S.); mmartinezp@aquatec.es (M.M.)

[2] Cetaqua, Water Technology Centre, Carretera d'Esplugues, 75, 08940 Cornellà de Llobregat, Barcelona, Spain; maria.guerrero@cetaqua.com (M.G.); eduardo.martinez@cetaqua.com (E.M.-G.)

[3] Flumen Research Institute, Universitat Politècnica de Catalunya, Jordi Girona 1-3, 08034 Barcelona, Spain

* Correspondence: luca.locatelli@aquatec.es

Received: 21 March 2020; Accepted: 29 April 2020; Published: 7 May 2020

Abstract: Green infrastructure (GI) contributes to improve urban drainage and also has other societal and environmental benefits that grey infrastructure usually does not have. Economic assessment for urban drainage planning and decision making often focuses on flood criteria. This study presents an economic assessment of GI based on a conventional cost-benefit analysis (CBA) that includes several benefits related to urban drainage (floods, combined sewer overflows and waste water treatment), environmental impacts (receiving water bodies) and additional societal and environmental benefits associated with GI (air quality improvements, aesthetic values, etc.). Benefits from flood damage reduction are monetized based on the widely used concept of Expected Annual Damage (EAD) that was calculated using a 1D/2D urban drainage model together with design storms and a damage model based on tailored flood depth–damage curves. Benefits from Combined Sewer Overflows (CSO) damage reduction were monetized using a 1D urban drainage model with continuous rainfall simulations and prices per cubic meter of spilled combined sewage water estimated from literature; other societal benefits were estimated using unit prices also estimated from literature. This economic assessment was applied to two different case studies: the Spanish cities of Barcelona and Badalona. The results are useful for decision making and also underline the relevancy of including not only flood damages in CBA of GI.

Keywords: urban flood; water quality; cost-benefit analysis; modelling; combined sewer overflows

---

1. Introduction

Green infrastructure (GI)—also recognized with the acronyms NBS (Nature-Based Solutions), SUDS (Sustainable Urban Drainage Systems), LID (Low Impact Development), BMP (Best Management Practices), WSUD (Water Sensitive Urban Design) and many others [1]—contributes to improve urban stormwater management and has several other societal benefits like air quality improvements, reduction of heat island effects, aesthetic and recreational values, and others [2]. Socio-economic assessment of GI is an important tool for urban drainage planning and decision making of climate change adaptation strategies [3].

Several studies have presented socio-economic assessments of different climate change adaptation options focusing on direct and indirect benefits derived from flood damage reduction capacity of GI. Velasco et al. [4] presented a cost-benefit analysis where only direct benefits were included in terms of avoided flood damages obtained by different adaptation scenarios in Barcelona: structural measures (pipe enlargement and stormwater tanks), GI, flood barriers for ground floor doors of businesses and private buildings and early-warning systems. Zhou et al. [3,5] presented a framework and its application to a Danish case study for economic assessment of different climate adaptation options focusing on flood impacts. The economical assessment was based on a cost-benefit analysis (CBA) with direct and indirect benefits derived from flood damage reduction that were monetized using flood models together with damage costs for houses, basements, sewers, roads, lakes and people health and also administrative and traffic delay costs. The damage costs were calculated using unit costs reported from case-specific literature. In these papers, flood adaptation options based on pipe enlargements were compared to stormwater infiltration through GI focusing on flood reduction benefits.

Further studies present socio-economic assessments including additional benefits not only related to direct or indirect flood damages [6]. Löwe et al. [7] presented a cost-benefit analysis (CBA) for comparing different flood adaptation options in Australia. The flood adaptation options consisting of pipe enlargement, flood zoning and rainwater harvesting through GI were compared including flood reduction benefits and also additional benefits derived from reduction of drinking water consumption. Zhou et al. [8] presented an integrated hydrological cost-benefit analysis for comparison of different climate adaptation options such as open urban drainage systems, pipe enlargement and local stormwater infiltration. Here, benefits derived from flood damage reduction were integrated with additional monetized benefits derived from increased property values in the areas where GI was planned and the consequent increase in property taxes. Finally, Cooper et al. [9] presented an integrated costs-benefits analysis of a berm (sea wall) to mitigate the effects of coastal flooding from sea storms. Here, the monetized benefits of the project included: avoided costs derived from building damages, management expenses, fatalities, debris removal, utility and municipal damages; benefits derived from recreational and health value and indirect costs derived from interruption of key transportation and commercial infrastructure located in the area. The recreational and health values were linked to the ecosystem services and health benefits to the surrounding community generated by the planned green areas along the berm.

Further studies underlined the importance of analyzing GI with a multidisciplinary approach. Venkataramanan et al. [10] presented a multidisciplinary literature review focusing on the interaction between human dimensions and socio-ecological-technical systems that are involved with GI in the context of flood risk management. Additionally, Wilkerson et al. [11] analyzed the role of socio-economic factors involved in the planning and management of urban ecosystem services.

The aim of this paper is to present a cost-benefit analysis that includes multiple benefits derived from green infrastructure in the context of urban drainage planning. The novelty of this study is the integration of water quantity and quality and other socio-economic benefits into CBA of GI in the context of urban drainage planning. The application of this analysis to two different case studies can also be considered as novel since the application of CBA is generally used for comparing different adaptation measures within the case study. GI benefits are calculated from direct and indirect flood impacts reduction, water quality related benefits and additional societal benefits. Benefits of flood damage reduction are calculated as avoided direct and indirect flood damage costs to buildings, vehicles, urban infrastructure and indirect costs. Flood damage costs are calculated using coupled 1D urban drainage and 2D surface runoff models together with tailored depth–damage and permeability coefficients functions. Water quality related benefits derived from CSO and waste water treatment cost reduction are calculated using a 1D urban drainage model and costs of wastewater treatment and CSO spills obtained from literature. Finally, additional societal benefits like increased aesthetic value, air quality improvement, habitat provision and reduced urban heat island effect and energy consumption, are calculated based on unit costs from literature. The socio-economic assessment is applied to two different case studies: the Spanish municipalities of Barcelona and Badalona. These two case studies were part of the two European H2020 research projects: BINGO (Bringing Innovation to onGOing water management. www.projectbingo.eu) and RESCCUE (Resilience to Cope with Climate Change in Urban Areas. www.resccue.eu). The aim of presenting two cases is mainly to show that the methodology can be applied to different cases. Nevertheless, the comparison can also bring new points of view in the discussion of GI in the context of urban drainage planning. The methodology proposed can be considered generally applicable to other cities in the context of green infrastructure and urban drainage planning.

2. Materials and Methods

2.1. The Two Case Studies

2.1.1. The Case Study of Barcelona

Barcelona (Figure 1) has an extension of approximately 100 km^2, 1,619,000 inhabitants and it is highly urbanized. An important part of its urban development lies in a flat area up to few tens of meters above mean sea level. The city faces the Mediterranean Sea and approximately half of its coast line is occupied by the harbor and the remaining by sandy beaches. In the opposite side of the sea, there are hills with significant slopes towards the urban area. The great majority of the drainage system is a combined one and Barcelona experiences urban pluvial floods due to intense rainfalls, steep slopes towards the flat urban area, high degree of imperviousness and, in recent years, expansion of new urban areas draining into an older drainage system. The mean annual rainfall is 612 mm/y, the degree of imperviousness is estimated to be approximately 70% of the whole municipal area even though it can reach much higher percentages in the urban areas (see for instance the two zoom-in areas in Figure 1). The city also experiences Combined Sewer Overflows (CSO) that generally occur during rainfall events larger than a few millimeters. CSOs pollute the river Besos (that coincides with the north-eastern boundary of the municipal area shown in Figure 1) and the sea water both in front of the beaches and in the harbor. Figure 1 also shows the planned GI that will be described in Section 2.2.1.

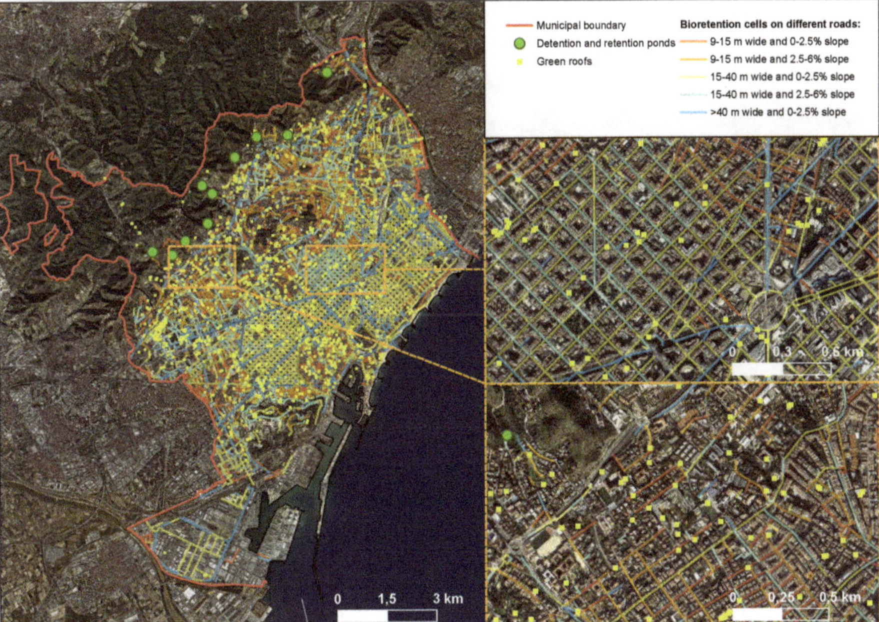

Figure 1. Plan view of Barcelona with all the planned GI: ponds, green roofs and bioretention cells. The colored lines show the classification of five different kind of streets where bioretention cells are planned (a different spatial allocation of bioretention cells was proposed as a function of the different street slope and width).

2.1.2. The Case Study of Badalona

Badalona (Figure 2), within the Barcelona Metropolitan Area, has an extension of approximately 21 km^2, 215,000 inhabitants (the fourth most populated city in Catalonia) and it is highly urbanized. An important part of its urban development lies in a flat area up to few tens of meters above mean sea level. In the north and north-western part of the municipality there are hills with significant slopes towards the urban area. On the opposite side the city has approximately 5 km of sandy beaches facing the Mediterranean Sea. Badalona experiences urban pluvial floods due to intense rainfalls, steep slopes towards the flat urban areas, high degree of imperviousness and, in recent years, expansion of new urban areas draining into an older drainage system. The mean annual rainfall is 568 mm/y, the degree of imperviousness is estimated to be approximately 57% of the whole municipal area even though it can reach much higher percentages in the urban areas (see for instance the two zoom-in areas in Figure 2). Almost all the drainage system is a combined one and CSOs that generally occur during rainfall events larger than a few millimeters pollute the sea water. Figure 2 also shows the planned GI that will be described in Section 2.2.3.

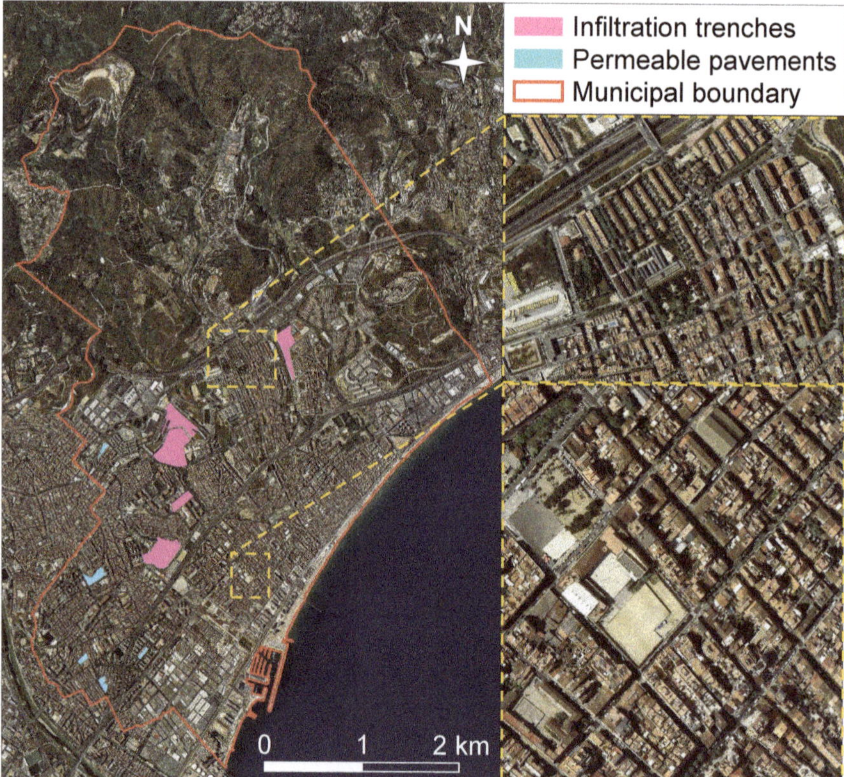

Figure 2. Plan view of Badalona with the planned green infrastructure (green roofs are not shown) and two zoom-in areas for better visualization of the urban environment.

2.2. The Climate Change Adaptation Scenarios

Green infrastructure was one out of the several climate change adaptation options (do nothing, pipe enlargement, new pipes and detention storages and early-warning systems) proposed and analyzed in agreement with the different local stakeholders. In this study, two different adaptation scenarios with future rainfalls are presented: the business as usual (BAU) scenario where no adaptation is considered, and the GI scenario. The BAU scenario is used as a reference scenario when calculating benefits as part of the cost-benefit analysis. Both BAU and GI scenarios were based on future simulated rainfalls. Two kinds of future rainfalls were estimated for each of the two case studies: a future design storm event relevant for single event flood simulations and a future continuous rainfall time series relevant for continuous urban water simulations that aimed at stimulating combined sewer overflows, water quality impacts on the Mediterranean Sea and annual combined sewer water fluxes at waste water treatment plants. The future design storm events were calculated by applying climate factors (CF) to current design storm events according to Arnbjerg-Nielsen et al. [12]. It is noted that significantly different approaches were used in Barcelona and Badalona in order to derive CF. In Barcelona the 50th percentiles of all Representative Concentration Pathways (RCP) 8.5 and 4.5 scenarios were used to compute CF as a function of both different return periods and rainfall durations. Instead, in Badalona the average values of RCP 8.5 were used to compute CF as a function of different return periods. Nevertheless, the obtained climate values are in both cases within the range proposed in other local studies [13]. Details on the derivation of CF and future rainfall time series are provided in the following. Further future climate variables like temperature, sea level rise, wind, solar radiation, etc., were not considered in the current climate change adaptation scenarios even though they likely impact the future urban drainage systems and GI performances [14].

2.2.1. Green Infrastructure in Barcelona

The proposed GI in Barcelona was agreed with local project stakeholders and it was mostly derived from a study of the Municipality of Barcelona [15] that aimed at increasing stormwater exploitation in the city. Three different types of GI were proposed: green roofs, bioretention cells and retention and detention basins. Figure 1 shows the location of GI throughout Barcelona. Extensive green roofs are assumed to be retrofitted to approximately 5% (143 ha) of all the roof area of Barcelona. This percentage was derived from a study for the Municipality of Barcelona [16] that analyzed the roof areas suitable for green roof retrofitting. Bioretention cells with a total area of approximately 181 ha are supposed to be implemented in almost all the streets of Barcelona as shown in Figure 1. The location and preliminary design of these bioretention systems were proposed in a study for the Municipality of Barcelona [15] that suggested five different spatial distribution and capacity of bioretention systems depending on street slope and width (the five street types classified were presented in Figure 1). The proposed systems are made of a top soil and vegetation layer and a deeper layer of more porous material for water detention and infiltration into the underlying soil. The bioretention cells are devised for managing stormwater runoff from part or the whole streets where they are built. Finally, ten retention and detention basins with a total volume of 128,700 m^3 are supposed to be located at the upstream parts of the urban area in order to collect stormwater runoff from the upstream rural areas for a 10-year return period design storm. Approximately half of the basin volume is allocated to retention with infiltration into the ground and the rest to detention and reduction of peak stormwater runoff. Other examples of the combination of retention and detention volumes can be found in the literature [17].

Overall, the GI implementation in Barcelona would reduce the total impervious area by approximately 14% for all the modelled area. Nevertheless, this reduction is higher in the city center reaching approximately 29%. It is noted that bioretention cells and retention and detention basins do manage stormwater runoff from their associated catchment areas (larger areas compared to their physical construction areas).

2.2.2. The Future Rainfalls in Barcelona

The future rainfalls in Barcelona were computed based on the results of CMIP5 climate models considering the RCP scenarios 8.5 and 4.5. Downscaling methods were then applied and verified using both the ERA-Interim re-analysis as a reference for reproducing the past climate variables and other statistical indicators. Future rainfalls were finally derived using both rainfall observations from local rain gauges and different atmospheric circulation models: ACCESS1, BCC-CSM1, CanESM2,

CNRM-CM5, GFDL-ESM2M, HADGEM2-CC, MIROC-ESM-CHEM, MPI-ESM-MR, MRI-CGCM3 and NorESM1. Each model provided past (1951–2005) and future (2021–2100) rainfall time series.

The CF used for flood simulations were computed for both different rainfall durations (5, 10, 15 min, etc.) and different return periods (T = 1, 10, 50, 100 and 500 y) by calculating the rainfall intensity ratio between the simulated future (2071–2010) and simulated historical period (1976–2005). The computed CF were in the range between 1.07 and 1.26 and corresponded to the 50th percentile of the predicted RCP 8.5 and 4.5 scenarios.

The future rainfall time series used for continuous urban water simulations was selected to be the same as the actual one. This choice came after analyzing the predicted future rainfall volume and annual number of rainfall events. The 50th percentile of the latter two variables did not show an increase in the future and therefore, together with the project stakeholders, it was decided to keep the current rainfall time series for continuous urban water simulations of the future climate change adaptation scenarios.

2.2.3. Green Infrastructure in Badalona

The proposed GI in Badalona was agreed together with local project stakeholders that spotted realistic near-future implementation areas. Three different types of GI were selected for the adaptation scenario: green roofs, permeable pavements and infiltration trenches (Figure 2). Extensive green roofs are assumed to be retrofitted to 5% of all the roof area of Badalona. Permeable pavements with a total area of 47,000 m^2 are supposed to be implemented in 7 different public squares and parks. Infiltration trenches are supposed to be implemented in 5 different public parks that have a total area of 298,372 m^2. These trenches are supposed to retain and infiltrate into the ground both the impervious and pervious stormwater runoff from the parks (mostly pervious areas) generated by a design storm of 10 years return period. A total trench volume of 1923 m^3 was estimated (assuming a 95% porosity of the trench filling material).

Overall, the planned GI implementation in Badalona would reduce the total impervious area by approximately 2%. It is noted that infiltration trenches do not reduce impervious areas; however, they do manage stormwater runoff from their associated catchment areas.

2.2.4. The Future Rainfalls in Badalona

Two different sources of future climate data were used in the case of Badalona. The future design storm events for flood simulations were obtained from climate projections results of the EURO-CORDEX project (www.euro-cordex.net) while the future rainfall time series for continuous urban water simulations were obtained from the decadal climate predictions of the Miklip project (www.fona-miklip.de) that were derived from the model MPI-ESM (www.mpimet.mpg.de/en/science/models/mpi-esm/).

The CF used for flood simulations were obtained by calculating the 24 h rainfall intensity ratio between future projections (2051–2100) and historical simulated rainfall (1951–2005). Three different RCP scenarios were analyzed: 8.5, 4.5 and 2.6. The CF obtained with average rainfall intensities from RCP 8.5 scenarios were the ones selected together with the project stakeholders for flood simulations. A CF of 1.15 for the 2-year return period design storm was obtained, 1.07 for the 10-year, 1.02 for the 100-year and 1.01 for the 500-year. In this case, the same climate factor is applied to all rainfall durations. Calculating climate factors from 24 h rainfall intensity ratio can be a limitation [18].

The future rainfall time series used for continuous urban water simulations were obtained in two steps: first, the daily rainfall was obtained using the Daily Spatio-Temporal Stochastic Precipitation Generator [19]; then, disaggregation of daily rainfall into 5 min values was made using a stochastic method that combined both the Bartlett–Lewis process [20] and further procedures (included into the R package 'HyetosMinute') in order to reproduce the 5 min rainfall observations from local rain gauges. This procedure provided an ensemble of 10 different time series with both historical and future rainfall. Only a single time series representing average future rainfall conditions was selected and used for continuous simulations with the urban drainage and the sea water quality model (presented in Section 2.3.2).

2.3. The Cost-Benefit Analysis

2.3.1. Costs

The capital (CAPEX) and operation and maintenance (OPEX) costs of the planned green infrastructure are based on unit costs obtained from both literature and local experience. The costs ranges found in provider websites, unpublished documents and literature have generally a large spread. In this cost-benefit analysis (CBA) the costs were derived partly from literature [21] and partly from unpublished documents and internal research projects. The different costs were converted into the same year value using consumer price indices. The CAPEX of extensive green roofs are assumed to be 80 €/m^2 and the OPEX 2.33 €/m^2/y. Bianchini et al. [22] reported a CAPEX range of 120–152 €/m^2 and an OPEX one of 1–12 €/m^2 for extensive green roofs. The CAPEX of bioretention cells are 45 €/m^2 plus 2.25 €/m^2 for plant implementation and the OPEX 0.45 €/m^3/y. The CAPEX of detention and retention ponds are 100 €/m^2 and the annual OPEX is 1.49% of the CAPEX. The CAPEX of permeable pavements are 49.5 €/m^2 and the OPEX 1.375 €/m^3/y. The CAPEX of infiltration trenches are assumed to be 185 €/m^3 and additional 742 €/m^3 the OPEX 50 €/m^3/y. The additional CAPEX of infiltration trenches in this case include the costs of additional manholes, inlets and pipes that need to be constructed since these systems are supposed to be constructed into a public park area where existing drainage connections are limited. The estimated CAPEX of the infiltration trenches proposed in Badalona are similar to the costs paid by the municipality for an executed project. Zhou et al. [8,23] used investment costs of infiltration trenches in the range between 16 and 91 €/m^2. Alves et al. [6] estimated annual OPEX as 3% of CAPEX costs.

The lifetime of an infrastructure can vary depending on Its maintenance: the higher the maintenance costs the longer the lifetime [24]. In Badalona extraordinary maintenance was assumed to be carried out every 20 years with a cost equal to the 23% of the CAPEX at each intervention. Similarly, in Barcelona it was assumed every 20 years for bioretention cells and 50 years for green roofs and retention and detention ponds with a cost equal to the 50% of the CAPEX at each intervention.

Residual GI value at the end of the project evaluation period was also considered according to European recommendations for evaluations of investments [25]. This reflects the value of the remaining potential use of GI since its services will be provided further beyond the end of the CBA evaluation period [25]. In this study, it was considered as a negative cost but it could also be considered as a benefit as the choice does not affect the net present value Equation (1) (it only affects the graphical presentation of cost and benefits).

2.3.2. Benefits

Several benefits can be included into CBA of green infrastructure [5,9,14]. Benefits can be direct and indirect, tangible (i.e., that can be quantified in monetary values) and intangibles [26]. In this study, direct and indirect tangible benefits are taken into consideration. The benefits of the GI scenario were calculated as avoided damages (or added values) compared to the BAU scenario that is considered to be the reference as typically done in similar CBA [3,5]. In this study, the benefits were organized into 3 different categories for a better representation and discussion of the results:

- Benefits derived from flood damage reduction. Benefits are defined as avoided direct and indirect flood damage costs. Flood damage costs were quantified in terms of Expected Annual Damage (EAD) using a 1D/2D urban drainage model together with design storms and a damage model based on tailored flood depth–damage curves [27]. The direct flood damages were quantified for infrastructure, vehicles, buildings and assets, while the indirect damages for business interruption.
- Benefits derived from water quality improvements. Benefits are defined as avoided direct and indirect damage costs. The direct damages are quantified as environmental costs produced by

CSO spills to receiving water bodies and for avoided costs of combined sewage treatment. Indirect damages are monetized for coastal economies that are affected by the polluted water.

- Additional benefits. Additional indirect benefits are monetized considering: increased aesthetic value, air quality improvement, reduction of the urban heat island effect and energy consumption, and habitat provision [22,24].

Direct flood damages in both Barcelona and Badalona were quantified using coupled 1D/2D (urban drainage/overland flow) models and damage models based on tailored flood damage curves (developed for indoor flood water levels) and permeability coefficient curves that were developed together with flood insurance experts [27]. The damage model takes as inputs the deterministic and spatially distributed values of maximum flood depth simulated with the 1D/2D urban drainage model. The simulated flood depth from the 1D/2D model (considered as outdoor flood depth) is converted into building indoor water levels using the permeability coefficient curves and then the flood damage curves are applied to indoor water levels. In the case of buildings with basements further model parameters control the indoor flood water exchange from ground floor to lower floors. Both the flood models and the damage models (of Barcelona and Badalona) were calibrated and validated using historical data. The flood models used water level data in the drainage network, rain gauge data and photos of urban floods during different past rain events. The damage models used flood insurance compensation data from different flood events during the last few decades [27]. The most influential model parameters of the 1D/2D model were the roughness coefficients of pipes and urban surfaces and of the damage model the parameters controlling the indoor flood water level exchange from ground floor to lower floors [27].

The 1D/2D model provides the maximum simulated flood depth for different design storms of different return periods between 1 and 500 years: 1, 10, 50, 100 and 500 years for Barcelona and 2, 10, 100 and 500 for Badalona. For each return period, the total flood damages at the urban scale were calculated by multiplying the maximum simulated flood depth at each cadastral parcel by permeability coefficients and flood depth–damage curves that were specifically tailored for Badalona and Barcelona for different land uses (hotels, warehouses, restaurants, dwellings, car parks, etc.) and vehicles [28]. The permeability coefficient curves were used to transform the 2D simulated flood levels on the urban floodable area into indoor water levels. Finally, Expected Annual Damage (EAD) was calculated including both direct and indirect damages as detailed in a previous study of Badalona [27]. Indirect flood damages due to business interruptions were estimated at 29% of the total direct damages using an input–output model [27]. This percentage is in the range of other studies that proposed 19–39% [29,30].

The 1D/2D hydrodynamic models were developed with InfoWorks ICM (www.innovyze.com) and calibrated and validated using local rainfall and water level data. The 1D sewer model of Badalona includes approximately 368 km of pipes, 11,338 manholes, 11,954 sub-catchments, 62 weirs, 4 sluice gates, and 1 detention tank of 30,000 m^3. The 2D model has 199,338 cells that form an unstructured mesh generated from a digital terrain model (DTM) of 2 m^2 resolution obtained by a LIDAR with a precision of approximately 15 cm for the altitudes. The size of the 2D cells is in the range of 16–64 m^2 in the urban areas where most of the flood damages occur. The 1D sewer model of Barcelona includes approximately 2041 km of pipes, 85,834 manholes, 980 weirs, 44 sluice gates, 75 pumps and 285 storage nodes representing different kinds of chambers and 10 detention tanks with a total volume of more than 400,000 m^3. The 2D model has 1,361,324 cells that form an unstructured mesh generated from a digital terrain model (DTM) of 2 m^2 resolution obtained from a LIDAR. The size of the 2D cells is in the range of 25–100 m^2.

Direct and indirect water quality benefits were computed using continuous simulation of a 1D urban drainage model to estimate annual volumes of CSO and combined sewage water sent at the treatment plant. The urban drainage models used were the 1D/2D models presented earlier but without the 2D overland flow model. The urban drainage models were then coupled to a sea water quality model [31] to simulate the sea water contamination from CSOs and to estimate the average duration of insufficient bathing water quality. The duration of insufficient bathing water quality was used as an input to a coastal economy model that estimates indirect damages to coastal economies caused by pollution of bathing waters and the consequent reduction of sea related leisure, sport and restoration activities. The coastal economy model includes different contributions. First, the daily direct added value of the coastal economy was calculated by selecting the expected business sectors affected by a beach closure (restaurants, small retails and maritime sector). This selection was based on the results of a field study based on surveys to beach goers and personal interviews to coastal business owners carried out in Barcelona and Badalona (see both H2020 BINGO and RESCCUE projects). Second, based on data from Barcelona's economic annual report [32] a 50% share of the annual coastal economic added value was assumed to come from the bathing season [33], which lasts approximately 3–4 months in Badalona and Barcelona. Only the direct added value of coastal districts (identified by comparing the CSO spill points with the districts maps) affected by CSO spills were included. Furthermore, assumptions of the magnitude of the impact per sector were made based on the results of the local surveys: 50% impact to restaurants, 25% to retails and 25% to maritime sector (water sport and private fishing). The daily economic impact obtained by dividing the value added by the number of days of the bathing season, was finally multiplied by the average number of sea water pollution days (where the beaches could potentially be closed to bathing) to estimate the potential annual indirect damages to the coastal economy.

The different GI systems were simulated in both the 1D and the 1D/2D drainage models by converting the planned GI areas from impervious areas into pervious areas with hydrological losses. This simplified approach was also used by Velasco et al. [4]. However, to the knowledge of the authors, this method was not validated with hydrological data and can be a limitation.

The direct damages produced by CSO spills to receiving water bodies were calculated using a reparation cost method, which assumes that the value of the damage is equal to the cost of repairing it [34]. The direct damage produced by CSO spills was obtained multiplying the average annual CSO volume by the unit CSO damage cost of 0.7 €/m^3 in Badalona. Instead, in the case of Barcelona different values were used: 2.69 €/m^3 for CSOs to the sea and 1.50 €/m^3 to the river and the harbor according to a Spanish regional normative devised for industrial spills [35]. Another benefit considered was the reduction of the sewage water to be treated by the wastewater treatment plants (WWTPs). The monetization of this benefit was calculated as the avoided costs of combined sewer water treatment that were estimated by multiplying the average treated annual volumes from the urban drainage model with a selected unit treatment cost of 0.12 €/m^3 that is considered reasonable for local WWTPs based on local expertise. The tangible indirect damages (and the consequent benefits calculated as avoided damages) to coastal economies were estimated using the pollution time from the sea water quality model and the coastal economy model explained before.

The additional indirect benefits considered are based on four contributions. The first is aesthetic value which is monetized as the willingness to pay for properties nearby, or that include green infrastructure, is measured through the increase of the value of these properties. This value could also include the increased property taxes acquired by the taxation authorities [8]. In this case, the benefits were estimated with a benefit transfer method to be the 3% of the CAPEX of GI [22]. The benefits derived from the reduction of energetic consumption (for indoor heating and cooling) and heat island effect are quantified using 0.049 €/m^2/y per green roof unit surface [24]. Benefits derived from urban heat island reduction obtained with bioretention cells (that in Barcelona are planned) were not included and this can be a limitation. The air quality benefits are derived from both emission reduction (of CO_2 and Nox) capacity of GI that was estimated to be 0.072 ton/ha and the cost of emissions of 3051 €/ton [24,36]. The habitat provision was based on the potential increase of urban ecosystems that support wildlife and it was estimated to be 2.8 €/m^2 for both case studies. This was estimated using a benefit transfer method from a study that assumed the value of habitat creation could be estimated at 15% of the value of natural land [22].

2.3.3. Net Benefits

The net present value (NPV) is calculated using Equation (1)).

$$NPV = \sum_{t=1}^{T} \frac{B_t - C_t}{(1+i)^t} \tag{1}$$

where B_t and C_t are the benefits and costs at each year t, i is the discount rate, T is the project evaluation period.

3. Results

3.1. Costs

Table 1 summarizes CAPEX and OPEX of the green infrastructure proposed in Badalona and Barcelona. The table shows that in the case of Badalona the total costs are approximately an order of magnitude lower compared to Barcelona. Barcelona has a bigger area and a much more ambitious implementation plan compared to Badalona. Further, the total GI costs of Badalona are dominated by green roofs. This is because green roofs are assumed to be retrofitted onto 5% of the total roof area of Badalona, whereas infiltration trenches are placed only on 7 different parks and infiltration pavements on 5 different parks and public squares.

Table 1. CAPEX and OPEX of the analyzed green infrastructure.

	Badalona		Barcelona	
	CAPEX [€]	**OPEX [€/y]**	**CAPEX [€]**	**OPEX [€/y]**
Green roofs	14,534,788	405,157	114,752,240	3,342,159
Infiltration trenches	1,783,561	96,150		
Permeable pavements	1,739,183	48,311		
Bioretention cells			85,509,743	1,357,298
Detention and retention ponds			12,870,000	191,763
TOTAL	18,057,531	549,618	213,131,983	4,891,220

3.2. Benefits

The first step in order to estimate benefits derived from flood damage reduction obtained by GI implementation is the estimation of EAD for both the BAU and the GI scenarios. Table 2 shows the EAD results. Generally, the EAD of these two BAU scenarios are considered to be overestimated, particularly in the case of Barcelona (see the Discussion section). Figure 3 shows the flood damage costs simulated as a function of different exceedance probabilities for the two case studies. The EAD that is the area below the curve of Figure 3 was calculated using simple trapezoidal contributions adopting the linear interpolation between the discrete points represented Figure 3.

Table 2. Flood Expected Annual Damage including both direct and indirect damages.

		M€/y
	EAD. BAU	62.65
Barcelona	EAD. GI	33.90
	Flood damage reduction	28.75
	EAD. BAU	1.93
Badalona	EAD. GI	1.86
	Flood damage reduction	0.07

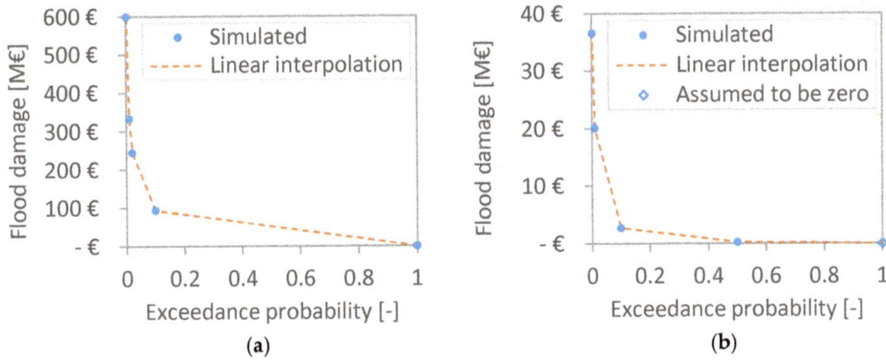

Figure 3. Flood damage as a function of the exceedance probability for Barcelona (**a**) and Badalona (**b**).

Table 3 shows the details of the monetized annual (not discounted) benefits for each of the three categories proposed and their percentage contribution to the total benefits in Barcelona and Badalona. The table shows that the benefits derived from reduced combined sewage treatment costs; from reduced indirect damages to coastal economies; from air quality improvement and from reduction of the urban heat island effect and energetic consumption are in the range of 0–1%.

Table 3. Annual value of benefits (not discounted).

Benefit Category	Description	Barcelona			Badalona		
		Value [€]	Percentage	Aggregated Percentages	Value [€]	Percentage	Aggregated Percentages
Benefits derived from flood damage reduction	Avoided direct and indirect flood damage costs	28,745,795	56%	56%	66,536	6%	6%
Benefits derived from water quality improvements	Avoided environmental damage due to CSO to receiving waters	11,876,496	23%	24%	44,306	4%	5%
	Avoided cost of combined waste water treatment	274,985	1%		945	0%	
	Avoided indirect damages to coastal economies	270,474	1%		9043	1%	
Additional benefits	Added aesthetic value	6,393,959	12%	20%	436,044	40%	89%
	Air quality improvement	71,272	0%		3992	0%	
	Habitat provision	4,016,328	8%		508,718	47%	
	Reduction of urban heat island effect and energy consumption	85,031	0%		10,770	1%	
TOTAL		51,734,342			1,080,354		

Figure 4 provides a graphical representation of the contribution of each of the three benefit categories proposed to the total benefits. Overall, significant differences are shown in the percentages of Barcelona and Badalona. In the case of Barcelona, the benefits derived from flood damage reduction are 56% of the total and in Badalona 6%. Additionally, water quality benefits have a larger share in Barcelona compared to Badalona. This is probably because of the widespread GI implementation of Barcelona compared to Badalona where a significantly less ambitious GI implementation plan was considered. A different GI location in Badalona could result in higher water related benefits. Cooper et al. [9] also looked into the contribution of multiple benefits associated to green infrastructure (considered as a coastal flood adaptation measure) showing that benefits from reduced residential damages were 69% of the total, recreational and health benefits 12% and avoided commercial damages 12%.

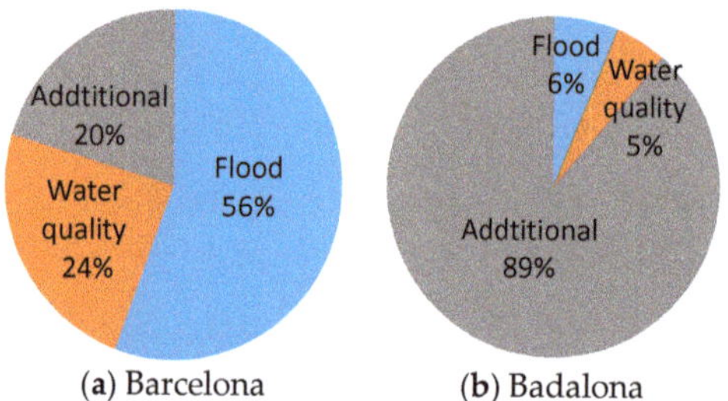

Figure 4. Contribution of each of the three different benefit categories to the total green infrastructure benefits. (**a**) Barcelona and (**b**) Badalona.

Figure 5 shows the discounted benefits during the study evaluation period (80 years, from 2020 to 2100 for the considered scenarios). The results show that the benefits reach their maximum when all GI are implemented: after 20 years in Barcelona and 5 years in Badalona.

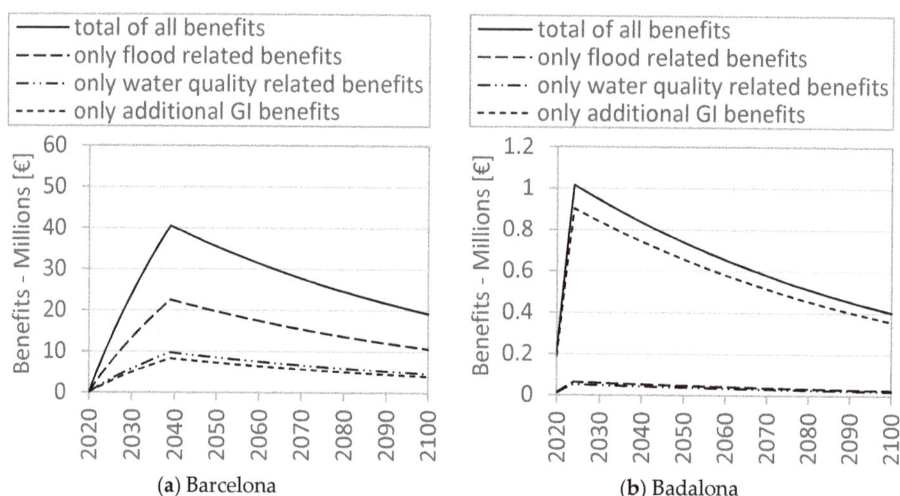

Figure 5. Contributions of the different benefit categories to the total green infrastructure benefits (Discount rate = 1.23%). (**a**) Barcelona and (**b**) Badalona.

3.3. Net benefits

Figure 6 shows the discounted (rate of 1.23%) costs and benefits. Note that the y-axes of Barcelona is approximately an order of magnitude higher than the Badalona one. This figure helps visualizing that the ratio between benefits and costs is generally higher in Barcelona compared to Badalona.

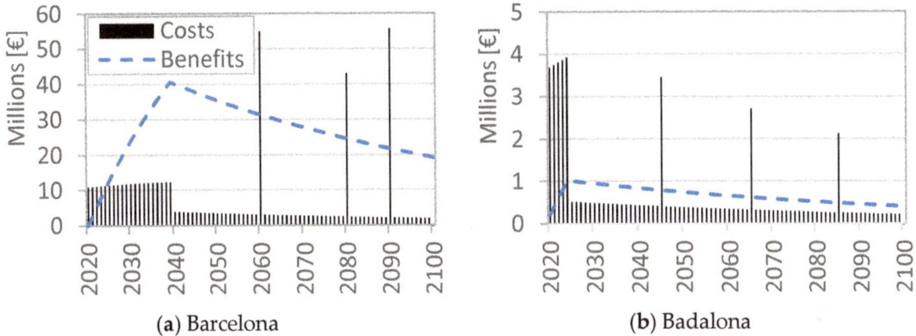

Figure 6. Discounted (i = 1.23%) costs and benefits of green infrastructure. (**a**) Barcelona and (**b**) Badalona.

The evaluation period T in this study was selected to be 80 years in both cases. Similar studies performing CBA for climate change adaptation measures in the context of urban drainage planning used 90 years [23]; 50 years [7] and 35 years [4].

Figure 7 shows the discounted marginal benefits during the project evaluation period. Both Barcelona and Badalona and two different discount rates (1.23% and 4%) were applied. Additionally, three different combinations of benefits are included and shown. First, only benefits derived from flood damage reduction are included, then flood together with water quality benefits and finally all benefits from the three categories proposed: flood, water quality and additional benefits. These three different combinations show that including multiple GI benefits significantly affects the results and this is relevant for decision making of urban drainage planning. In the case of Barcelona, the NPV obtained considering only benefits derived from flood damage reduction and a discount rate of 1.23% (Figure 7a) increases by a factor of 1.74 when including flood and water quality benefits and by 2.37 when including all the three benefits categories. Instead, with a discount rate of 4.00% (Figure 7b), the NPV obtained considering only benefits derived from flood damage reduction increases by a factor of 1.95 when including flood and water quality benefits and by 2.76 when including all the three benefits categories. In the case of Badalona, the NPV also increases significantly by including multiple benefits. In this case, factors of increase of NPV are considered misleading because the NPV is mostly negative. The NPV in Badalona becomes positive only at the last five years of the study evaluation period (Figure 7c).

In the two cases analyzed in this study two different discount rates were considered: the 1.23% that was recommended for climate change adaptation projects in the region of Catalonia [37] where the two considered cities are located and the 4% that was used in another CBA of climate change adaptation measures of Barcelona [4]. The discount rate is a controversial topic in economic valuation of policies, in particular in the context of climate change as it involves intergenerational and social valuation issues (Atkinson et al., 2018). In addition CBA results are very sensitive to the discount rate, particularly for projects with a long time horizon, where small changes of discount rate can influence the suggested decisions [38]. High discount rates imply that future economic impacts would have a lower weight compared to today's value, and could lead to an underestimation of future benefits derived from damage reduction measures [39,40]. A CBA of GI for the case study of Melbourne used 1.4% [7] and the range of 1 to 4% for GI Danish case studies of [3,5]. Some literature also proposed a 1% discount rate (Aaheim, 2010; Lopez, 2008; Stern, 2007). Different public institutions propose different discount rates. For instance, in Denmark, the Danish Environmental Protection Agency (EPA) recommended a 3% for environmental projects while the Department of Finance suggested 5% [5]. The US EPA recommended 2–3% while the American office of management and budget proposed 7% [5]. For developing countries the World Bank recommend 10% because of the significant GDP growth [38,41]. The UK Government proposed 3.5% for project evaluation periods of 1–30 years, 3%

for 31–75 years, 2.5% for 76–125 years, 2% for 125–200 years, 1.5% for 201–300 years and 1% for larger periods [42]. Generally, it is recommendable to consider different discount rates in order to quantify favorable and unfavorable scenarios.

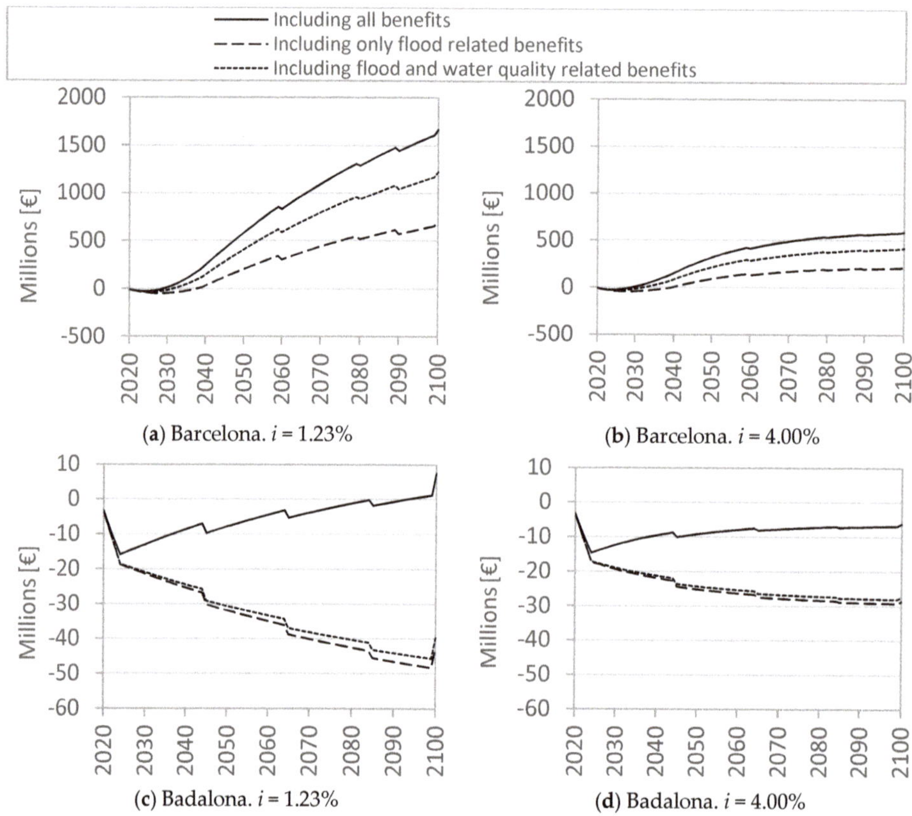

Figure 7. Accumulated marginal benefits of the proposed green infrastructure including different benefit categories and two different discount rates. (**a**) Barcelona with i = 1.23%; (**b**) Barcelona with i = 4.00%; (**c**) Badalona with i = 1.23%; (**d**) Badalona with i = 4.00%.

The cumulative NPV including all benefits in Barcelona becomes positive after 10 years with a discount rate of 1.23% (Figure 7a) and after 11 years with a discount rate of 4.0% (Figure 7b). In Badalona it becomes positive after 75 years with a discount rate of 1.23% (Figure 7c) and it remains negative with a discount rate of 4.0% (Figure 7d). Overall, the GI planning scenario of Barcelona seems to be a better socio-economical option compared to inaction. Instead, the GI planning scenario in Badalona seems to be a worse socio-economical option compared to inaction. Similarly, Zhou et al. [5] presented several stormwater infiltration scenarios that can be considered as GI scenarios showing positive NPV at discount rates of 1% and both positive and negative NPV at discount rates of both 3 and 5%. Zhou et al. [23] presented a negative 50th percentile of the NPV of their stormwater infiltration adaptation scenario. Alves et al. [6] obtained negative NPV for both a green roof and a permeable pavements adaptation scenario and a positive NPV for rainwater harvesting. Zhou et al. [3] reported a positive NPV for their stormwater infiltration scenario.

4. Discussion

This study presented a CBA to evaluate the socio-economic viability of selected GI applied to two different case studies: Barcelona and Badalona. The results are significantly different among the two cases: Barcelona has higher NPV compared to Badalona. Additionally, the accumulated marginal benefits (Figure 7) of Barcelona are mostly positive and become positive after tens of years compared to Badalona where they are mostly negative during the 80 years project evaluation period. The dominating GI benefits (Figure 4) in Barcelona are from flood damage reduction while in Badalona from additional benefits (mostly aesthetic and habitat provision). Direct comparison between the two case studies is difficult for several reasons: the scale difference (Barcelona is much bigger than Badalona), the current situation (Barcelona has much higher flood damage costs than Badalona), the different approaches used to derive CF and the differences of GI planning (Barcelona has an intensive GI implementation plan while Badalona has a sparse one). The fact of having a sparse implementation plan that was not devised to solve particular urban water problems might result in the lower socio-economic performance of the case of Badalona compared to Barcelona. Further comparison between the results obtained in Badalona and Barcelona is considered out of the scope of this study.

Generally, CBA are sensitive to parameter uncertainty and model assumptions. Therefore, quantifying uncertainty of NPV estimations of climate change adaptation options is relevant and in this study only uncertainty related to discount rate was addressed as also done in other studies [9]. Uncertainty is often quantified by analyzing different present and future climate scenarios [4,5,7] and by analyzing different investment options [3]. Zhou et al. [23], instead of using a scenario approach where variables are changed individually, quantified the NPV uncertainties using a Monte Carlo approach to fully explore the propagation of uncertainty from different models and variables choices to the final NPV. A significant source of uncertainty also comes from the hydrological performance of GI [43–45].

Additionally, in this study the EAD is considered to be overestimated, particularly for the case of Barcelona (Table 2) where EAD seems to be high when compared to flood damage compensations data. From 1996 to 2018, pluvial floods, only in the city of Barcelona, have caused more than EUR 34 million in compensations, for industries, offices, dwellings, vehicles and civil works, according to the classification adopted by the Spanish Insurance Compensation Consortium (CCS). In 2018, damages caused by four heavy rainfalls amounted to around EUR 5.5 million. It was the third most damaging year in terms of insurance indemnifications within the last 22 years. The first two years were 1999 and 2002, which compensations amounted to EUR 7.3 million and EUR 6.5 million respectively. Such values only include compensations that the CCS paid. Therefore, total damages (including also indirect damages) are usually higher. Three main contributions were identified to produce the EAD overestimation:

(a) The hyetographs design storms (for all the considered return periods) were obtained from few rain gauges and uniformly applied to the whole catchment area in Barcelona. When calculating flood damages, it can be relevant to use design storms obtained by spatially averaged (over the catchment area) Intensity Duration Frequency (IDF) curves or multiply the rainfall intensity by a reduction coefficient that is a function of the catchment area: the larger the area, the lower the coefficient. In the case of Badalona, the project storm hyetographs presented blocks with maximum rainfall intensity corresponding to different return periods in order to take into account the correspondence of the project storms intensities with the observed rainfall data for extreme events [46].

(b) The discretization used in the flood damage vs exceedance probability curve and the integration method used to compute EAD (the integral of the flood damage curve over the exceedance probability domain) introduced a significant numerical error with consequent overestimation, particularly for the case of Barcelona. Figure 3 showed the flood damage vs exceedance probability for the case of Barcelona together with the linear interpolation lines that were used for the calculation of the area below the curve that corresponds to the EAD. The figure shows that the selected simulated points might not be enough to properly describe the non-linear relation between flood damages and exceedance probability, particularly for the case of Barcelona (Figure 3a) for the range of exceedance probability between 0.1 (10 year return period) and 1 (1 year return period). By introducing new simulation points (for instance at 0.2 exceedance probability) the EAD (the area below the curve) might significantly reduce [4,47].

Even though the EAD is considered overestimated and two main different causes were identified, the EADs for Barcelona were not re-calculated as these values were included in the latest drainage master plan of Barcelona.

5. Conclusions

This study presented a cost-benefit analysis to evaluate the socio-economic viability of GI, which was considered as a climate change adaptation option in the cities of Badalona and Barcelona. The GI planning of the two cities is significantly different: Barcelona proposed a widespread GI implementation plan while Badalona proposed a much lower degree of implementation. CBA is relevant for decision making of urban drainage planning and is useful for comparing different scenarios: in this case a business as usual (BAU) and the GI scenario were compared. Multiple benefits derived from GI implementation were considered and they were grouped into three different categories: benefits derived from flood damage reduction, from water quality improvements and from additional benefits. For each categories both direct and indirect tangible (that can be monetized) benefits were defined and quantified. The largest share of GI benefits in Barcelona was from reduced flood damages (56%), while in Badalona was from additional benefits like added value of properties and habitat provision (89%). The GI benefits derived from reduced sewage treatment costs; from reduced indirect damages to coastal economies; from air quality improvement and from reduction of the heat island effect and energy consumption resulted in the range of 0–1% playing an insignificant role in the socio-economic assessment. The calculated cumulative net present value (NPV) in Barcelona became positive after 10–11 years considering all benefits, whereas in Badalona was mostly negative. Overall, this study presented and quantified how different multiple benefits that can contribute to net present value as part of CBA. The details provided in this paper guarantee the replicability of the presented CBA to other case studies.

Author Contributions: Conceptualization, L.L. and M.G.; methodology, L.L., M.G., E.M.-G. and B.R.; software, L.L. and D.S.; validation, L.L. and D.S.; investigation, L.L. and M.G.; resources, M.M.; data curation, L.L., E.M.-G. and B.R.; writing—original draft preparation, L.L.; writing—review and editing, L.L., M.G., B.R. and E.M.-G.; visualization, L.L.; project administration, L.L., M.M. and B.R.; funding acquisition, M.M. and B.R. All authors have read and agreed to the published version of the manuscript.

Funding: This research was funded by the BINGO European H2020 project, Grant Agreement No.641739 and the RESCCUE European H2020 project, Grant Agreement No. 700174.

Acknowledgments: The authors thank AMB, BCASA and the Municipalities of Badalona and Barcelona for their collaboration.

Conflicts of Interest: The authors declare no conflict of interest.

References

1. Fletcher, T.D.; Shuster, W.; Hunt, W.F.; Ashley, R.; Butler, D.; Arthur, S.; Trowsdale, S.; Barraud, S.; Semadeni-Davies, A.; Bertrand-Krajewski, J.-L.; et al. SUDS, LID, BMPs, WSUD and more—The evolution and application of terminology surrounding urban drainage. *Urban Water J.* **2014**, *9006*, 1–18. [CrossRef]

2. Yu, H.Z. Complicated structure modeling in front-zone of Hala'alate Mountain of northwestern margin, Junggar Basin. *Nat. Gas Geosci.* **2014**, *25*, 91–97.

3. Zhou, Q.; Halsnæs, K.; Arnbjerg-Nielsen, K. Economic assessment of climate adaptation options for urban drainage design in Odense, Denmark. *Water Sci. Technol.* **2012**, *66*, 1812–1820. [CrossRef]

4. Velasco, M.; Russo, B.; Cabello Termes, M.; Sunyer, D.; Malgrat, P. Assessment of the effectiveness of structural and nonstructural measures to cope with global change impacts in Barcelona. *J. Flood Risk Manag.* **2018**, *11*, S55–S68. [CrossRef]

5. Zhou, Q.; Mikkelsen, P.S.; Halsnæs, K.; Arnbjerg-Nielsen, K. Framework for economic pluvial flood risk assessment considering climate change effects and adaptation benefits. *J. Hydrol.* **2012**, *414–415*, 539–549. [CrossRef]

6. Alves, A.; Gersonius, B.; Kapelan, Z.; Vojinovic, Z.; Sanchez, A. Assessing the Co-Benefits of green-blue-grey infrastructure for sustainable urban flood risk management. *J. Environ. Manag.* **2019**, *239*, 244–254. [CrossRef] [PubMed]

7. Löwe, R.; Urich, C.; Sto. Domingo, N.; Mark, O.; Deletic, A.; Arnbjerg-Nielsen, K. Assessment of urban pluvial flood risk and efficiency of adaptation options through simulations—A new generation of urban planning tools. *J. Hydrol.* **2017**, *550*, 355–367.

8. Zhou, Q.; Panduro, T.E.; Thorsen, B.J.; Arnbjerg-Nielsen, K. Adaption to extreme rainfall with open urban drainage system: An integrated hydrological cost-benefit analysis. *Environ. Manag.* **2013**, *51*, 586–601. [CrossRef]

9. Cooper, W.; Garcia, F.; Pape, D.; Ryder, D.; Witherell, B. Climate Change Adaptation Case Study: Benefit-Cost Analysis of Coastal Flooding Hazard Mitigation. *J. Ocean Coast. Econ.* **2016**, *3*. [CrossRef]

10. Venkataramanan, V.; Lopez, D.; McCuskey, D.J.; Kiefus, D.; McDonald, R.I.; Miller, W.M.; Packman, A.I.; Young, S.L. Knowledge, attitudes, intentions, and behavior related to green infrastructure for flood management: A systematic literature review. *Sci. Total Environ.* **2020**, *720*, 137606. [CrossRef]

11. Wilkerson, M.L.; Mitchell, M.G.E.; Shanahan, D.; Wilson, K.A.; Ives, C.D.; Lovelock, C.E.; Rhodes, J.R. The role of socio-economic factors in planning and managing urban ecosystem services. *Ecosyst. Serv.* **2018**, *31*, 102–110. [CrossRef]

12. Arnbjerg-Nielsen, K. Quantification of climate change effects on extreme precipitation used for high resolution hydrologic design. *Urban Water J.* **2012**, *9*, 57–65. [CrossRef]

13. Rodr, R.; Navarro, X.; Casas, M.C.; Ribalaygua, J.; Russo, B.; Pouget, L.; Reda, A. Influence of climate change on IDF curves for the metropolitan area of Barcelona (Spain). *Int. J. Climatol.* **2014**, *654*, 643–654.

14. Löwe, R.; Urich, C.; Kulahci, M.; Radhakrishnan, M.; Deletic, A.; Arnbjerg-Nielsen, K. Simulating flood risk under non-stationary climate and urban development conditions—Experimental setup for multiple hazards and a variety of scenarios. *Env. Model. Softw.* **2018**, *102*, 155–171. [CrossRef]

15. Àrea de Medi Ambient. *Pla Tècnic per L'aprofitament dels Recursos Hídrics Alternatius a Barcelona: Vol I/III Doc. Núm. 1: Memòria i Annexos*; Repositori Obert de Coneixement de l'Ajuntament de Barcelona: Barcelona, Spain, 2009.

16. BCN. Ecologia. Cobertes i Murs Verds a Barcelona: Estudi sobre les existents, el potencial i les estretègies d'implantació. 2010. Available online: http://bcnecologia.net/sites/default/files/proyectos/doc_cobertes_i_murs_verds_01m2010_0.pdf (accessed on 1 April 2020).

17. Locatelli, L.; Gabriel, S.; Mark, O.; Mikkelsen, P.S.; Arnbjerg-Nielsen, K.; Taylor, H.; Bockhorn, B.; Larsen, H.; Kjølby, M.J.; Blicher, A.S.; et al. Modelling the impact of retention-detention units on sewer surcharge and peak and annual runoff reduction. *Water Sci. Technol.* **2015**, *71*, 898–903. [CrossRef] [PubMed]

18. Monjo, R.; Gaitán, E.; Pórtoles, J.; Ribalaygua, J.; Torres, L. Changes in extreme precipitation over Spain using statistical downscaling of CMIP5 projections. *Int. J. Clim.* **2016**, *36*, 757–769. [CrossRef]

19. Rust, H.W.; Richling, A.; Meredith, E.; Fischer, M.; Vagenas, C.; Kadow, C.; Ulbrich, U. Climate Predictions and Downscaling to Extreme Weather. 2018. Available online: http://www.projectbingo.eu/output/climate-predictions (accessed on 1 April 2020).

20. Rodriguez-Iturbe, I.; Cox, D.R.; Isham, V. Some models for rainfall based on stochastic point processes. *Proc. R. Soc. Lond. A Math. Phys. Sci.* **1987**, *410*, 269–288.

21. Narayanan, A.; Pitt, R. *Costs of Urban Stormwater Control Practices*; Department of Civil, Construction, and Environmental Engineering, The University of Alabama: Tuscaloosa, AL, USA, 2006.

22. Bianchini, F.; Hewage, K. Probabilistic social cost-benefit analysis for green roofs: A lifecycle approach. *Build. Environ.* **2012**, *58*, 152–162. [CrossRef]

23. Zhou, Q.; Arnbjerg-Nielsen, K. Uncertainty assessment of climate change adaptation options using an economic pluvial flood risk framework. *Water* **2018**, *10*, 1877. [CrossRef]

24. Feng, H.; Hewage, K.N. Economic Benefits and Costs of Green Roofs. In *Nature Based Strategies for Urban and Building Sustainability*; Elsevier Inc.: Amsterdam, The Netherlands, 2018; pp. 307–318. ISBN 9780128123249.

25. Sartori, D.; Catalano, G.; Genco, M.; Pancotti, C.; Sirtori, E.; Vignetti, S.; Del Bo, C. Guide to Cost-benefit Analysis of Investment Projects: Economic appraisal tool for Cohesion Policy 2014–2020. *Eur. Commision* **2014**, 349. [CrossRef]

26. Smith, K.; Ward, R. *Floods: Physical Processes and Human Impacts*; Wiley: Hoboken, NJ, USA, 1998; ISBN 978-0-471-95248-0.

27. Martínez-Gomariz, E.; Locatelli, L.; Guerrero, M.; Russo, B.; Martínez, M. Socio-Economic Potential Impacts Due to Urban Pluvial Floods in Badalona (Spain) in a Context of Climate Change. *Water* **2019**, *11*, 2658. [CrossRef]

28. Martínez-Gomariz, E.; Gómez, M.; Russo, B.; Sánchez, P.; Montes, J.A. Methodology for the damage assessment of vehicles exposed to flooding in urban areas. *J. Flood Risk Manag.* **2018**, 1–15. [CrossRef]

29. Hallegatte, S. An adaptive regional input-output model and its application to the assessment of the economic cost of Katrina. *Risk Anal.* **2008**, *28*, 779–799. [CrossRef] [PubMed]

30. Lorenzo, C.; Standardi, G.; Bosello, F.; Mysiak, J. *Assessing Direct and Indirect Economic Impacts of a Flood Event through the Integration of Spatial and Computable General Equilibrium Modelling*; Research Papers Issue RP0202 December 2013; Centro Euro-Mediterraneo sui Cambiamenti Climatici: Lecce, Italy, 2013.

31. Locatelli, L.; Russo, B.; Martinez, M. Evaluating health hazard of bathing waters affected by combined sewer overflows. *Nat. Hazards Earth Syst. Sci. Discuss.* **2019**, 1–19. [CrossRef]

32. Barcelona, D. Pla estratègic dels espais litorals de la ciutat. Available online: https://www.decidim.barcelona/processes/PlaLitoralBCN (accessed on 1 April 2020).

33. Turrión-Prats, J.; Duro, J.A. Tourist seasonality in Catalonia. *Tour. Econ.* **2017**, *23*, 846–853. [CrossRef]

34. De Bruin, K.; Dellink, R.; Agrawala, S.; Dellink, R. Economic Aspects of Adaptation to Climate Change: Integrated Assessment Modelling of Adaptation Costs and Benefits. *Oecd Environ. Work. Pap.* **2009**, *22*, 36–38.

35. Decreto 459/2013. *de 10 de Diciembre, Sobre los Vertidos Efectuados Desde Tierra al Mar. Boletín Oficial del País Vasco*; Departamento de Medio Ambiente y Politica Territorial: Bilbao, Spain, 2014.

36. Holt, A. *Ex Post Evaluation of Cohesion Policy Interventions 2000-2006 Financed by the Cohesion Fund. WP C Cost Benefit Analysis of Environment Projects*; Publications Office of Eurepean Union: Luxembourg, 2011.

37. Chiabai, A.; Galarraga, I. *Determining Discount Rates: An Application of the Equivalency Principle*; EU: Brussels, Belgium, 2016.

38. Griffin, R.C. *Water Resource Economics: The Analysis of Scarcity, Policies, and Projects*; MIT Press: Cambridge, MA, USA, 2006; ISBN 9780262072670.

39. Aaheim, A. The determination of optimal climate policy. *Ecol. Econ.* **2010**, *69*, 562–568. [CrossRef]

40. Stern, N. *The Economics of Climate Change: The Stern Review*; Cambridge University Press: Cambridge, UK, 2007; ISBN 9780511817434.

41. Pearce, D.; Atkinson, G.; Mourato, S. *Cost-Benefit Analysis and the Environment: Recent Developments*; Organisation for Economic Cooperation and Development (OECD): Paris, France, 2006; ISBN 9789264010055.

42. Oxera a Social Time Preference for Use in Long-Term Discounting. Available online: https://www.oxera.com/publications/a-social-time-preference-for-use-in-long-term-discounting/ (accessed on 3 March 2020).

43. Locatelli, L.; Mark, O.; Mikkelsen, P.S.; Arnbjerg-Nielsen, K.; Wong, T.; Binning, P.J. Determining the extent of groundwater interference on the performance of infiltration trenches. *J. Hydrol.* **2015**, *529*, 1360–1372. [CrossRef]

44. Locatelli, L.; Mark, O.; Mikkelsen, P.S.; Arnbjerg-Nielsen, K.; Bergen Jensen, M.; Binning, P.J. Modelling of green roof hydrological performance for urban drainage applications. *J. Hydrol.* **2014**, *519*, 3237–3248. [CrossRef]

45. Bockhorn, B.; Klint, K.E.S.; Locatelli, L.; Park, Y.-J.; Binning, P.J.; Sudicky, E.; Bergen Jensen, M. Factors affecting the hydraulic performance of infiltration based SUDS in clay. *Urban Water J.* **2017**, *14*, 125–133. [CrossRef]

46. Raso, J.; Malgrat, P.; Castillo, F. Improvement in the selection of design storms for the New Master Drainage Plan of Barcelona. *Water Sci. Technol.* **1995**, *32*, 217–224. [CrossRef]

47. Olsen, A.S.; Zhou, Q.; Linde, J.J.; Arnbjerg-Nielsen, K. Comparing methods of calculating expected annual damage in urban pluvial flood risk assessments. *Water* **2015**, *7*, 255–270. [CrossRef]

Article

The Contribution of NBS to Urban Resilience in Stormwater Management and Control: A Framework with Stakeholder Validation

Paula Beceiro [1,2,*], Rita Salgado Brito [1] and Ana Galvão [2]

[1] Urban Water Unit, National Civil Engineering Laboratory, LNEC, Av. Brasil 101, 1700–066 Lisbon, Portugal; rsbrito@lnec.pt

[2] CERIS, Instituto Superior Técnico, Universidade de Lisboa, Av. Rovisco Pais, 1049–001 Lisbon, Portugal; ana.galvao@tecnico.ulisboa.pt

* Correspondence: pbeceiro@lnec.pt

Received: 31 January 2020; Accepted: 16 March 2020; Published: 24 March 2020

Abstract: Urban waters represent a crucial component for the enhancement of urban resilience due to their importance in cities. Nature-based solutions (NBS) have emerged as sustainable solutions to contribute to urban resilience in order to meet the challenges of climate change. In order to promote the use of NBS for increasing urban resilience, tools that demonstrate the value of this type of solutions over the long-term are required. A performance assessment system provides an adequate basis for demonstrating this value, as well as for diagnosing the current city situation, selecting and monitoring the implementation of solutions. Regarding NBS management, some assessment approaches have been published, focusing on assessing the effectiveness of NBS in the face of climate change and supporting their design and impact assessment. Nevertheless, an integrated approach to assess the NBS contribution for urban resilience has not been published. This paper presents a comprehensive resilience assessment framework (RAF) to evaluate the NBS contribution for urban resilience, focused on solutions for stormwater management and control. Furthermore, details on stakeholders' validation, with focus on the metrics' relevance and applicability to cities, is also presented.

Keywords: Ecosystem Services (ES); Nature-Based Solutions (NBS); Resilience Assessment Framework (RAF); stakeholders' validation; stormwater management and control; urban resilience

---

1. Introduction

Climate change has fostered the need to develop and improve urban resilience by promoting a resilient city's capabilities to absorb disruption, learn from the past, adapt, transform and prepare for the future. Resilience emerged as an interesting concept on cities, often theorized as highly complex adaptive systems [1,2]. Resilience is commonly understood as the capacity of a system to absorb disturbance and re-organize while undergoing change so as to still retain essentially the same function, structure, identity and feedbacks [3]. The concept of resilience emerged in the 1960s from the growing interest in ecology to determine population stability between communities. Resilience emerged as an interesting perspective on cities, often theorized as highly complex adaptive systems [4–6].

In the urban water cycle, the evolution of the drainage systems followed the evolution of the resilience concept, from one single point of view (e.g., economic or social resilience) to a broader and more inclusive definition, encompassing the multiple dimensions of urban resilience. Urban water services are of fundamental importance in the promotion of urban resilience. Urban waters are essential to support nature-based solutions (NBS) functions and to ensure the provision of ecosystem services from NBS, such as air quality improvement or urban heat island mitigation. Evaluating and enhancing resilience in the urban water cycle is a crucial step toward more sustainable urban water management [7].

Historically, the main objectives of urban drainage systems were to ensure efficient management of peak flows and adequate treatment of polluted waters, aiming to ensure public health and prevent flooding. More recently, integrated approaches for urban water management emerged and other key issues were identified for sustainable water management, such as surface and ground water quality, ecological concerns, and recreational uses [8].

The European Commission defines NBS as actions that aim to helps societies address a variety of environmental, social and economic challenges in a sustainable way [9]. In essence, NBS can be defined as living solutions inspired by, continuously supported by and using nature, which are designed to address several societal challenges in a resource-efficient perspective and to provide simultaneously economic and environmental benefits [9]. NBS involve actions for conservation or rehabilitation of natural ecosystems and improvement or creation of natural processes in modified or artificial ecosystems [10]. Some examples of NBS for stormwater management are infiltration basins, green roofs, constructed wetlands or swales with vegetation cover, among others.

In the water sector, the NBS concept can be found in technologies such as sustainable urban drainage systems (SUDSs), which mimic nature to manage stormwater runoff and provide other services to the urban environment. SUDSs are recognized as one of the main NBS techniques to improve urban resilience regarding stormwater management. These techniques can also be found with different designations, such as low impact development (LID), best management practices (BMPs), and green infrastructures. In this sense, an integrated approach of urban water management that incorporates a SUDS as a fundamental component is the Water Sensitive Urban Design, originate and widely applied in Australia. This approach to urban planning and design aims to integrate in the urban design the various disciplines of engineering and environmental sciences associated with the provision of water services [11].

To date, several studies were developed focused on the analysis of the resilience and sustainability enhancement based on SUDS implementation [12,13]. In the UK water sector, SUDSs are increasingly promoted in order to enhance flood resilience in urbanized areas and its application to increase resilience has also been studied [14,15]. For example, an approach adopted to quantify the cost-effectiveness of resilience measures and integrative and adaptable flood management plans was proposed [16].

The European Research and Development Program promotes a large number of projects related to NBS to increase knowledge and to create technical, political and other conditions for cities renaturalization. Improving risk management and resilience, using nature-based solutions, represents one of the mains goals of the EU Research and Innovation agenda [9]. These R&D projects will analyse several objectives and perspectives, as the improvement of regulatory instruments, the increase of the natural capital through NBS or the capacity to obtain a more sustainable and resilient urban ecosystem. In the context of urban resilience, some NBS studies were carried out focusing on some ES enhancement or on specific challenges, such as urban heat island mitigation [17,18], air quality improvement, climate mitigation and adaptation [19,20], and water quality improvement [21], among others.

Public participation is becoming increasingly embedded in the decision-making processes [22]. In this sense, several studies highlight the need to ensure a broader stakeholders' engagement in the development and implementation of assessment tools [23]. The development of the assessment process with stakeholders' collaboration promotes their empowerment and enhance their role in decision-making processes [24]. Regarding to the assessment process, stakeholders help to highlight weaknesses, to prioritize interventions and to identify the assessment tools adequacy to diverse locations. Stakeholders networks on NBS design, planning, and implementation are essential to ensure the transference of successful approaches between countries, communities and case studies [25]. Several available assessment frameworks include the stakeholders' participation in the development and implementation processes [26–28].

Currently, there is a need to analyse the NBS contribution to urban resilience and to develop tools that demonstrate the long term value of these solutions [29]. Several frameworks to assess resilience are being developed, such as the RESCCUE Resilience Assessment Framework (RESCCUE RAF) [26], the Disaster Resilience scorecard for cities [27], the Arup and Rockefeller city resilience framework [28] and the USEPA framework [30], among others.

Based on the review of the available resilience assessment frameworks (RAFs), the relevant *attributes* for resilience assessment were identified [31]. From this analysis, an RAF needs to i) propose a multi-dimension methodology that includes subjective and objective information, allowing us to measure urban resilience in one scale; ii) identify resilience objectives and criteria; iii) use qualitative and quantitative metrics addressing performance, cost and risk; iv) define reference values and a final resilience assessment; and v) identify the urban resilience capabilities associated with the proposed metrics. Additionally, there is a need for the RAF to consider and to allow the assessment of short- and long-term changes [30].

Regarding specifically the NBS for stormwater management and control, some assessment frameworks have already been developed, focusing on assessing NBS effectiveness in the face of climate change [32], supporting the design and impact assessment of the NBS for climate resilience [25] and to specific urban challenges, such as green space management or air quality [33]. Even though the NBS assessment frameworks are not directly focused on urban resilience, they may support a specific assessment framework focused on NBS contribution to urban resilience. To date, a comprehensive assessment approach has not been published to assess the contribution of NBS to urban resilience.

In this sense, among the several projects analysed, common concerns and knowledge gaps relevant to the development of a specific RAF for NBS were identified. Based on this analysis, an appropriate RAF for NBS should assess the following *aspects*: (i) social, environmental, economic, and governance dimensions; (ii) spatial and land use planning at the city level; (iii) service and infrastructure management; (iv) potential capabilities to provide ecosystem services (ES) and to enhance natural capital and biodiversity; (v) impacts on the surrounding area; (vi) infrastructure implementation and design, including adequate monitoring and maintenance processes; (vii) infrastructure performance under normal and stressing condition, considering acute shocks and continuous stresses; and (viii) infrastructure interdependencies with other urban services.

The main objectives of this paper are to present the methodology adopted for the construction of an RAF for NBS, as a specific RAF to evaluate the NBS contribution for urban resilience, and the stakeholders' validation, with focus on the analysis of relevance and applicability of metrics to cities. The main innovative contribution of this work is to propose a multidimensional and comprehensive RAF to assess the NBS contribution for urban resilience, focused on NBS for stormwater management and control. The proposed framework, driven by resilience objectives and assessment criteria, aims to integrate the *attributes* identified for urban resilience assessment and the relevant *aspects* for the NBS evaluation.

2. Methodology

2.1. Construction of a Resilience Assessment Framework for NBS

A new methodology for the construction of a specific assessment framework to evaluate the NBS contribution to urban resilience, with focus on solutions for stormwater management and control, is presented. The specific NBS considered in the RAF are the following: infiltration basins, green roofs and walls, vegetated swales, infiltration trenches, and porous pavements. This methodology allows to develop a multidimensional and comprehensive RAF that considers a broader definition of urban resilience. The methodology considers objective and subjective information and allows resilience to be measured on a single scale. As previously mentioned, urban resilience is defined as a city's ability to absorb disturbances, learn from the past, adapt, transform and prepare for the future. In this sense, the RAF considers five *capabilities* of a resilient city—namely, absorb, learn, adapt, transform, and prepare.

The proposed methodology is based on the performance assessment structure proposed by the ISO 24500 standards [34–36] for water supply and wastewater system management. The ISO 24500 structure is grounded on the definition of objectives, criteria and metrics. In the proposed methodology, resilience objectives aim to consider the several NBS contributions to urban resilience and the criteria allow to evaluate several aspects or points of view of the RAF objectives. Metrics are parameters or functions used to assess the criteria. In order to facilitate the RAF application, objectives were further grouped into two main dimensions.

The RAF seeks also to ensure the alignment with asset management, taking into consideration the fundamentals of value of the assets, leadership in service provision, assurance of alignment in the organization and of resources for effective implementation of a plan, along with the RAF application in the short and long term [37]. These conditions are incorporated in metric´s definition. The main *attributes* for an adequate urban resilience assessment (e.g., metric´s reference values ought to be defined) and the main *aspects* to be evaluated in the NBS (e.g., infrastructure management), identified in the literature review [31], are also considered.

The development of the RAF was performed in five main steps—i) identification of the urban resilience dimensions in the RAF; ii) definition of objectives, criteria, and metrics (O-C-M); iii) validation of the O-C-M of the RAF; iv) definition of reference values for each metrics; and v) consolidation of the RAF. The focus of this paper is on the methodology's first four steps.

2.2. Step 1: Identification of the Urban Resilience Dimensions in the RAF

Based on the existing literature review, the main areas that contribute to urban resilience (e.g., social and environmental areas) and the several aspects for an appropriate NBS assessment were identified (e.g., impact on the surrounding area). The main areas identified as well as *aspects* for NBS assessment were incorporated into resilience dimensions. Dimensions were also defined in order to combine similar levels of assessment. Two resilience dimensions were identified in the RAF. The first dimension addresses the assessment of resilience at the city level, and the second dimension is focused on the assessment of resilience at NBS level.

2.3. Step 2: Definition of Objectives, Criteria and Metrics

In the second step, urban resilience objectives, criteria and metrics were identified. Resilience has to be a tangible concept that cities are able to understand and measure, in order to build robust strategies and prioritize investments. The assessment of the RAF framework considers NBS performance by evaluating the contribution of these solutions at city, service, and infrastructure levels. In this way, it will be possible to identify how, when and where to act first in case of incipient resilience.

The resilience objectives highlight the several NBS contributions to urban resilience. The resilience criteria cover the aspects or points of view that evaluate the achievement of the objectives. The proposed metrics allow a clear assessment of the criteria, supporting the definition of explicit targets and monitoring of results. The use of quantitative and qualitative metrics allows the incorporation and evaluation of objective and subjective information, covering a more comprehensive definition of urban resilience.

The RAF includes metrics that assess performance, cost and risk of the NBS in accordance with the standard EN 752:2008 [38]. In metrics' definition, the related urban resilience capabilities (e.g., to be prepared) are identified. The metrics determination can resort to data from different sources and complexity, allowing the RAF application by cities with different information maturity. The RAF includes metrics with three levels of complexity—based on the existing data in the city (data based), based on a procedure defined for specific metrics (procedure based), or based on results from a mathematical model (model based). The method for metric determination and the specification of the required information was defined in this step.

2.4. Step 3: Validation of the Objectives, Criteria, and Metrics of the RAF

In this step, the RAF was submitted to stakeholders to identify knowledge gaps and improvement opportunities. The RAF validation includes the involvement and participation of stakeholders with different contexts, in a different resilience development level and with diverse NBS in the city, as water utilities or municipalities. The proposed dimensions, resilience objectives, criteria, and metrics were analysed with the stakeholders during working sessions. Metrics determination for participating cites and case studies is also carried in this step, taking into consideration the available information. In addition, a survey was conducted to determine the RAF metrics' relevance and the feasibility of application to their own cities.

The validation step includes (i) RAF submission to stakeholders to identify improvement opportunities and gaps, (ii) RAF application to participating cities, (iii) determination of metrics relevance and feasibility of application to cities, and (iv) complete RAF application to case studies.

Seven participating cities have contributed to the RAF validation steps—namely, Almada (Portugal), Coimbra (Portugal), Lisbon (Portugal), Porto (Portugal), Barcelona (Spain), Bristol (the United Kingdom), and Vancouver (Canada). The participating cities present different challenges regarding urban resilience and NBS. The participation of cities allows to validate the RAF taking into consideration different international and urban context, city dimension and management, NBS management (e.g., private, public), social involvement, and awareness, among other factors. Stakeholders from water utilities, municipal council and private organisation participated in the working sessions. A total of eight organizations validated the RAF.

2.5. Step 4: Definition of Reference Values for Each Metric

Reference values for each metric were defined to assess the NBS contribution to urban resilience on a normalized scale. The definition of reference values was based on existing literature review, regulations, standards and available assessment frameworks. An overall assessment will be proposed to measure the NBS contribution to urban resilience.

Reference values allow metric´s classifications by means of a judgment (e.g., satisfactory or unsatisfactory performance). Comparing the results of a metric with its reference values allow to pinpoint the existing problems and to monitor the implemented solutions.

2.6. Complementary Profile

A complementary profile of the city needs to be established prior to the application of the methodology. This profile is intended to collect the characteristics of the city, of the NBS management service, and of the existing NBS under assessment. Specific information (e.g., urban context) is also detailed.

3. Resilience Assessment Framework for NBS

3.1. Overall RAF Structure

The RAF is structured in two resilience dimensions, namely "Integration of NBS in the city" (Dimension I) and "Operation and services of NBS" (Dimension II). A set of 10 resilience objectives is proposed in the RAF. Dimension I considers four objectives and 12 criteria. Dimension II considers six objectives and 13 criteria. A set of 71 metrics is proposed, divided in 34 metrics for Dimension I and 37 metrics for Dimension II. In Dimension I, 24 data based and 10 procedure based metrics are proposed. In Dimension II, 25 data based, three procedure based, and nine model based metrics are considered. Figure 1 presents the RAF dimensions, objectives, and criteria.

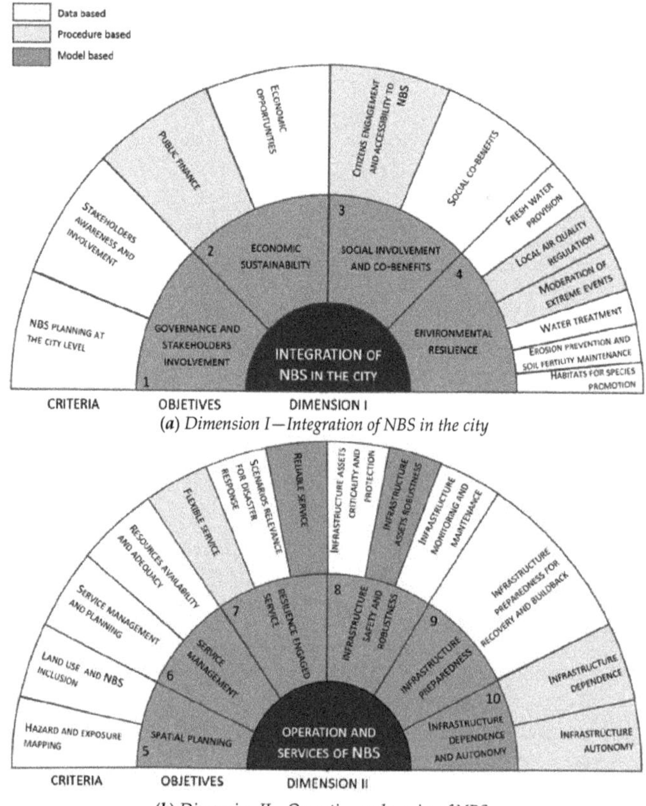

(**a**) *Dimension I—Integration of NBS in the city*

(**b**) *Dimension II—Operation and service of NBS*

Figure 1. Resilience assessment framework (RAF) objectives and criteria for (**a**) first and (**b**) second dimension, specifying the source of information required for metrics determination, by criterion.

3.2. Urban Resilience Dimensions (Step 1)

The resilience assessment proposal aims to evaluate the NBS contribution for urban resilience, focused on NBS for stormwater management and control. This RAF aims to support the diagnosis, decision-making, implementation, planning and management of the NBS and to identify solutions with potential to contribute to city resilience. Based on the identified *attributes* for resilience assessment and *aspects* for NBS evaluation, two dimensions are defined in the RAF.

Dimension I aims to assess the integration of NBS in the city governance and stakeholder involvement, economic sustainability and social involvement, as well as NBS contribution to the environmental resilience. The contribution of the NBS for urban resilience is assessed at the city level in this dimension.

Dimension II aims to assess the adequacy of the NBS to urban planning, the NBS functioning, regarding service management (e.g., service articulation between entities, allocation of financial and technical resources), its consequences in the surrounding area (e.g., ES improvement, flooded area, affected buildings), and the performance of the infrastructure. In this dimension, the NBS adequacy regarding urban, functional and physical components is assessed at the NBS level. Some objectives of the functional and physical components can be evaluated considering all NBS in the city or just some specific NBS.

3.3. Objectives, Criteria, and Metrics (Step 2)

The objectives aim to consider the several NBS contributions to urban resilience [20]. In this sense, the proposed resilience objectives consider the relevant governance, environmental, social and economic concerns and the main aspects of the city, service and infrastructure required to assess this contribution.

The resilience objectives of the Dimension I are focused on the NBS contribution to the social, environmental, economic, and governance aspects at the city level. These objectives aim to ensure city preparedness for governance, planning and financial aspects of the NBS. Also, these objectives aim to guarantee the NBS capabilities to promote green jobs, social co-benefits and ES, preparing the city for future impacts. The proposed objectives for this dimension are detailed as follows:

- Objective 1—Governance and stakeholders' involvement aims to ensure NBS planning at city level and the stakeholders' awareness and involvement. The criteria associated with this objective are the NBS planning at the city level and the stakeholder awareness and involvement. The criteria assess the governance component at city level, evaluating the adequacy of the NBS planning, the identification of the risk, identification of ES and protective infrastructure and the NBS alignment with ES. The proposed criteria also assess the stakeholders' engagement, the community involvement in the NBS processes (e.g., planning, decision making) and the existence of awareness campaigns.
- Objective 2—Economic sustainability aims to ensure financial capacity related to NBS and potential economic opportunities. The proposed criteria for this objective are the public finance and the economic opportunities. The criteria assess the existence of a specific budget for NBS, identify the monitoring and maintenance annual costs and assess the development of initiatives to promote the NBS in households. Concerning the economic opportunities, the criteria identify the creation of new green jobs, business and activities and tourism enhancement by NBS.
- Objective 3—Social involvement and co-benefits aims to ensure the citizen involvement, accessibility to NBS and social co-benefits. The proposed criteria for this objective are the citizens' engagement, NBS accessibility and the social co-benefits. The proposed criteria assess the citizens' engagement to NBS, the public accessibility and the NBS distribution. The proposed criteria also assess the main ES provision related to social co-benefits (e.g., urban heat island mitigation, health and well-being co-benefits).
- Objective 4—Environmental resilience aims to ensure the ES provision from NBS, relating to the environmental component. The proposed criteria for this objective are the fresh water provision, the local air quality regulation, the moderation of extreme events, the water treatment, the erosion prevention and maintenance of soil fertility and the habitats for species promotion. The capabilities of the NBS to provide ES are evaluated.

The resilience objectives of the Dimension II pretend to ensure the adequacy of the spatial planning and both service and infrastructure management at the NBS level. These objectives question the preparedness and adaptation of the urban planning and the integrated service management to NBS. In this sense, an adequate integration of these solutions in the risk identification, land use planning, and city policy are aimed at. Furthermore, these objectives aim to assure the capability of the city to absorb disturbance, transform itself and prepare for future scenarios, based on the existing NBS. The proposed objectives for this dimension are detailed as follows:

- Objective 5—Spatial planning aims to ensure hazard and exposure mapping and NBS identification in land use planning and risk areas at city level. The proposed criteria for this objective are hazard and exposure mapping and land use and NBS inclusion. These criteria assess the existence of updated hazard maps, the NBS identification on risk areas and their inclusion on the land use planning. Also, the NBS inclusion on major urban development and projects by policy is assessed.
- Objective 6—Service management Stent aims to ensure the integrated management of the service and its articulation and the adequacy of competences and resources. The proposed criteria for this

objective are the service management and planning and the resources availability and adequacy. These criteria assess the existence of an integrated management for NBS and of an articulation and exchange of information between entities. Regarding available resources, the existence of adequate competences and of a specific entity in charge, with appropriate financial and technical resources, are also assessed.

- Objective 7—Resilience engaged service aims to ensure service flexibility, disaster response and service reliability. The proposed criteria for this objective are the flexible service, the scenarios relevance for disaster response and the reliable service. As regards to flexible service, this criterion assesses the ES improvement and water reuse. Moreover, an adequate disaster response is assessed through the definition of relevant scenarios for heat wave, flooding and droughts. These criteria also assess the reliability of the service by minimizing the impact in the surrounding areas (e.g., flooding, critical location).

- Objective 8—Infrastructure safety and robustness aims to identify the criticality of the infrastructure and ensure the infrastructure assets' robustness, monitoring and maintenance. The proposed criteria for this objective are the infrastructure assets criticality and protection, the infrastructure assets robustness and the infrastructure monitoring and maintenance. The proposed criteria assess the identification of the critical components and the implementation of protective buffers. Regarding the infrastructure assets robustness, this criterion assesses the hydraulic and water quality performance regarding the infrastructure design conditions. Also, criteria assess the development and implementation of monitoring and maintenance plans, identified variables and other relevance aspects to be monitored and controlled.

- Objective 9—Infrastructure preparedness aims to ensure infrastructure preparedness to recover and buildback after a disruptive event. The proposed criterion for this objective is the infrastructure preparedness for recovery and buildback, which assess the infrastructure preparedness in the face of short and long-term changes, by addressing the impacts related to acute shocks and to continuous or chronic stresses.

- Objective 10—Infrastructure dependence and autonomy aims to identify the dependencies between other urban services and NBS infrastructure and the NBS autonomy. The proposed criteria for this objective are the infrastructure dependence and the infrastructure autonomy. In this sense, the criteria assess NBS dependencies from other services, the infrastructure of other services dependent on NBS, and the identification of the NBS infrastructure´s autonomy.

In order to allow for an objective assessment of each criterion, specific metrics were defined, including both quantitative and qualitative metrics. Metric selection seeks to properly evaluate the proposed criteria, taking into consideration that the metrics are interrelated. It is necessary to understand how they provide comprehensive information on the resilience maturity. The determination of the metrics in the Dimension I presents a higher feasibility of application, since most are data based and only some metrics are procedure based. The Dimension II presents greater complexity because it is necessary to develop a mathematical model of the NBS's hydraulic behaviour for proposed model-based metrics. Due to the high number of proposed metrics, they cannot be detailed in this manuscript. All objectives, criteria and corresponding metrics are supplied in the supplementary material (Table S1).

3.4. Validation of the RAF (Step 3)

RAF validation by the stakeholders provide the opportunity to contribute to adjust metrics' definition, to identify relevant sources of information for metrics' determination and test the assessment approach adequacy to different development levels of urban resilience. With this approach, stakeholders' contributions allow to consolidate the RAF, particularly, the proposed metrics and reference values.

This section presents an analysis of the metrics' relevance and feasibility of application to cities based on the stakeholders' opinion. According to [39], metrics' relevance was classified considering three levels—(i) essential, which should be integrated in the assessment of any city (essential for all cities); (ii) complementary, when the assessment corresponds to an intermediate level; and (iii) comprehensive, when the purpose is to make an in-depth assessment of the city. According to the feasibility of application, metrics were classified as high, medium, and low feasibility. Figure 2 presents the responses to metrics relevance and feasibility of application to cities, aggregated at the criterion level. Detailed responses for each metric are presented in Appendix A.

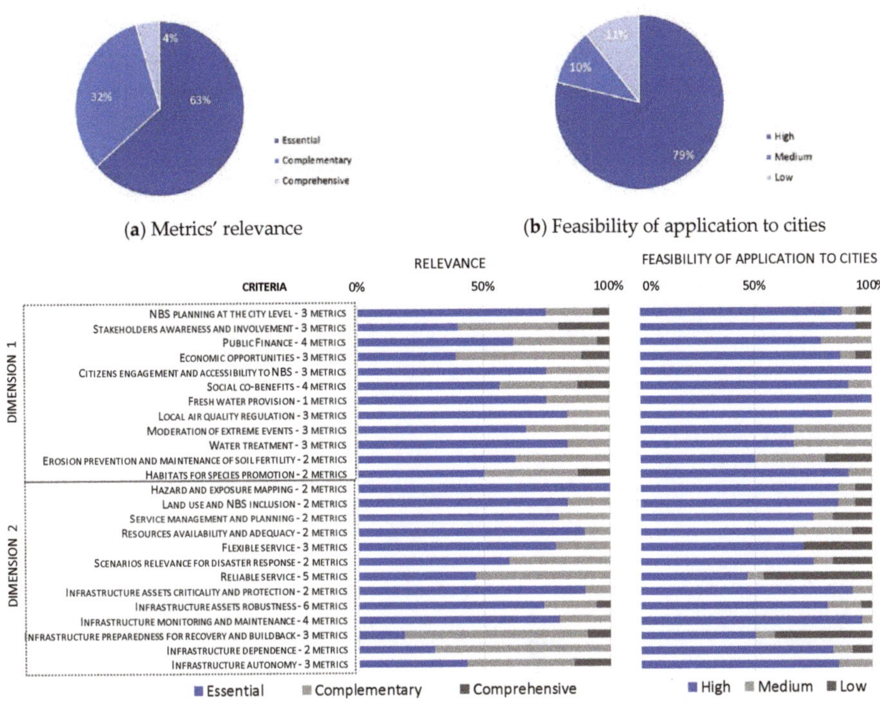

(a) Metrics' relevance

(b) Feasibility of application to cities

(c) Responses to metrics relevance and feasibility of application to cities, aggregated at criterion level

Figure 2. Stakeholders' opinion regarding metrics' relevance and feasibility of application to cities.

Considering an overview of the metrics relevance, 63% and 32% are considered essential and complementary, respectively (Figure 2a). Only 4% of the metrics are classified as comprehensive. In terms of the feasibility of application, 79% and 10% of the metrics are considered to have high and medium feasibility application to cities, respectively (Figure 2b). Nevertheless, 11% of the RAF metrics are considered with low feasibility of application. The responses to metrics relevance and feasibility of application to cities, aggregated at the criterion level (Figure 2), confirm the conclusions obtained for the overall results, aggregated for the whole RAF (Figure 2a,b).

These results highlight the relevance of the selected metrics and, consequently, the assessment criteria defined in the resilience assessment proposal. In addition, the stakeholders' opinion regarding the feasibility of application provided the opportunity to identify which metrics are more suitable for a city depending on its resilience development level. In this sense, the stakeholders' opinion supported the selection of metrics to be considered in the RAF according to different resilience development levels. Recommendations will be further proposed in setting a tailored roadmap for the RAF application to the city with a preselection of metrics depending on the city´s resilience development level.

3.5. Reference Values (Step 4)

The classification of the metrics' result is made by associating each answer to a resilience development level, related to the reference values. The metric results are classified as (i) incipient (non-existent or at early stage of development); (ii) progressing (significant steps have already been taken); and (iii) advanced (consolidated results). The assessment of each metric is made according to reference values, defined from an extensive literature review on each metric. A resilience development level between 0 and 3 is then assigned, based on reference values, namely (i) incipient [0, 1], progressing [1, 2], and advanced [2, 3, 26].

Given the RAF structure (Figure 1), the results of the metrics contained in each criterion might be averaged to a criterion resilience development level, and further on, upwards to an objective and then to a dimension development level. Table 1 presents examples of three metrics, their reference values and the set of references used to support them.

Table 1. Examples of data, procedure, and model-based metrics and corresponding reference values proposed in the RAF for nature-based solutions (NBS).

Objective	Criterion (Metric Type)	Metric	Reference values	References
Social involvement and co-benefits	Citizens engagement and accessibility to NBS (DB*)	NBS distribution Are NBS scattered in the city?	3) Yes, NBS are scattered in the city, existing one or more NBS in each neighbourhood; 2) Yes, NBS are partially scattered in the city but they don't exist in every neighbourhood; 1) No, a significant number of NBS (with an area higher than 0.25ha) are concentrated in a few location or 50% of NBS area corresponds to one NBS;	[25] [40] [41] [42]
Environmental resilience	Local air quality regulation (PB)	Carbon sequestration and storage Is a carbon sequestration and storage increase expected due to NBS implementation?	3) Yes, above 60 t/ha; 2) Yes, between 10 and 60 t/ha; 1) Yes, less than 10 t/ha; 0) No.	[43] [44] [45] [46] [47] [48]
Resilience engaged service	Reliable service (MB)	Flooded area [Maximum flooding area, related to stormwater drainage problems / area of NBS urban catchment] x 100	3) No flooded areas; 2) Less than or equal to 2.5% area is flooded; 1) More than 2.5% and less than 5% area is flooded; (0) More than or equal to 5% and less than 10% area is flooded.	[27] [49] [50]

* DB: Data based, PB: Procedure based, MB: Model based.

4. Conclusions

This paper presents the methodology adopted for the construction of a RAF to assess NBS contribution to urban resilience and the developed structure of the RAF for NBS, focused on solutions for stormwater management and control. This resilience assessment proposal ensures the evaluation of the main *attributes* of the urban resilience and the relevant *aspects* for the NBS evaluation. The RAF aims to support NBS diagnosis and to assist decision-making in its planning, implementation, and management. Also, this framework allows to identify NBS with potential to contribute to city resilience.

The determination of the metrics relevance and feasibility of application to cities is a fundamental step in the validation step. Considering the metrics relevance, the stakeholders' opinion allowed to conclude that most metrics are considered essential for the assessment of the NBS contribution for urban resilience. Regarding the feasibility of application, a higher variability in the stakeholders' responses was obtained. As expected, procedure and model based metrics were labelled as medium or low feasibility of application, mainly at the service and infrastructure level.

In this sense, the criteria identified as with lower feasibility of application correspond to the following criteria: (i) flexible service, (ii) reliable service, and (iii) infrastructure preparedness for recovery and buildback. Regarding the flexible service criterion, the lower feasibility of application is due to the lack of awareness of ES and the difficulties related to the use of the NBS retained water for other purposes, at city level. From another hand, the lower feasibility of application of the reliable service and infrastructure preparedness for recovery and buildback criteria is because metric determination is carried out based on a mathematical model.

The stakeholders' participation highlighted the relevance of each metric and criteria defined in the RAF. Stakeholders' opinion allowed to identify the RAF adequacy to diverse city maturity levels and helps to select adequate metrics for the cities according to the urban resilience development level. Based on this analysis, the consolidation of metrics definition and required data will be carried out.

The RAF consolidation will be carried out after its complete application to the case study. In this step, the metrics' definition and the required information will be verified. In this sense, future work will focus on the consolidation of the RAF and on the proposal of a roadmap for the RAF application to any city.

The RAF application should follow the proposed roadmap and consider pre-selected metrics. The pre-selected metrics and the required information depend of the urban resilience development (incipient, progressing, advanced). For example, only for cites with advanced urban resilience development the model based metrics will be determined, and, consequently, the information provided by the mathematical model, which is more detailed and difficult to obtain, will be required.

The determination of the RAF will be carried out by a multidisciplinary team composed by human resources of the entities in charge of NBS management, stormwater management services and green space management. The RAF can be applied to assess the contribution to urban resilience of all existing NBS in the city, a group of NBS or a specific NBS. For this reason, the NBS under assessment should be identified in the complementary profile.

Supplementary Materials: The following are available online at http://www.mdpi.com/2071-1050/12/6/2537/s1. Table S1. Objective, criteria, and metric of Dimension I "Integration of NBS in the city".; Table S2. Objective, criteria, and metric of Dimension II "Operation and service of NBS".

Author Contributions: The conceptualization of the work and of the methodology were carried out by P.B. Original draft preparation was developed by P.B. and review and editing was accomplished by P.B., R.S.B., and A.G. All authors have read and agreed to the published version of the manuscript.

Funding: This research was funded by Portuguese Foundation for Science and Technology (FCT) through the PhD fellowship PD/BD/135216/2017.

Acknowledgments: The authors gratefully acknowledge the support of the Portuguese Foundation for Science and Technology (FCT), through the PhD fellowship PD/BD/135216/2017. The authors acknowledge the RESCCUE project for the opportunity to participate in this special issue. The authors would like to thank the following organizations for the collaboration and all constructive comments during this work and for the collaboration in the RAF validation step: Águas de Coimbra, Águas do Porto, Ajuntament de Barcelona, Bristol City Council, Câmara Municipal de Lisboa, Câmara Municipal do Porto, City of Vancouver, and SMAS de Almada.

Conflicts of Interest: The authors declare no conflict of interest. The funders had no role in the design of the study; in the collection, analyses, or interpretation of data; in the writing of the manuscript, or in the decision to publish the results.

Appendix A

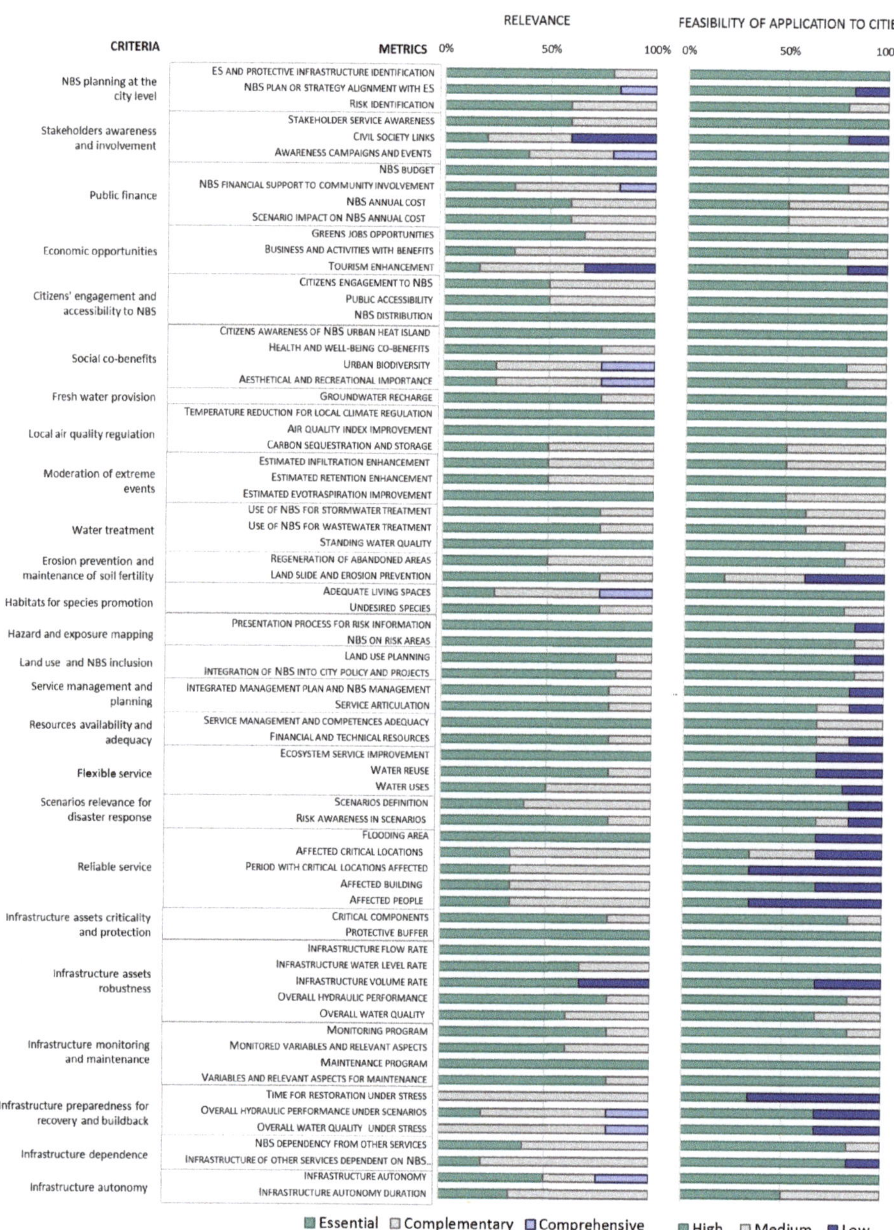

Figure A1. Stakeholders' opinion regarding the metrics relevance and feasibility of application to their own city at the metric level.

References

1. Meerow, S.; Newell, J.P.; Stults, M. Defining urban resilience: A review. *Landsc. Urban Plan.* **2016**, *147*, 38–49. [CrossRef]
2. Jagt, A.P.V.; Smith, M.; Ambrose-Oji, B.; Konijnendijk, C.; Giannico, V.; Haase, D.; Lafortezza, R.; Nastran, M.; Pintari, M.; Železnikari, S.; et al. Co-creating urban green infrastructure connecting people and nature: A guiding framework and approach. *J. Environ. Manag.* **2019**, *233*, 757–767. [CrossRef] [PubMed]
3. Folke, C. Resilience: The emergence of a perspective for social-ecological systems analyses. *Glob. Environ. Chang.* **2006**, *16*, 253–267. [CrossRef]
4. Cardoso, M.A.; Almeida, M.C.; Telhado, M.J.; Morais, M.; Brito, R. Assessing contribution of climate change adaptation measures to build resilience in urban areas. Application to Lisbon. In Proceedings of the 8th International Conference on Building Resilience—ICBR Lisbon'2018 Risk and Resilience in Practice: Vulnerabilities, Displaced People, Local Communities and Heritages, Lisbon, Portugal, 14–16 November 2018.
5. Eggermont, H.; Balian, E.; Azevedo, J.M.N.; Beumer, V.; Brodin, T.; Claudet, J.; Fady, B.; Grube, M.; Keune, H.; Lamarque, P.; et al. Nature-based solutions: new influence for environmental management and research in Europe. *GAIA-Ecol. Perspect. Sci. Soc.* **2015**, *24*, 243–248. [CrossRef]
6. OECD. *Resilient Cities*; Preliminary Version; OECD Publishing: Paris, France; Available online: http://www.oecd.org/cfe/regional-policy/resilient-cities.htm (accessed on 12 December 2016).
7. Diao, K.; Sweetapple, C.; Farmani, R.; Fu, G.; Ward, S.; Butler, D. Global resilience analysis of water distribution systems. *Water Res.* **2016**, *106*, 383–393. [CrossRef]
8. Shutes, B.; Raggatt, L. *Deliverable 2.2. 5 Development of Generic Best Management Practice (BMP) Principles for the Management of Stormwater as Part of an Integrated Urban Water Resource Management Strategy*; SWITCH Project Deliverable (www.switchurbanwater.eu); Urban Pollution Research Centre: London, UK, 2010.
9. European Commission. *Towards an EU Research and Innovation Policy Agenda for Nature-Based Solutions and Re-Naturing Cities*, Final Report of the Horizon 2020 expert group on nature-based solutions and re-naturing cities (Full version); European Commission: Brussels, Belgium, 2015.
10. UNESCO. *2018 UN World Water Development Report, Nature-Based Solutions for Water*; UNESCO: Paris, France, 2016.
11. Wong, T.H.; Brown, R.R. The water sensitive city: Principles for practice. *Water Sci. Technol.* **2009**, *60*, 673–682. [CrossRef]
12. Ellis, J.B.; Lundy, L.; Revitt, M. An integrated decision support approach to the selection of Sustainable Urban Drainage Systems (SUDS). In Proceedings of the SWITCH Conference: The Future of Urban Water, Paris, France, 24–26 January 2011.
13. Viavattene, C.; Ellis, J.B. The management of urban surface water flood risks: SUDS performance in flood reduction from extreme events. *Water Sci. Technol.* **2013**, *67*, 99–108. [CrossRef]
14. Mugume, S.N.; Gomez, D.E.; Fu, G.; Farmani, R.; Butler, D. A global analysis approach for investigating structural resilience in urban drainage systems. *Water Res.* **2015**, *81*, 15–26. [CrossRef]
15. Potter, K.; Vilcan, T. Managing urban flood resilience through the English planning system: Insights from the 'SuDS-face'. *Phil. Trans. R. Soc. A* **2020**, *378*, 20190206. [CrossRef]
16. Djordjević, S.; Butler, D.; Gourbesville, P.; Mark, O.; Pasche, E. New Policies to Deal with Climate Change and Other Drivers Impacting on Resilience to Flooding in Urban Areas: The CORFU Approach. *Environ. Sci. Policy* **2011**, *14*, 864–873.
17. Panno, A.; Carrus, G.; Lafortezza, R.; Mariani, L.; Sanesi, G. Nature-based solutions to promote human resilience and wellbeing in cities during increasingly hot summers. *Environ. Res.* **2017**, *159*, 249–256. [CrossRef] [PubMed]
18. Zölch, T.; Henze, L.; Keilholz, P.; Pauleit, S. Regulating urban surface runoff through nature-based solutions–An assessment at the micro-scale. *Environ. Res.* **2017**, *157*, 135–144. [CrossRef] [PubMed]
19. Calliari, E.; Staccione, A.; Mysiak, J. An assessment framework for climate-proof nature-based solutions. *Sci. Total Environ.* **2019**, *656*, 691–700. [CrossRef] [PubMed]
20. Naumann, S.; Kaphengst, T.; McFarland, K.; Stadler, J. *Nature-Based Approaches for Climate Change Mitigation and Adaptation. The Challenges of Climate Change–Partnering with Nature*; German Federal Agency for Nature Conservation (BfN), Ecologic Institute: Bonn, Germany, 2014.

21. Hancz, G.; Biró, J.; Biró, B. Estimation of potential runoff quality improvement as a result of applied green infrastructure measures in a Hungarian town. *J. Int. Sci. Publ. Ecol. Saf.* **2018**, *12*, 2015.

22. Reed, M.S.; Graves, A.; Dandy, N.; Posthumus, H.; Hubacek, K.; Morris, J.; Prell, C.; Quinn, C.H.; Stringer, L.C. Who's in and why? A typology of stakeholder analysis methods for natural resource management. *J. Environ. Manag.* **2009**, *90*, 1933–1949. [CrossRef] [PubMed]

23. Larkin, S.; Fox-Lent, C.; Eisenberg, D.A.; Trump, B.D.; Wallace, S.; Chadderton, C.; Linkov, I. Benchmarking agency and organizational practices in resilience decision making. *Environ. Syst. Decis.* **2015**, *35*, 185–195. [CrossRef]

24. Cox, R.S.; Hamlen, M. Community disaster resilience and the rural resilience index. *Am. Behav. Sci.* **2015**, *59*, 220–237. [CrossRef]

25. Raymond, C.M.; Berry, P.; Breil, M.; Nita, M.R.; Kabisch, N.; de Bel, M.; Enzi, V.; Frantzeskaki, N.; Geneletti, D.; Cardinaletti, M.; et al. *An Impact Evaluation Framework to Support Planning and Evaluation of Nature-Based Solutions Projects. Report prepared by the EKLIPSE Expert Working Group on Nature-Based Solutions to Promote Climate Resilience in Urban Areas*; Centre for Ecology & Hydrology: Wallingford, UK, 2017.

26. Cardoso, M.A.; Brito, R.; Pereira, C.; David, L.M. Avaliação da resiliência dos serviços urbanos de águas face às alterações climáticas. In Proceedings of the XVI Seminário Ibero-Americano sobre Sistemas de Abastecimento e Drenagem, SEREA19, Lisbon, Portugal , 15–17 July 2019.

27. UNISDR. *Disaster Resilience Scorecard for Cities. Preliminary Level Assessment*; United Nations Office for Disaster Reduction: Geneva, Switzerland, 2017.

28. ARUP. *City Resilience Framework. 100 Resilient Cities*; The Rockefeller Foundation; ARUP: New York, NY, USA, 2015.

29. Dhakal, K.P.; Chevalier, L.R. Managing urban stormwater for urban sustainability: Barriers and policy solutions for green infrastructure application. *Environ. Manag.* **2017**, *203*, 171–181. [CrossRef]

30. USEPA. *Evaluating Urban Resilience to Climate Change: A Multi-Sector Approach. United States Environmental Protection Agency*; EPA/600/R-16/365F; Office of Research and Development: Washington, DC, USA, 2017.

31. Beceiro, P.; Brito, R.S.; Galvão, A. Contribution of Nature-Based Solutions (NBS) for resilience in cities. In Proceedings of the 18° Encontro de Engenharia Sanitária e Ambiental (ENASB) and 18° Simpósio Luso-Brasileiro de Engenharia Sanitária e Ambiental (SILUBESA), Porto, Portugal, 10–12 October 2018.

32. Kabisch, N.; Frantzeskaki, N.; Pauleit, S.; Naumann, S.; Davis, M.; Artmann, M.; Haase, D.; Knapp, S.; Korn, H.; Stadler, J.; et al. Nature-based solutions to climate change mitigation and adaptation in urban areas: Perspectives on indicators, knowledge gaps, barriers, and opportunities for action. *Ecol. Soc.* **2016**, *21*, 2. [CrossRef]

33. NATURE4CITIES. NATURE4CITIES—D2.1—System of Integrated Multi-Scale and Multi-Thematic Performance Indicators for the Assessment of Urban Challenges and NBS. Available online: https://docs.wixstatic.com/ugd/55d29d_3b17947e40034c168796bfc9a9117109.pdf (accessed on 1 March 2019).

34. ISO. ISO 24510:2007. *Activities Relating to Drinking Water and Wastewater Services—Guidelines for the Assessment and for the Improvement of the Service to Users*; International Organization for Standardization: Geneva, Switzerland, 2007.

35. ISO. ISO 24511:2007. *Activities Relating to Drinking Water and Wastewater Services—Guidelines for the Management of Wastewater Utilities and for the Assessment of Drinking Water Services*; International Organization for Standardization: Geneva, Switzerland, 2007.

36. ISO. ISO 24512:2007. *Service Activities Relating to Drinking Water and Wastewater—Guidelines for the Management of Drinking Water Utilities and for the Assessment of Drinking Water Services*; International Organization for Standardization: Geneva, Switzerland, 2007.

37. ISO. ISO 55000:2014. *Asset Management—Overview, Principles and Terminology*; International Organization for Standardization: Geneva, Switzerland, 2014.

38. CEN. EN 752:2008. *Drain and Sewer Systems Outside Buildings*; European Committee for Standardization: Brussels, Belgium, 2008.

39. Cardoso, M.A.; Brito, R. Approach to Develop a Climate Change Resilience Assessment Framework. In Proceedings of the IWA's 2019 LESAM and PI Conferences, Vancouver, BC, Canada, 23–27 September 2019.

40. Oliveira, S.; Andrade, H.; Vaz, T. The cooling effect of green spaces as a contribution to the mitigation of urban heat: A case study in Lisbon. *Build. Environ.* **2011**, *46*, 2186–2194. [CrossRef]

41. Pafi, M.; Siragusa, A.; Ferri, S.; Halkia, M. *Measuring the Accessibility of Urban Green Areas. A comparison of the Green ESM with Other Datasets in Four European Cities*; EUR 28068 EN; Publications Office of the European Union: Luxembourg, 2016. [CrossRef]

42. Van den Bosch, A.; Mudu, M.; Uscila, P.; Barrdahl, V.; Kulinkina, A.; Staatsen, B.; Staatsen, B.; Swart, W.; Kruize, H.; Zurlyte, I.; et al. Development of an urban green space indicator and the public health rationale. *Scand. J. Public Health* **2016**, *44*, 159–167. [CrossRef]

43. Besir, A.B.; Cuce, E. Green roof and facades. A comprehensive review. *Renew. Sustain. Energy Rev.* **2018**, *82*, 915–939. [CrossRef]

44. Chen, W.Y. The role of urban green infrastructure in offsetting carbon emissions in 35 major Chinese cities: A nationwide estimate. *Cities* **2015**, *44*, 112–120. [CrossRef]

45. Davies, Z.G.; Edmondson, J.L.; Heinemeyer, A.; Leake, J.R.; Gaston, K.J. Mapping an urban ecosystem service: Quantifying above-ground carbon storage at a city-wide scale. *J. Appl. Ecol.* **2011**, *48*, 1125–1134. [CrossRef]

46. Hossain, M.A.; Shams, S.; Amin, M.; Reza, M.S.; Chowdhury, T.U. Perception and barriers to implementation of intensive and extensive green roofs in Dhaka, Bangladesh. *Buildings* **2019**, *9*, 79. [CrossRef]

47. Marchi, M.; Pulselli, R.M.; Marchettini, N.; Pulselli, F.M.; Bastianoni, S. Carbon dioxide sequestration model of a vertical greenery system. *Ecol. Model.* **2015**, *306*, 46–56. [CrossRef]

48. Novak, D.J. Atmospheric carbon dioxide reduction by Chicago's urban forest. In *Chicago's Urban Forest Ecosystem: Results of the Chicago Urban Forest Climate Project*; Gen. Tech. Rep. NE-186; Department of Agriculture, Forest Service, Northeastern Forest: Radnor, PA, USA, 1994.

49. Beceiro, P. 1D/2D Integrated Modelling and Performance Assessment to Support Floods Management. Application to Stormwater Urban Drainage System in Estuarine Areas. Master's Thesis, Technical University of Lisbon, Lisbon, Portugal, 2016.

50. Cardoso, M.A.; Brito, R.S.; Pereira, R.; David, L.; Almeida, M.C. Resilience Assessment Framework – RAF. Description and implementation. Deliverable 6.4, RESCCUE project. 105 pp. (under publication). 2020. Available online: http://www.resccue.eu/resccue-project (accessed on 1 November 2019).

Article

Methodology to Prioritize Climate Adaptation Measures in Urban Areas. Barcelona and Bristol Case Studies

María Guerrero-Hidalga [1,*], Eduardo Martínez-Gomariz [1,2], Barry Evans [3,4], James Webber [3], Montserrat Termes-Rifé [1,5], Beniamino Russo [6,7] and Luca Locatelli [6]

1   Cetaqua, Water Technology Centre, Cornellà de Llobregat, 08940 Barcelona, Spain; eduardo.martinez@cetaqua.com (E.M.-G.); mtermes@ub.edu (M.T.-R.)
2   Flumen Research Institute, Universitat Politècnica de Catalunya, 08034 Barcelona, Spain
3   Centre for Water Systems, University of Exeter, Exeter EX4 4QF, UK; b.evans@exeter.ac.uk (B.E.); J.Webber2@exeter.ac.uk (J.W.)
4   College of Sciences, School of Built Environment, Massey University, Wellington 4442, New Zealand
5   Economics Department, Universitat de Barcelona, 08034 Barcelona, Spain
6   AQUATEC—Suez Advanced Solutions, 08038 Barcelona, Spain; brusso@aquatec.es (B.R.); luca.locatelli@aquatec.es (L.L.)
7   Grupo de Ingeniería Hidráulica y Ambiental (GIHA), Escuela Politécnica de La Almunia (EUPLA), Universidad de Zaragoza, La Almunia de Doña Godina, 50100 Zaragoza, Spain
*   Correspondence: maria.guerrero@cetaqua.com; Tel.: +34-9-3312-4844

Received: 22 May 2020; Accepted: 8 June 2020; Published: 12 June 2020

Abstract: In the current context of fast innovation in the field of urban resilience against extreme weather events, it is becoming more challenging for decision-makers to recognize the most beneficial adaptation measures for their cities. Detailed assessment of multiple measures is resource-consuming and requires specific expertise, which is not always available. To tackle these issues, in the context of the H2020 project RESCCUE (RESilience to cope with Climate Change in Urban arEas), a methodology to effectively prioritize adaptation measures against extreme rainfall-related hazards in urban areas has been developed. It follows a multi-phase structure to progressively narrow down the list of potential measures. It begins using less resource-intensive techniques, to finally focus on the in-depth analysis on a narrower selection of measures. It involves evaluation of risks, costs, and welfare impacts, with strong focus on stakeholders' participation through the entire process. The methodology is adaptable to different contexts and objectives and has been tested in two case studies across Europe, namely Barcelona and Bristol.

Keywords: climate change adaptation; climate risk; socio-economic assessment; urban resilience

---

1. Introduction

There is an increasing movement of local policy-makers to act against climate change, urging to adapt their cities and improve their resilience with suitable policies [1]. Designing and implementing general local-scale policies requires multidisciplinary studies and taking broad assumptions to address uncertainties, in order to maximize the welfare associated with a desirable outcome, such as improving public transport [2]. Climate change adaptation policy making has additional complexities. In the design phase, which is the scope of this paper, there are uncertainties related to climate predictions, vulnerability, and risk impacts, as well as socio-economic, technological, and environmental future trends that complicate the understanding of the potential outcome of a measure [3]. In addition, they have a cross-sectoral nature and diversity in terms of typology (e.g., structural, nature-based, or digital), scales (from a building to national scale), and timescales (short, medium, and long term) [3]. These complexities undermine measures' assessment results that consider only one criterion. Therefore, a systematic prioritization assessment is required in order to support the selection of the most suitable set of measures for each city under a changing climate context.

The purpose of the assessment of climate change adaptation measures is to minimize the degree of uncertainty when selecting policies that reduce the impacts of the changing climate [2]. For this purpose, technical and economic assessment approaches are usually considered (or a combination of them), as they provide relevant information about the costs and expected physical effects of the measures. However, in general, the degree of uncertainty of the results is linked to data availability and specific technical expertise to carry out detailed assessments. Therefore, when selecting a method, such limiting factors (i.e., data and skills availability) should be considered.

The present work has been developed in the context of the H2020 project RESCCUE [4], where a number of innovative models and tools were developed in order to help urban areas to become more resilient to climate change. One of the tools is the present methodology that aims to facilitate the ranking of all available adaptation measures considering the particularities of each city. Results are presented for Bristol and Barcelona (Figure 1), two of the three case study cities, to display the different approaches taken by each city.

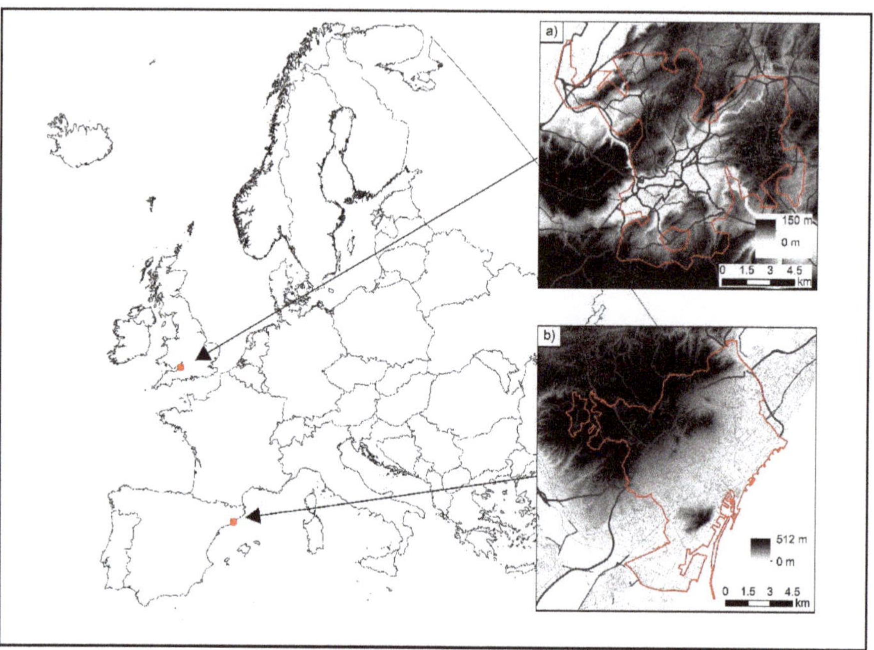

Figure 1. Location of the two cities where the methodology has been applied: (**a**) Bristol, United Kingdom and (**b**) Barcelona, Spain.

Section 2 summarizes the main methods used to evaluate adaptation measures. Section 3 starts with a brief description of the key terminology, and it is followed by the description of the methodological approach, presented by stages. Main methodological differences between Barcelona and Bristol studies are presented in order to compare the different approaches taken. The results for both case studies—Section 4—are also divided by stages. In the discussion, results are commented, as well as the main challenges found when applying the methodology and recommendations for further application of the methodology. The conclusion summarizes the main results.

2. Literature Review

A deep analysis of the most relevant methods in the field was carried out to understand the potential applicability to the context of the project RESCCUE [4]. A brief description of them is presented in order to provide context.

- Cost-effectiveness analysis (CEA) is a simple method that offers guidance to rank different alternatives, providing a ratio of necessary investment to achieve a unit of effectiveness (to be selected based on a pre-set objective) [5]. It is attractive as it can quickly scan and rank different options whose benefits are not easily measurable. Although it has limitations on assessing whether the policies are "worth doing", in the sense that it does not determine whether benefits are greater than costs [6] and lacks the capacity to identify non-direct impacts [7].
- Multi-criteria analysis (MCA) is a multi-step method based on the synthesis of already existing assessment studies [3]. It has diverse forms, but the general MCA method uses a scoring system to determine the potential accomplishment of a policy objective, giving weights (allocated by experts) to the different indicators included, previously normalized [8]. The major benefit is the possibility to assess variables of different nature and scales into the same framework (i.e., monetary, quantitative, or qualitative data). This is at the same time its major disadvantage, as trying to simplify and normalize different units and criteria can lead to a loss of accuracy [9]. Another characteristic of MCA is the consideration of multiple stakeholders in the evaluation process of indicators that are not measurable with quantitative data. If participants are impartial experts in relevant fields, results will be relevant, otherwise the scores risk being biased [6].
- Risks reduction assessment involves assessing either the health or environmental risks (or both) attached to a policy or project [6]. It is a valuable method for urban adaptation and mitigation appraisals as it is based in the concept of the disaster risk triangle (hazard, vulnerability, and exposure) [10]. It provides detailed results in terms of probability of damage for the selected return periods for the design storms, which facilitate the estimation of the potential damage reduction indicators compared to the do-nothing option [11]. A significant disadvantage of the method is its high time resources and specialized personnel requirements.
- Cost-benefit analysis (CBA) is a popular method to appraise the expected net results of different investment or policy options. It considers all costs and resulting benefits through the project or policy life, including economic (actual revenues and costs), social, and environmental changes derived from their implementation, through different available monetization methods [6]. In addition, in the context of natural hazards, this monetization usually implies consideration of risk reduction efficiency.

In order to address these challenges, the methodology was designed taking the most relevant parts of each method. It follows an MCA approach on its structure, to accomplish a gradual down-selection process. The initial "wish list" of measures is assessed and ranked in a first stage using the CEA method and co-benefits scoring, since they require less resources (time and expertise). After a first selection based on the initial results, most of the efforts are put in the last steps where only a small number of measures are studied in detail. The detailed assessment consists in exhaustive risk assessment and CBA.

3. Methodology

The proposed prioritization methodology was developed to offer a flexible approach, able to adapt to different urban contexts. The balance between expertise, resources requirements, accuracy, and replicability of results was an important consideration during the design process. The capacity of the methodology to allow for different levels of detail was also considered, due to the diverse data availability, which normally limits the assessment potential. Meeting these requirements was made possible by developing a method that followed the principles of MCA, in the sense that it (i) gives relevance to stakeholder decisions; (ii) uses normalized quantitative and qualitative indicators through a scoring system; (iii) is able to rank options with different goals [9]; and (iv) offers a multi-phase analysis approach. The phases are composed by combinations of assessment methods, ordered from coarser to more detailed assessments. The first stage includes a CEA and co-benefits scoring assessment, whereas the following phase is based on more detailed assessments—risks reduction assessment and CBA. In addition, the methodology proposes several variables to rank results that help decision-makers to downselect the most suitable measures for their specific policy goals. An introductory diagram of the methodology is available in Figure 2, which will be further explained in this section. In addition, the key terminology employed in the present methodology and the stages proposed to apply it is presented.

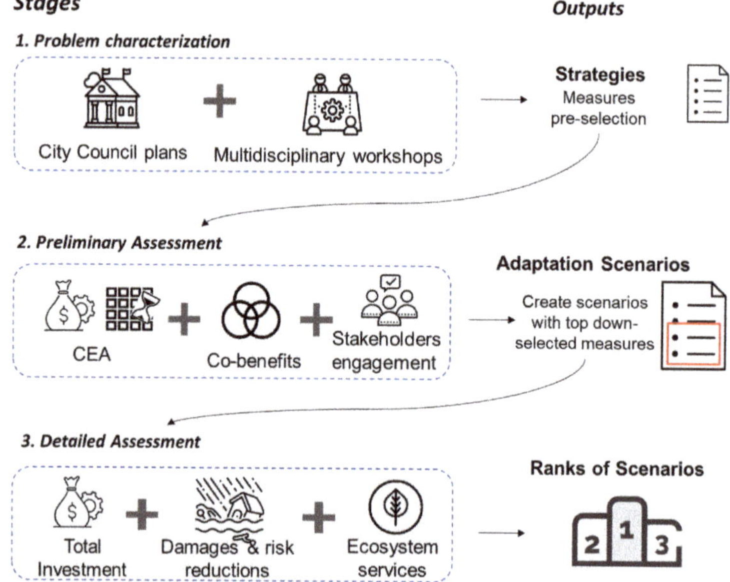

Figure 2. Outline of the methodology stages and outputs.

3.1. Key Terminology

The main terminology comprised in the proposed prioritization method could be summarized as follows:

1. Adaptation strategies are understood as sets of measures that aim to tackle one concrete issue related to climate change adaptation. Measures contain specific actions targeted to address a problem within the scope of the strategy they belong to. It means that a measure can be displayed in more than one strategy, because it tackles more than one problem (e.g., green areas could support flooding control and air pollution strategies).

2. Initial and recurrent investments of each measure are required. As it is an ad-hoc analysis, costs are expected to be estimates transferred from similar actions in different locations, scaled to the new location's size. It can be done using unitary values ($€/m^2$), and GDP to adjust through the purchasing power parity index [12] and exchange rate if necessary. For accuracy purposes, researchers should always try to find budget references from sites as similar to the research site as possible.

3. Economic, social, and environmental co-benefits are those benefits or positive effects generated in parallel to the main objective of the policy [13], understood as the specific climate change adaptation goal. Specific indicators for each co-benefit category, are presented in Table 1 below.

The quantification and monetization of co-benefits is surrounded by uncertainty [14], thus co-benefits were accounted for using a specific scoring system evaluated by experts from diverse disciplines in each case study. This method also involves uncertainty, but the method is less time-consuming and considers local knowledge.

4. An effectiveness indicator helps to assess the success of the resources used in achieving the objective of each measure [15]. It is important to select one that is valid across measures of different characteristics, and that requires available information. The effectiveness indicator selected in the present study is the reduction of downtime of urban services after an extreme weather event—recovery time reduction. This indicator provides information about the duration of modeled floods scenarios, and if scenarios with and without measures are compared, the variation of recovery time can be used as an effectiveness indicator.

5. Climate risks reduction is understood as a percentage of high-risk area reduced, may be used as an indicator of the adaptation effectiveness. The higher the number of risks assessed, the more comprehensive the prioritization of adaptation measures would be. Risks, such as the stability of pedestrians or vehicles exposed to water flows and damages caused to properties and vehicles, have been assessed among the case studies presented herein. Results are sought for both the business as usual scenario and the scenarios where adaptation measures are implemented.

Table 1. Representation of the categories of co-benefits and their indicators scored from 0 to 10 with expert judgement.

Economic	Social	Environmental
Cost savings	Reduced mortality impacts	Improved air quality
Reduced energy losses	Reduced health impacts	Improved water quantity
Job creation	Reduced mortality from diseases	Reduced aquifer depletion
Possible reduction in prices	Enhanced public amenity	Reduced water pollution
Increased labor productivity	Reduced impacts on vulnerable groups	Reduced land contamination
Increased economic production	Reduced number of householder/business forced from home/workplace	Improved biodiversity and ecosystems
Increased property values	Social inclusion	Maintained and increased green space
		Reduced environmental impacts through associated awareness
		Increased biodiversity and ecosystem services
		Effective/uninterrupted water collection and security
		Erosion control
Average (economic) = $\sum$ Scores/number of indicators ($n = 7$)	Average (social) = $\sum$ Scores/number of indicators ($n = 7$)	Average (environmental) = $\sum$ Scores/number of indicators ($n = 11$)

3.2. Stage 1: Problem Characterization

The first step was to identify the most pressing bio-physical and socio-economic issues [16]. Climate change scenarios and their impacts on the cities under study were required. In most large European cities, climate change predictions are available [17] and climate plans that address their major concerns with regards to climate change adaptation and/or mitigation [1].

In the current case studies, the downscaled climate-model projections used were developed by the FIC (Climate Research Foundation) [18], while Bristol used also climate data derived from UK Climate Change Projections 2009 (UKCP09) [19]. Local climate action plans were considered, as well as the conclusions from local workshops conducted to address the most pressing issues identified by stakeholder engagement in each site. A common workshop structure was proposed and followed by each city, adapting details according to their specific needs, such as the number and background of members. The main structure followed in the first workshop included context introduction, a discussion among partners regarding the approaches taken, strategies to be included, and the prioritization method to follow. The second workshop focused on reviewing the expected outcomes in terms of hazard reduction and the presentation and discussion of detailed strategies already identified by the city council.

The outcome, presented in the results section, was a selection of the extreme events and hazards to address and the development of a longlist of measures that are expected to alleviate the impacts of the former).

The second step consisted in selecting and applying a categorizing criterion to create strategies that are aligned with the objectives sought. In the context of urban services, the following options were proposed:

- By type of hazard (e.g., flooding events and droughts);
- By specific urban service targeted (e.g., transport, energy, and water supply);
- By type of measure (e.g., engineering, nature-based, or technological).

Researchers of the Barcelona case study decided to categorize strategies by the type of hazard they targeted, namely pluvial flooding and Combined Sewer Overflow (CSO), according to the objective of the city's stakeholders to focus on the reduction of those impacts. Similarly, the Bristol case study chose to categorize strategies by the predominant hazards identified in the city, which were pluvial and fluvial flooding and CSO events. These issues were forecasted to cause significant disruption to many aspects of the communities, in particular to the transport network, power supply, and properties located in high-risk areas.

This classification permits comparison of measures that "compete" against each other to alleviate one specific hazard instead of an overall prioritization of measures earmarked for different issues.

3.3. Stage 2: Preliminary Rank of Adaptation Measures

The preliminary assessment phase served as an overall screening of the extensive list of adaptation measures considering cost, effectiveness, and welfare aspects. The process involved the ranking of the weighted output of the application of cost-effectiveness analysis (CEA) and the assessment of co-benefits for each measure within a given strategy.

First, the CEA served as comparative assessment of different options that aim to achieve a given objective not measurable in monetary terms [20]. It did so by assessing alternatives in terms of the cost per unit of benefit delivered. In the RESCCUE context, the objective was to increase a city's resilience through the reduction of the floods and CSO spills impacts. The effectiveness of the consecution of this objective was measured through the variation of the recovery time (VRT) from a modeled climate-related event. The recovery time was based on a 1D/2D hydrodynamic model developed to simulate floods in the assessed cities related to a range of return periods. For both case studies, the 1D/2D model was carried out to obtain the time to recover from a flood episode under different measures. An additional 1D drainage model was developed and applied to simulate CSO spills into water bodies and to estimate the average duration of insufficient water quality for the different scenarios modeled. The detailed methodology of both models can be found in [21,22]. The variation of the recovery time for a measure i and event e ($VRT_{i,e}$) was calculated by subtracting the time obtained in the business as usual (BAU) scenario for the same event ($RT_{BAU,e}$) with the one obtained with the measure modeled ($RT_{i,e}$) (Equation (1)):

$$VRT_{i,e} = RT_{BAU,e} - RT_{i,e}. \tag{1}$$

Hydrodynamic modeling is the preferred option to assess the effectiveness of adaptation actions in urban services. However, if a 1D/2D modeling software is not available, a 1D model would also offer an alternative, as it provides the time during which the drainage network is working under surcharged conditions. This surcharged time could be used alternatively as an effectiveness indicator.

The cost was included as the equivalent annual cost (EAC) (Equation (2)), which is the annual estimated cash flow over the lifespan of the project, considering discounting [23]. This allows harmonization of all costs for comparison of measures through the time horizon of the study, set at 2100. They consider initial investment, annual costs, reinvestment (if necessary), and residual value at the end of the assessment time period:

$$EAC_j = \frac{NPV_j}{A_{(t,i)}}, \quad A_{(t,i)} = \sum_{t=1}^{T} \frac{1 - \frac{1}{(1+i)^T}}{i} = \frac{1 - (1+i)^{-T}}{i}, \tag{2}$$

where $A_{(t,i)}$ is the annuity factor which is the sum of all discount factors for the duration of the project; T is the time horizon, and i is the discount rate. The discount rate selected for Barcelona and Bristol case studies was 1.23%. It was based on research on the most suitable long-term rate for both regions (Catalonia, Spain and West Country, England) carried out within the European project EconAdapt [24]. This is aligned with the Stern economic school of thought that considers that climate change impact's increases in the long-term future should be accounted for through low or decreasing discount rates [25,26]. Both sites considered the lower range of the discount rate in the scenario modeled with economic growth.

In the case that costs of measures were not available or not accurate enough, a literature review could be carried out in order to develop a scoring system for expected costs of implementation and maintenance of the measures proposed. In the preliminary assessment, costs accuracy of measures is not essential, but actual relative differences between measures is required in order to obtain a realistic preliminary ranking. However, the measures selected to undergo the detailed assessment (next stage) required more accurate results, as their outputs were expected to be more precise.

The equivalent annual costs (EAC) of each measure i divided by its VRT, resulted in the cost-effectiveness assessment (CEA) ratio indicator (Equation (3)). In the framework of increasing urban resilience, the reduction of the city recovery time is an important indicator. Therefore, in the CEA indicator, a "penalty" is levied to those measures that do not reduce it at all. Results were ranked from the most (smaller result) to the least (larger result) preferred option:

$$CEA_i = \left\{ \begin{array}{l|l} \frac{EAC_i}{VRT_i} & if\ VRT_i > 0 \\ EAC_i \times 2 & if\ VRT_i \leq 0 \end{array} \right\}. \tag{3}$$

In parallel, co-benefits were included in order to assess the indirect effects of measures. In both case studies, there was a lack of accessible quantitative information to assess the co-benefits. Although, there was also a strong interest in including them in the prioritization exercise. The solution was found in assessing co-benefits through a site-specific semi-qualitative scoring system, using local and technical experts from each site. Based on the MCA method, stakeholders from both city councils and utilities and technical experts (engineers, economists, and natural scientists) contributed in participatory processes in order to assess the co-benefits [13,27].

In a first round of workshops, there was a selection of indicators from the co-benefits standardization framework in the C40 context by the London School of Economics [13]. From their extensive framework, a multidisciplinary group of experts selected indicators relevant to resilience and urban services. They were classified by economic, social, and environmental co-benefits (Table 1). In a second workshop, the experts working group were asked to score every measure under the selected indicators using a 0 to 10 scoring system. There was a discussion and voting exercise for each indicator and measure, and consensus was found through a session facilitator. Average values were estimated for each category of co-benefit, in order to include average values per category for each measure in the ranking exercise. One should note the caveat related to the quantification of co-benefits related to their extremely context-driven nature [7]. The sign and size of their impact on welfare depend heavily on local circumstances. Therefore, the co-benefits scores assessed for Barcelona and Bristol are unique for those case studies and measures, and new studies should internally assess their own potential co-benefits impacts.

At the end of the second workshop, the working group agreed to assign a weighted percentage to each variable (i.e., CEA (25%) and economic (25%), social (25%), and environmental (25%) co-benefits) in order to align the importance given to the variables with the overall aim of the decision-makers. To calculate the overall ranking, the results of each variable were normalized in order to use a common scale for the different data types on a common scale. Each data point was given a score between 0 and 1, based on its relative position within all values of the variable. This allowed to calculate the weighted scores of each measure per strategy and produce the rankings.

The aim of this phase was to offer coarse results that facilitate decision-makers to shortlist the measures that deserve further analysis. Thus, this stage is only relevant if there is a large number of measures (e.g., more than ten) to be screened. A low number of measures to assess can be considered affordable in terms of finding the resources to carry out a detailed damage assessment for all measures.

3.4. Stage 3: Detailed Assessment of Adaptation Scenarios

3.4.1. Stakeholders' Selection of Sets of Measures for Detailed Assessment

The two groups of stakeholders—city councils and utilities representatives and researchers from the RESCCUE project—held a third workshop with the aim of deciding which measures were worth an in-depth assessment. Using the preliminary ranking results, all members reached a consensus to select the measures, following the MCA method that relies on subjective expert judgement or stakeholder opinion [3]. In these workshops, it was decided that, in the detailed assessment measures, would be grouped by adaptation scenarios, as decision makers sought to understand the impact of implementing combination of measures, rather than individual measures results.

3.4.2. Technical Detailed Assessment: Risks Modeling

Different climate impacts have been assessed for the two case studies of the project RESCCUE. Although other hazards have been addressed, only floods, both pluvial and fluvial, and CSO spills are considered here for comparative purposes of the two cities. Damages due to floods and CSO spills can be many, and these can be classified as direct or indirect, and in turn tangible or intangible [28]. A variety of damages caused by these hazards have been assessed in these two cities, such as flood damages to properties and vehicles, traffic disruption or damages to the electrical grid. Table 2 summarizes the different hazards and risks assessed for each city.

Table 2. Summary of detailed assessments conducted in Barcelona and Bristol.

Case Study	Hazard	Risk Assessment
Barcelona	Pluvial floods	Damage impact to properties and vehicles Intangible damages to pedestrians and vehicles Street waste containers instabilities
	Water quality	CSO spills in bathing areas
Bristol	Pluvial and fluvial floods	Damage impacts to general infrastructures Traffic disruption Energy sector damage
	Water quality	CSO spills

Various methodologies to assess these risks have been employed in the two case studies. In the Barcelona case study, the following detailed assessments were carried out to evaluate the efficiency of the proposed adaptation scenarios within the pluvial flooding strategy:

- Economic damage assessment: a detailed estimation of pluvial flood damages to properties and vehicles was conducted. A 1D/2D hydrodynamic model of the entire city permitted estimation of the flow parameters (i.e., water depth and velocity) on the surface (i.e., city streets) for different return periods of rainfall (i.e., 1, 10, 50, 100, and 100) [29]. Therefore, this urban drainage model output has been used as an input for the flood damage models. The model proposed by Martínez-Gomariz et al. [30] was applied to estimate damages to vehicles, based on the depth-damage curves developed by the US Army Corps of Engineers [31]. A new model to estimate damage to properties in dense urban environments has been developed within the framework of the RESCCUE project. The model was constructed according to the suggestions of an insurance surveyor expert in flood damage appraisals [32]. This model relies on the accuracy of depth-damage curves that were constructed specifically for the city of Barcelona. These curves were developed based on damage claims of previous flood events together with the expert opinion when there was a lack of data [33]. Both models provided a total amount of direct economic damage for properties and vehicles that were aggregated per each return period. With this aggregation, the expected annual damage (EAD) [34] could be determined and used as a risk indicator. Therefore, the difference between the EAD before and after measures implementation provides an indicator of effectiveness of adaptation measures in terms of economic damages reduction.

- Intangible damage assessment: social impacts focused on safety of pedestrians and vehicles exposed to extreme pluvial flood events. Risk was defined as the combination of hazard and vulnerability by Turner et al. [35]. According to this approach, implemented in other previous studies [36,37], hazard assessment is based on the severity and frequency of the surface hydrodynamic variables and is classified based on specific flood hazard experimental criteria regarding pedestrian and vehicular stability in urban flooded areas [38–40]. On the other hand, flood vulnerability for pedestrian was assessed through several indicators like demographic density, percentage of people with critical age, and foreign inhabitants, and the number of critical infrastructures [36,37]. Vulnerability for vehicular circulation was assessed based on the vehicular daily intensity. Finally, a risk matrix combined hazard and vulnerability indexes to express flood risk.

- Stability of street waste containers: flood impacts over the waste collection system were assessed. A specific measure about fixation of waste containers was included in 2 scenarios [29]. The indicator displays results in terms of percentage of reduction of unstable containers.

For the CSO spills reduction strategy, two detailed assessments were carried out to understand the impacts of the adaptation scenarios:

- Direct impacts on human health: insufficient bathing quality time (in hours) during the bathing season was carried, modeled for the coastal areas of Barcelona where CSO events occur [22]. Specifically, pollutant hazard was assessed through a coupled urban drainage and seawater quality model that was developed, calibrated, and validated based on local observations. The study quantified the health hazard of bathing waters affected by CSOs based on two novel indicators: the mean duration of insufficient bathing water quality (1) per bathing season and (2) after single CSO/rain events. More information about the proposed technique to assess human health hazard produced by CSO could be found in [41].

- Indirect impacts on business: potential economic losses as a consequence of closing related businesses (water sports, restaurants by the seaside, and fishing activities) due to bathing waters contamination [42]. Estimations were obtained using revenues of the affected sectors and neighborhoods where CSO spills occur. The relative damage proportion to the total revenue was assigned using the results of a survey and questionnaire carried out to citizens and business owners of the area. Results were expressed in monetary terms.

In the Bristol case study, the evaluation of adaptation measures for pluvial and fluvial flooding was carried out under the following methodology:

- Damage assessment: flood models were developed evaluating property damage through application of the damage assessment tool developed in EU project CORFU (Collaborative Research on Flood Resilience in Urban Areas) [43] and designing intervention scenarios representing property level protection. Flood models included fluvial and pluvial events at a range of return periods (T20 to T1000 and T10 to T100, respectively). Analysis of peak flood depth mapping for a baseline representing current day conditions and a climate change scenario derived from UKCP09 [19], assuming BAU emissions up to the year 2115 was carried out.

A damage cost assessment per scenario was generated by combining flood depths with building classifications [44] and depth-damage curves [45]. Intervention scenarios designed to protect properties within the 20-year flood outlines, either for all residential buildings in the three worst affected areas (zonal target) or for the same number of worst-impacted properties distributed across the study area (individual target), were compared. Interventions were represented by adapting the damage curve to prevent damages below a 600 mm water-depth threshold, reflective of a likely effective level of protection [46].

- Traffic impact assessment: The flood maps that were used for the damage assessment were loosely coupled with micro-simulation traffic model SUMO (Simulation of urban Mobility) [47]. A modified approach outlined by [48] was applied to simulate the effects of flooding on traffic, whereby the speed limits on specific sections of roads were reduced or road sections were close temporarily during a thirty-minute flood event depending on the maximum-recorded flood-depths. For the adaptation scenario, the bridges that cross the central river sections within the city were assumed to be locally protected from flooding in order to observe how—by keeping these specific roads open and unflooded—the flows of traffic within the network could be greatly improved. The impact costs values were derived via the use of speed versus cost table from the multicolored manual (MCM) [45].
- Electricity system: within the Bristol case study, the effect of localized improvements to infrastructure protection were analyzed [49]. Detailed assessment results were achieved by carrying out a sensitivity analysis that altered the fragility (depth vs. infrastructure failure) curves of the electrical substations.

These methodologies have been applied for the BAU scenario, in which no adaptation is considered and for a scenario in which measures are implemented. For instance, once a certain flood is modeled (i.e., 10-year return period design storm), the risk for pedestrians, based on a person's stability threshold, is assessed first without measures (BAU). Afterwards, a measure is implemented in the flood model, which yields a lower flood, and thus the risk is re-assessed. The comparison of both model results yields the variation in the high-risk area for pedestrians, considered an effectiveness indicator of risk reduction measures. Similar procedures have been conducted for the hazards and risks assessments listed in Table 2 for the two cities.

3.4.3. Economic Assessment: Environmental Cost-Benefit Analysis

Cost-benefit analysis (CBA) was the method selected for the final comparison of the potential adaptation scenarios. This method allowed to integrate all previous assessment results involving direct or indirect changes in environmental, economic, and societal variables, translated into monetary terms [6]. It provided a comparative overview of the potential effects of the different scenarios in terms understood by all stakeholders.

The net sum of all relevant positive and negative outcomes of a scenario is known as the total economic value (TEV) [6] and is typically divided between use and non-use (passive) value [6]. In this study, the focus was set on use values, which relate to the actual use of the good in question, as for example, the use of green areas in cities for recreation [6]. Whereas, non-use values are those related to their existence, altruistic or bequest value [6]. When possible, market prices were used to value the changes provided by the measures' implementation (or related goods or services), while benefit transfer was used when direct values were not available. Benefit transfer relies on unit values obtained in previous studies to estimate the value in the study site, adapting them to the characteristics of the new site [50]. Ecosystem services are understood as the direct and indirect contributions of ecosystems to humans [51]. They play an important role, since their principles have been accounted for in the sustainable urban drainage (SUDS) measures [52], in an attempt to demonstrate the benefits of "greening" urban areas. This followed the trend in the urban planning sector of putting emphasis on accounting for the co-benefits provided by nature-based solutions [52–56].

The time horizon selected for the analysis was 2020–2100, aligned with the timespan of the general assessment for the project RESCCUE. Costs included all required investment efforts of each city council, as well as the operating costs for the lifespan of the analysis. Benefits came from two sources. First, the avoided costs were estimated through the difference between the estimated economic damage assessment in the BAU scenario and in each of the alternative scenarios [11]. In addition, monetary values were included to account for the benefits of the improvement in the provision of ecosystem services of SUDS measures. They were adapted using the benefit transfer method [57]. These benefits include the reduction of the heat island effect, which implies reductions on electricity consumption [58,59]; air quality improvements [58]; habitat creation and aesthetic value, related to the increase in willingness to pay for properties with surrounding green areas [59]. Net benefits (Equation (4)) aggregate benefits and costs to determine the TEV or complete impact of the scenarios (*j*) proposed:

$$\text{Net benefit}_j = \text{Benefits}_j - \text{Costs}_j. \tag{4}$$

The net benefits expected through the lifetime of the project (years from t = 1 to T = 80) were discounted to reflect future values in present terms, obtaining the net present value (NPV) (Equation (5)), using the same discount factor as the CEA. Similarly, the results were annualized for comparative reasons, thus presented using the annual equivalent present value (AEPV) (Equations (6) and (7)), for each scenario *j*:

$$NPV_j = \sum_{t=1}^{T} \frac{Benefits_{j,t} - Costs_{j,t}}{(1+i)^t}, \tag{5}$$

$$AEPV_j = \frac{NPV_j}{A_{(t,i)}}, \tag{6}$$

$$A_{(t,i)} = \frac{1 - (1+i)^{-T}}{i}, \tag{7}$$

where $A_{(t,i)}$ is the annuity factor of the present value.

3.5. Stage 4: Final Ranking

The multiple results obtained through the detailed assessment allowed to rank adaptation scenarios under various criteria. In the Barcelona case study, scenarios were prioritized by: (1) area of risk reduction, (2) by avoided damage, (3) by costs, and (4) by net benefit criteria. In the Bristol case study, scenarios were ranked based on total damage.

4. Results: Application in Two European Cities

The application of the methodology is presented in stages, comparing the two case studies in order to display the different approaches taken.

4.1. Problem Characterization in Practice

The outcome of the first round of workshops held in each city was the definition of the problem characterization (Table 3). The second workshop identified an extensive list of measures and the adaptation strategies to be assessed (Table 4).

Table 3. Description of hazards selected as outcome of the problem characterization.

Case Study	Extreme Event under Assessment	Relevance
Barcelona	Pluvial flooding CSO spills	Climate forecasts predict increases in the frequency of extreme weather events: Increase of 20% for 100 years return period rain (T100) is expected for the period 2041–2070, + 40% expected by the end of the century [18].
Bristol	Fluvial flooding Pluvial flooding CSO spills	Climate forecasts indicate that fluvial and pluvial flood events are likely to worsen in response to an increasing likelihood of extreme rainfall. For example, between 2041 and 2070, the 1 year extreme daily rainfall is predicted to increase from 33 to 58 mm (UKCP09 median value) [19]. The fluvial system is also particularly vulnerable to tidal interference increasing river levels in the city, with UKCP18 projections indicating a 10 cm sea level rise across the 2041 to 2070 horizon, leading to significant areas of the city facing a future threat [60].

Table 4. Description of selected strategies and their measures to be prioritized in the two case studies.

Case Study	Selected Strategies and Their Measures
Barcelona	**Pluvial flooding** 1. Improvements of surface drainage system 2. Increase of sewer system capacity—New pipes (I) 3. Increase of sewer system capacity—New detention tanks (II) 4. SUDS scheme (increased area of green roofs, infiltration trenches, and detention basins) 5. Early Warning System 6. Self-healing algorithm implemented in the electrical distribution grid 7. Ensure the stability of waste containers **Combined sewer overflows (CSO) spills** 1. SUDS scheme 2. Early Warning System (EWS) 3. Detention tanks for CSO prevention 4. Improvements of the capacity of sewer interceptor and WWTP 5. End of pipe CSO treatment

Table 4. *Cont.*

Case Study	Selected Strategies and Their Measures
	Pluvial flooding
	1. Property level protection of crucial infrastructure
	2. Demountable flood protection barriers
	3. Identify high risk areas by carrying out studies of flood modeling analysis
	Fluvial flooding
	1. Flood proof crucial infrastructures
	2. Build riverside flood defense walls
	3. Demountable flood protection barriers
	4. Identify high risk areas by conducting studies involving flood modeling analysis
Bristol	CSO events (in the Ashton Vale area) [1]
	1. Inlet diameters increase
	2. Disconnecting paved surfaces from the combined system
	3. Full surface water separation
	4. SUDS scheme (swales, filter trenches, permeable paving, detention basins, mixed schemes)
	5. Raising curb heights
	6. Increasing surface water sewer system capacity
	7. Tide isolation of drainage systems
	8. Improvements to watercourse capacities

[1] CSO events strategy is presented in Appendix A, because it was assessed in only one area of the city and thus not comparable with the other two strategies embracing the entire city.

The longlist of adaptation actions related to urban services of Barcelona contained 4 strategies and 27 measures, collected from the local climate action plan and workshops. Stakeholders of the Barcelona case study agreed to focus the efforts on improving the resilience of the city against pluvial flooding and CSO spills events. Therefore, the 2 strategies under assessment are related to those hazards and contained structural, nature-based solutions and technological measures aimed to reduce damages related to those hazards.

In the Bristol case study, 3 strategies and 14 measures related to urban-services were identified initially during the workshops and from local action plans. Stakeholders decided to focus on the improvement of Bristol's resilience against pluvial and fluvial flooding events and CSO spills.

4.2. Preliminary Assessment of Adaptation Measures in Practice

After defining the strategies and their respective measures, CEA and co-benefits were estimated, following the methodology described above. The variables used as an input in the CEA were the results of the hydrodynamic models and annualized costs estimations, whereas the co-benefits scores were obtained by participatory processes of multidisciplinary experts in Barcelona and Bristol. The ranking results of the preliminary assessment in Barcelona and Bristol are presented in Tables 5 and 6 respectively. Weights given to each indicator were determined under consensus during the second workshop, following stakeholders' judgement.

Table 5. Preliminary ranking results for the adaptation measures included in the flood and CSO strategies assessed for the Barcelona case study.

	Weights Given	25%	25%	25%	25%
Rank	Pluvial Flooding Strategy Measures	CEA (€/h)	Economic	Social	Environmental
1	SUDS (green roofs, infiltration trenches, detention basins for rural catchments)	16,466,678	56%	69%	85%
2	Ensure the stability of waste containers	24,010	13%	13%	5%
3	Early Warning System	29,478	13%	13%	5%
4	Increase of sewer system capacity—new pipes (I)	5,597,181	13%	60%	35%
5	Increase of sewer system capacity—new detention tanks (II)	18,687,976	1%	71%	34%
6	Improvements of surface drainage system	4,072,897	11%	41%	9%
7	Self-healing algorithm in the electrical distribution grid	127,304	10%	0%	0%
Rank	CSO Strategy Measures	CEA (€/h)	Economic	Social	Environmental
1	SUDS	584,465	56%	69%	85%
2	Early Warning System	1528	13%	13%	5%
3	Detention tanks	452,461	1%	71%	34%
4	End of pipe CSO treatment	4,687,158	9%	29%	8%
5	Improvements of the capacity of sewer interceptor	59,500	1%	17%	8%

Table 6. Preliminary ranking results for the adaptation measures included in the pluvial and fluvial strategies within the central area of Bristol.

	Weights Given	25%	25%	25%	25%
Rank	Pluvial Flood Strategy Measures	CEA (€/h)	Economic	Social	Environmental
1	Demountable flood protection barriers	7243	39%	81%	39%
2	Identify high risk areas (flood modeling analysis studies)	60,790	23%	57%	14%
3	Flood proof crucial infrastructures	134,608	7%	3%	14%
Rank	Fluvial/tidal Flood Strategy Measures	CEA (€/h)	Economic	Social	Environmental
1	Demountable flood protection barriers	8450	39%	81%	39%
2	Build riverside flood defense walls	3,749,280	64%	87%	41%
3	Identify high risk areas by conducting studies involving flood modeling analysis	56,509	23%	57%	14%
4	Flood proof crucial infrastructure	177,683	7%	3%	14%

Table 5 presents SUDS measures as the preferred option, followed by structural measures for both strategies. The equal distribution of weights between indicators implies that those with larger indirect benefits are prioritized over the ones that are just more cost-effective. The reinforcement of the stability of waste containers is the second option for the floods strategy, and the early warning system for the CSO strategy.

Table 6 shows that demountable flood protection barriers are the most preferred measure for protection against pluvial flooding and for fluvial and tidal flooding, combined in this instance with riverside defense walls that offer protection up to a 1 in 200-year event (ranked second). For the Ashton Vale region (Appendix A), a wider range of measures were selected to offer improved protection against pluvial flooding and reduce CSO spills events.

4.3. Detailed Assessment in Practice

4.3.1. Selection of Adaptation Scenarios

Results from the first ranking exercise gave stakeholders insights on the advantages and disadvantages of the measures proposed. This facilitated the discussion regarding the selection of the most relevant measures to include in the adaptation scenarios, aimed to answer different policy questions. For example, in the case of Barcelona's strategy for pluvial flood impacts reduction, there was a clear interest from policy-makers in the co-benefits provided the SUDS measures, while technical experts highlighted the potential reduction of risks and recovery time offered by the structural measures, although the costs of the latter were high and therefore difficult to meet. Therefore, they decided to assess the potential impact of implementation of SUDS across the entire city, while the structural measures were divided by zones. Adaptation scenarios were created to obtain results for those requests (Table 7). Regarding the stability of waste containers, detailed assessment was not found relevant, as the preliminary assessment already displayed a low CEA result, which was enough to be included in the new Climate Action Plan for Barcelona.

Table 7. Description of adaptation scenarios selected for Barcelona for the two strategies assessed.

Strategy Name	Adaptation Scenarios (AS)
Pluvial flooding impacts reduction	1. Flood_AS1. SUDS emplaced through the entire city 2. Flood_AS2. SUDS and structural measures (SM) through the entire city 3. Flood_AS3. SUDS (entire city) and SM within Zone 1 (Z1) 4. Flood_AS4. SUDS (entire city) and SM within Zone 2 (Z2) 5. Flood_AS5. SUDS (entire city) and SM within Zone 3 (Z3) 6. Flood_AS6. SUDS (entire city) and SM within Zone 4 (Z4) 7. Flood_AS7. SUDS (entire city) and SM within Zone 5 (Z5) 8. Flood_AS8. SUDS (entire city) and SM within Zone 6 (Z6)
CSO spills reduction	1. CSO_AS1. SUDS emplaced through the entire city 2. CSO_AS2. SUDS (entire city) and detention tanks

In the Bristol case study, the needs of stakeholders guided the way adaptation strategies were categorized (Table 8). City council representatives were most concerned with fluvial flooding linked to the River Avon, its associated watercourses and tidal interactions of the Severn Estuary, alongside pluvial flooding originating from extreme rainfall across the urban area. The interest was again on gaining knowledge on the zonal assessment of the city, although structural measures were identified as the most promising interventions. Due to restrictions in budgets, the Bristol case study chose to investigate only the use of property level protection, since it allowed adaptation measure assessment using previous flood modeling.

4.3.2. Detailed Assessment Results

In the Barcelona case study, the technical and economic assessments were carried out for the proposed adaptation scenarios related to flooding and CSO reduction strategies. The economic damage assessment provided the expected annual damages (EAD) for each scenario. Comparing the scenarios against the BAU, the avoided damage was estimated for each adaptation scenario. These were accounted as benefits in the CBA. Ecosystem services were included as benefits provided by the SUDS measures thus present in all adaptation scenarios. Costs were estimated for the adaptation scenarios, adjusting the previous estimates for individual measures of the CEA to the new adaptation scenarios. Net benefits were obtained in the CBA for each scenario.

Furthermore, for the flooding reduction strategy, estimates from the intangible damage assessment gave the percentage of high-risk area reduction for pedestrians and vehicles for each scenario.

In the Bristol case study, a detailed analysis of the proposed adaptation scenarios was undertaken using a different approach. Stakeholders were more interested in the potential impacts of different intensities events for two scenarios and compare them to the BAU. Measures were ranked based on the total damage costs expected during a range of extreme flood events in the city. The assessment was calculated by developing a baseline of expected flood damages during a flood event in the future, and comparing them to the expected flood damages with property level protection applied to targeting buildings or strategic zones.

Table 8. Adaptation scenarios for each strategy in Bristol case study.

Strategy Name	Adaptation Scenarios (AS)
Pluvial flooding impacts reduction	1. Pluvial_AS1. BAU with climate change 2. Pluvial_AS2. BAU CC with zonally targeted interventions 3. Pluvial_AS3. BAU CC with individually targeted interventions
Fluvial flooding impacts reduction	1. Fluvial_AS1. BAU with climate change 2. Fluvial_AS2. BAU CC with zonally targeted interventions 3. Fluvial_AS3. BAU CC with individually targeted interventions

4.4. Final Ranking: Results of the Detailed Assessment

The results from the detailed assessment provided information to generate four prioritization rankings relevant for stakeholders (Tables 9 and 10), three of them contained monetary criteria—avoided damage, net benefits and costs, and one was in percentage terms, representing the high-risk area reduction.

Table 9. Prioritization results under the 3 monetary criteria for pluvial flooding in Barcelona.

Rank	Avoided Damage		
	ID	**Scenario**	**AEPV (Million €/Year)**
1	Flood_AS2	SUDS + Structural measures entire city (SM)	49.0
2	Flood_AS7	SUDS + SM in Z5	43.1
3	Flood_AS4	SUDS + SM in Z2	41.9
4	Flood_AS5	SUDS + SM in Z3	41.7
5	Flood_AS8	SUDS + SM in Z6	41.2
6	Flood_AS6	SUDS + SM in Z4	40.2
7	Flood_AS3	SUDS + SM in Z1	40.0
8	Flood_AS1	SUDS entire city	23.7
Rank	**Costs**		
	ID	**Scenario**	**AEPV (Million €/Year)**
1	Flood_AS1	SUDS entire city	9.9
2	Flood_AS6	SUDS + SM in Z4	10.9
3	Flood_AS8	SUDS + SM in Z6	11.0
4	Flood_AS5	SUDS + SM in Z3	11.6
5	Flood_AS7	SUDS + SM in Z5	12.2
6	Flood_AS4	SUDS + SM in Z2	12.3
7	Flood_AS3	SUDS + SM in Z1	15.1
8	Flood_AS2	SUDS + SM entire city	22.4

Table 9. *Cont.*

Rank	Net Benefits		
	ID	**Scenario**	**AEPV (Million €/Year)**
1	Flood_AS7	SUDS + SM in Z5	39.6
2	Flood_AS8	SUDS + SM in Z6	38.9
3	Flood_AS5	SUDS + SM in Z3	38.7
4	Flood_AS4	SUDS + SM in Z2	38.3
5	Flood_AS6	SUDS + SM in Z4	38.1
6	Flood_AS2	SUDS + SM entire city	35.3
7	Flood_AS3	SUDS + SM in Z1	33.6
8	Flood_AS1	SUDS entire city	22.5

Table 10. Prioritization results under high-risk area reduction criterion for pluvial flooding in Barcelona.

Rank	Risk Reduction (T10)					
	ID	**Scenario**	**Risk for Pedestrians**	**ID**	**Scenario**	**Risk for Vehicles**
1	Flood_AS2	SUDS + SM entire city	99%	AS2	Str. BCN + SUDS	99%
2	Flood_AS3	SUDS + SM in Z1	79%	AS7	Z5 + SUDS	90.4%
3	Flood_AS4	SUDS + SM in Z2	79%	AS4	Z2 + SUDS	90.2%
4	Flood_AS7	SUDS + SM in Z5	77%	AS3	Z1 + SUDS	87.2%
5	Flood_AS5	SUDS + SM in Z3	76%	AS5	Z3 + SUDS	86.6%
6	Flood_AS6	SUDS + SM in Z4	74%	AS8	Z6 + SUDS	85.4%
7	Flood_AS8	SUDS + SM in Z6	65%	AS6	Z4 + SUDS	84.1%
8	Flood_AS1	SUDS entire city	34%	AS1	SUDS	45%

The aim was to provide as much information as possible to solve the concerns of decision makers regarding the different aspects of the policy making process, i.e., budgetary, welfare, and risks. The BAU scenario is omitted from the ranking as the objective of the city council is to act against floods. For the avoided damage and risk reduction criteria, the preferred scenario is to implement SUDS and structural measures in the entire city of Barcelona; whereas for the net benefit criteria, the scenario with SUDS and structural measures in zone 5 ranks first. In the case of cost criteria, the SUDS measures are the preferred option. The zone 5 scenario also ranks high for the rest of the criteria. This is expected to support stakeholders by facilitating the decision-making process of implementing adaptation measures. More detailed results of the risk assessment can be found in Appendix B.

The prioritization exercise for the CSO strategy (Table 11) was carried out using 3 monetary criteria (i.e., avoided damage, costs, and net benefits), using the outputs from the CBA that used the damages analysis results.

Table 11. Results of ranking of adaptation scenarios selected for CSO under the three selected criteria in Barcelona.

Rank	Avoided Damage	
	Scenario	AEPV (€/year)
1	CSO_AS2. SUDS and detention tanks	18,352,567
2	CSO_AS1. SUDS (entire city)	10,041,691
Rank	**Costs**	
	Scenario	AEPV (€/year)
1	CSO_AS1. SUDS (entire city)	9,918,794
2	CSO_AS2. SUDS and detention tanks	16,145,628
Rank	**Net Benefits**	
	Scenario	AEPV (€/year)
1	CSO_AS2. SUDS and detention tanks	10,850,498
2	CSO_AS1. SUDS (entire city)	9,078,777

These results helped decision-makers to understand the most advantageous adaptation measures given their selected criteria, which might be subject to maximizing net welfare gains, or minimize costs or damages. If the avoided damage and net benefits criteria are considered, the scenario of combined SUDS and tanks was preferred, whereas if the criterion follows cost-efficiency, the SUDS scenario ranked first.

The results of the detailed assessment carried out in the Bristol case study provided a rank based on the total damage costs expected during a range of extreme flood events in the city for pluvial and tidal/fluvial flood scenarios (Tables 12 and 13).

Table 12. Estimated total flood damages resulting from extreme pluvial flood events to all building classes for different return periods (T*years*) (£) in Bristol.

Rank	Scenarios	T10	T20	T30	T100
1	BAU CC with individually targeted interventions	31,880,000	45,322,000	-	96,796,000
2	BAU CC with zonally targeted interventions	35,218,000	48,312,000	-	98,804,000
3	Business as usual, considering climate change impact (BAU-CC)	36,692,000	50,088,000	-	100,757,000

Table 13. Estimated total flood damages resulting from extreme tidal/fluvial flood events to all building classes for different return periods (T*years*) (£) in Bristol.

Rank	Scenarios	T20	T100	T200	T1000
1	BAU CC with individually targeted interventions	-	155,695,000	482,760,000	537,228,000
2	BAU CC with zonally targeted interventions	-	156,622,000	482,738,000	537,258,000
3	BAU-CC	-	160,006,000	483,009,000	537,446,000

The analysis identified that the most effective intervention scenario was individually targeted property level protection to reduce the impact of both pluvial and fluvial flooding hazards. The results of the total damage assessment, of the previous CEA and co-benefits results were detailed enough for stakeholders, and therefore a complete CBA was considered not necessary for Bristol's decision-makers. Instead, stakeholders were more interested in understanding the changes pertaining to the traffic and energy sectors when the selected scenario was applied. Therefore, in the traffic sector, the effectiveness of the adaptation scenario was evaluated by comparing the reduction in recovery time, vehicular cost estimations with respect to their average speed, and PMx emissions, under flooded conditions compared to dry weather conditions [61]. The energy sector's detailed assessment compared the BAU scenario to the selected scenario under different levels of fragility curves of the electrical substations, to provide further information to decision makers on the potential impacts reduction generated by the measures proposed [49].

5. Discussion

The present methodology is built upon existing knowledge of applied quantitative and qualitative methodologies related to urban climate change adaptation and resilience. The flexibility of the methodology was a key feature, and it has been achieved by introducing iterative stakeholder engagement in each stage. The fact that multidisciplinary research groups worked along with city councils and other stakeholders, gave the methodology a practical focus that included tools to face the most common barriers and opportunities met by most climate change adaptation working groups, such as lack of data sources. The methodology offers a modular system that is able to adapt to the realities of the different case studies, allowing to select the required steps for each case. Another advantage is the use of existing and proved methodologies, which are common to stakeholders, with known strengths and weaknesses.

The potential to combine technical, economic, social, and environmental results is a powerful policy instrument, as their decision-making process must consider all those aspects that affect the present and future of their citizens' welfare [2]. It is particularly relevant for those cases where there are many measures that need to be screened and not enough time or capital resources to do so with the necessary detail to support policy decision-making. In that sense, co-benefits assessment in climate adaptation projects can contribute substantially to policy decision-making [62]. First, it is suggested that most co-benefits display positive welfare outcomes in the short term—for example, air pollution improvement, which is the concern of policy makers; whereas direct climate change policy benefits—such as heavy flood damage reduction—may only be perceived in the long term. Second, the co-benefits are usually enjoyed at a local or regional scale and thus are closer to the citizens bearing the costs, they provide incentives for decision makers to act [62].

The methodology enabled the views and expertise of multiple stakeholders to be included through workshops. It has the benefit of fewer requirements from the CEA during the first stage, although the outcomes lack sound conclusions. The uncertainty of results decreases with the participatory process of multidisciplinary experts and further detailed assessment to the selected measures.

The involvement of policy-makers from the beginning of the analysis was regarded as positive, compared to traditional approach of just presenting final results through an assessment report. In addition, such an engagement captured local expertise, enabling intervention development to progress with the understanding of specific high-risk areas and ongoing organizational initiatives. The stakeholder engagement was also able to capture mixed priorities from different groups and integrate this understanding within the context of a wider RESCCUE analysis, drawing together a network of interconnected infrastructure, including the energy, transport, and waste-water sectors. If consultation is transparent and well-structured, it can give valuable insights of local knowledge [27].

Although the focus of the present work is on urban infrastructure and services, it does not neglect social-oriented policies. However, it recognizes that these "soft" policies, such as encouraging residents to change public behaviors [63], belong to a different field of study and policy-making process, out of the scope of this work. However, the final goal of both types of policies is similar: to increase wellbeing, either from a psychosocial perspective or by reducing physical risks and damages.

One of the main barriers for urban adaptation assessment is the general lack of standardized data and/or time resources. Adapting the approach to accommodate varying data availability was an additional challenge. Furthermore, researchers found difficult to align work across stakeholders and case studies, due to existing differences in policy approaches, priorities, and types of stakeholders and their involvement. Another drawback of the methodology is that the co-benefits are city-specific and subject to the expert criteria. Analyses based on experiences, data, and perspectives from stakeholders are based on historical hazard impacts, rather than providing a true representation of future risks. This potential bias can be minimized if robust climate change forecasting and risk and damage modeling is introduced in the assessment. Therefore, although the hydrological model (suggested in the methodology to assess the recovery time) increases the technical expertise demand in the preliminary stage, it is essential to secure informed decision-making throughout the process. In addition, it is an established technique, and the exercise is not as complex as the detailed damage and risk assessment proposed in the third stage, which is highly recommended for final decision-making.

A limitation of an approach grounded in stakeholder engagement is the need to manage conflicting priorities and adjust messages to an audience with varying levels of expertise. This is particularly relevant when managing organizational and community stakeholders, where costs and benefits of intervention actions may not be aligned. Unconscious bias, such as exposure to specific hazards or experience may also influence engagement, leading to a focus on specific issues at the expense of a broader analysis. These issues have been addressed in RESCCUE through engaging a diverse group of technical stakeholders who shared their expertise and results to multidisciplinary audiences.

Similarly, highly spatial hazards such as pluvial flooding can be very dependent on unknown or unpredictable factors which may not have occurred during past events; for example, aging infrastructure may block previously functioning drainage features and significantly alter the response of urban catchments rainfall. These issues cannot be fully mitigated, and for that reason, it was crucial that the methodology recognized those limits by including historic and experiential data. In fact, the consequences of a changing climate and the continuing paving of urban areas (increased permeability) lessen the validity of decision based on historical facts.

Future recommendations to expand the replicability potential is to create a web-based assessment tool to facilitate the implementation in other urban areas.

6. Conclusions

The methodology recognizes that there are several viable strategies with different contributions to society and the aim of researchers is not selecting the best for each case, but to provide results from different perspectives to support city planners to take better-informed decisions.

Combined assessments of technical, socio-economic, and environmental aspects provide added value to policy makers, compared to assessment results addressing only one feature. Their role is to consider multiple aspects that may affect the citizens they represent when selecting a policy or project.

The relevance of scientific research and multidisciplinary technical assessment is only revealed when it is coupled and in harmony with the needs of decision-makers and citizens. Certainly, that was the focus of the present methodology and it was highly valued during its application. Therefore, the results presented here are expected to provide relevant tools for stakeholders to take informed decisions regarding adaptation to climate change in urban areas.

Author Contributions: Conceptualization, M.G.-H. and E.M.-G.; methodology, M.G.-H., E.M.-G. and M.T.-R.; validation, M.T.-R., J.W. and B.E.; formal analysis, M.G.-H., E.M.-G., B.R. and L.L. (Barcelona case study); B.E. and J.W. (Bristol case study); investigation, M.G.-H.; data curation, M.G.H.; Writing—Original draft preparation, M.G.-H.; Writing—Review and editing, M.G.-H., E.M-G and M.T.-R.; supervision, E.M.-G.; project administration, E.M.-G.; funding acquisition, E.M.-G. and M.G.-H. All authors have read and agreed to the published version of the manuscript.

Funding: This research, under the RESCCUE Project, was funded by the European Commission Horizon2020 funding program. Grant Agreement No. 700174.

Acknowledgments: The authors are grateful to BCASA (from Barcelona City Council) and Bristol City Council for their contributions and insights to implement the methodology and fit it into their new Drainage Master Plans. Thank you also to all partners of the Project RESCCUE for their work during the 4 years of the project, which made this work possible.

Appendix A

Table A1. Ranking of pluvial flood related measures based on effectiveness and co-benefits within the Ashton Vale area of Bristol with respect to flooding.

	Proposed Weights	**25%**	**25%**	**25%**	**25%**
Rank	**CSO Proposed Measures**	**CEA (€/h)**	**Economic**	**Social**	**Environmental**
1	Inlet increase	8750	36%	60%	35%
2	Disconnecting paved surfaces from (combined) sewer system	6332	33%	10%	36%
3	Swale	9629	26%	40%	55%
4	Surface water separation	4197	6%	13%	23%
5	Filter trenches	9629	26%	40%	41%
6	Permeable paving	9629	23%	43%	40%

Table A1. *Cont.*

	Proposed Weights	25%	25%	25%	25%
Rank	CSO Proposed Measures	CEA (€/h)	Economic	Social	Environmental
7	Raise curb height	1319	4%	4%	0%
8	Increase of combined sewer system capacity	32,013	1%	71%	34%
9	Sustainable Urban Drainage systems (SUDS)	9629	14%	20%	40%
10	Detention basin	9629	6%	13%	9%
11	Increase surface water sewer system capacity	17,097	6%	0%	0%
12	Improvements to drainage system (watercourse)—isolation from high tide conditions in River Avon	334,056	10%	9%	5%
13	Improvements to drainage system (watercourse)—capacity	158,973	6%	0%	0%

Appendix B

This appendix is aimed to present more detailed results of the assessment carried out for the assessment of pedestrian and vehicles under different return periods in Barcelona. Table A2 displays the risk assessment results for the 5 different return periods assessed for all scenarios in the Barcelona case study. Table A3 displays results of expected annual damage (EAD) in monetary terms for the same cases. Whereas, Figures A1 and A2 represent the assessment of the stability of waste containers carried out.

Table A2. Results of pluvial flood risk reduction assessment for pedestrian and vehicles under different return periods in Barcelona.

	Adaptation Scenario	Risk Reduction for Pedestrians					Risk Reduction for Vehicles				
		T1	T10	T50	T100	T500	T1	T10	T50	T100	T500
1	SUDS Entire BCN	0%	34%	20%	17%	14%	0%	45%	25%	21%	14%
2	SUDS + Structural Measures Entire BCN	0%	99%	87%	76%	59%	0%	99%	94%	87%	66%
3	SUDS + Structural Measures Zone 1	0%	79%	65%	58%	48%	0%	87%	74%	65%	49%
4	SUDS + Structural Measures Zone 2	0%	79%	63%	56%	47%	0%	90%	76%	67%	51%
5	SUDS + Structural Measures Zone 3	0%	76%	63%	55%	45%	0%	87%	75%	66%	50%
6	SUDS + Structural Measures Zone 4	0%	74%	59%	53%	47%	0%	84%	69%	59%	48%
7	SUDS + Structural Measures Zone 5	0%	77%	62%	55%	48%	0%	90%	71%	61%	48%
8	SUDS + Structural Measures Zone 6	0%	65%	60%	55%	45%	0%	85%	68%	60%	45%

Table A3. Expected annual damage (EAD) from pluvial floods for properties and vehicles for all adaptation scenarios modeled in the Barcelona case study.

	Scenario	EAD to Properties	EAD to Vehicles	EAD Variation
0	Baseline (current rainfall)	31,150,112 €	3,275,292.05 €	-
1	BAU [1] (future rainfall)	44,494,008 €	4,482,544.03 €	+42% (vs. Baseline)
2	SUDS Entire city [2]	23,680,855 €	2,747,960.30 €	−46%
3	SUDS + Structural Measures (SM) Entire city	2,235,685 €	341,720.20 €	−95%
4	SUDS + SM Zone 1	10,270,756 €	825,019.44 €	−77%
5	SUDS + SM Zone 2	8,408,240 €	927,384.48 €	−81%
6	SUDS + SM Zone 3	8,509,119 €	970,750.84 €	−81%
7	SUDS + SM Zone 4	9,821,973 €	999,403.93 €	−78%
8	SUDS + SM Zone 5	7,403,553 €	771,070.23 €	−83%
9	SUDS + SM Zone 6	8,888,035 €	1,008,856.05 €	−80%

[1] BAU is compared to baseline and all adaptation scenarios are compared to BAU; [2] SUDS are assessed for the entire city of Barcelona in all scenarios.

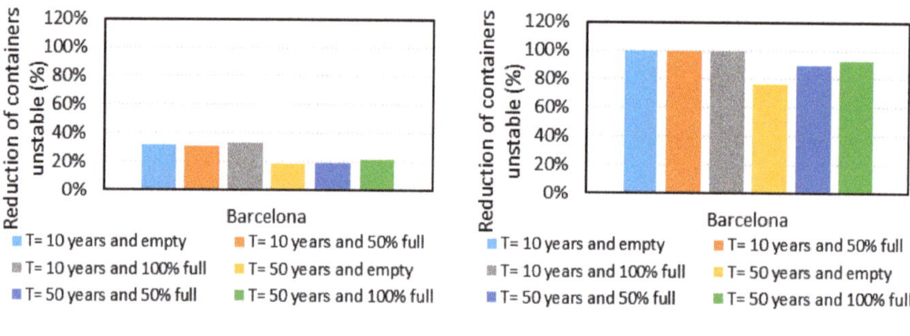

(**a**) Without fixation pieces (Flood_AS1)　　　　(**b**) With fixation pieces (Flood_AS10)

Figure A1. Reduction of the number containers potentially unstable at a city scale once applied (**a**) Flood_AS1 and (**b**) Flood_AS10.

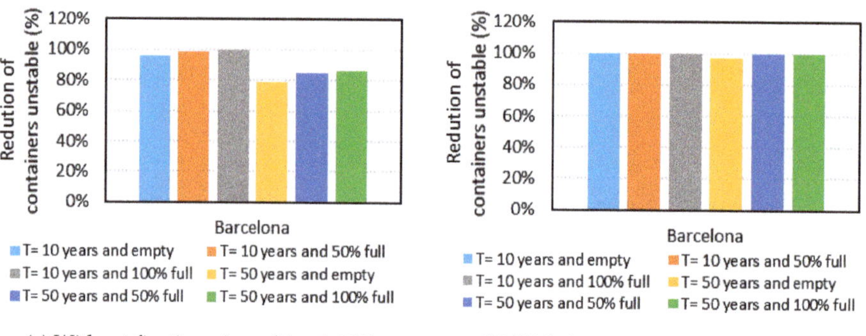

(**a**) Without fixation pieces (Flood_AS2)　　　　(**b**) With fixation pieces (Flood_AS11)

Figure A2. Reduction of the number containers potentially unstable at a city scale once applied (**a**) Flood_AS2 and (**b**) Flood_AS11.

References

1. Reckien, D.; Salvia, M.; Heidrich, O.; Church, J.M.; Pietrapertosa, F.; De Gregorio-Hurtado, S.; D'Alonzo, V.; Foley, A.; Simoes, S.G.; Krkoška Lorencová, E.; et al. How are cities planning to respond to climate change? Assessment of local climate plans from 885 cities in the EU-28. *J. Clean. Prod.* **2018**, *191*, 207–219. [CrossRef]
2. Jo, Y.C.H. Climate change adaptation measures for priority ranking. *Methodol. Adapt. Meas.* **2013**, *12*, 23–44.
3. Gianoli, A.; Grafakos, S.; Olivotto, V.; Haqua, A.N. Climate Change Adaptation Projects: Integrating Prioritization and Evaluation. In *The Sixth Urban Research and Knowledge Symposium*; The World Bank: Barcelona, Spain, 2012. Available online: https://www.researchgate.net/profile/Anika_Haque/publication/322363707_CLIMATE_CHANGE_ADAPTATION_PROJECTS_INTEGRATING_PRIORITIZATION_AND_EVALUATION/links/5a560702aca272bb6962d410/CLIMATE-CHANGE-ADAPTATION-PROJECTS-INTEGRATING-PRIORITIZATION-AND-EVALUATION.pdf. (accessed on 25 February 2020).
4. Velasco, M.; Russo, B.; Martínez, M.; Malgrat, P.; Monjo, R.; Djordjevic, S.; Fontanals, I.; Vela, S.; Cardoso, M.A.; Buskute, A. Resilience to cope with climate change in urban areas-A multisectorial approach focusing on water-The RESCCUE project. *Water (Switzerland)* **2018**, *10*, 1356. [CrossRef]
5. Brouwer, R.; Georgiou, S. *Animal Waste, Water Quality and Human Health*; Dufour, A., Bartram, J., Bos, R., Gannon, V., Eds.; IWA Publishing: London, UK, 2012; ISBN 9781780401232.
6. Atkinson, G.; Braathen, A.; Groom, B.; Mourato, S. *Cost-Benefit Analysis and the Environment: Further Developments and Policy Use*; OECD: Paris, France, 2018; Volume 2, ISBN 978-92-64-08516-9.
7. Ürge-Vorsatz, D.; Novikova, A.; Sharmina, M. Counting Good: Quantifying the Co-Benefits of Improved Efficiency in Buildings. In Proceedings of the European Council for an Energy Efficient Economy (ECEEE) Summer Study, La Colle sur Loup, France, 1–6 June 2009.
8. Noleppa, S. *Economic Approaches for Assessing Climate Change Adaptation Options under Uncertainty*; Deutsche Gesellschaft für Internationale Zusammenarbeit (GIZ) GmbH: Bonn, Germany, 2013.
9. Meyer, V.; Becker, N.; Markantonis, V.; Schwarze, R. *Costs of Natural Hazards—A Synthesis*; CONHAZ Project: London, UK, 2012.
10. Li, C.; Cheng, X.; Li, N.; Du, X.; Yu, Q.; Kan, G. A framework for flood risk analysis and benefit assessment of flood control measures in Urban Areas. *Int. J. Environ. Res. Public Health* **2016**, *13*, 787. [CrossRef]
11. Messner, F.; Penning-Rowsell, E.; Green, C.; Meyer, V.; Tunstall, S.; Van der Veen, A. *Guidelines for Socio-economic Flood Damage Evaluation*; FloodSite Project: Wallingford, UK, 2007.
12. Vachris, M.A.; James, T. International price comparisons based on purchasing power parity|Request PDF. *Mon. Labor. Rev.* **1999**, *122*, 3–12.
13. Floater, G.; Heeckt, C.; Ulterino, M.; Mackie, L.; Rode, P.; Bhardwaj, A.; Carvalho, M.; Gill, D.; Bailey, T.; Huxley, R. *Co-Benefits of Urban Climate Action: A Framework for Cities*; London School of Economics and Political Science: London, UK, 2016.
14. Azqueta, D.; Alviar, M.; Dominguez, L.; O'Ryan, R. *Introducción a la Economía Ambiental*; McGraw-Hill Interamericana de España: Madrid, Spain, 2007; ISBN 8448160584-9788448160586.
15. Mandl, U.; Dierx, A.; Ilzkovitz, F. The Effectiveness and Efficiency of Public Spending. In *Economic Papers*; DG Economic and Financial Affairs, European Commision: Brussels, Belgium, 2008; ISBN 9789279082269.
16. Máñez, M.; Cerdà, A. *Prioritisation Method for Adaptation Measures to Climate Change in the Water Sector. CSC Report 18*; Climate Service Center: Hamburg, Germany, 2014.
17. Guerreiro, S.B.; Dawson, R.J.; Kilsby, C.; Lewis, E.; Ford, A. Future heat-waves, droughts and floods in 571 European cities. *Environ. Res. Lett.* **2018**, *13*. [CrossRef]
18. Climate Research Foundation (FIC). Downscaled Climate Model Outputs of the RESCCUE Project. Available online: https://www.ficlima.org/intercambio/indexed/RESCCUE/ (accessed on 25 February 2020).
19. Met Office UKCP UK Climate Predictions (UKCP09). Available online: https://webarchive.nationalarchives.gov.uk/20181204111018/http://ukclimateprojections-ukcp09.metoffice.gov.uk/ (accessed on 26 February 2020).
20. Van den Berg, M.; van Gils, P.F.; de Wit, G.A.; Schuit, A.J. *Economic Evaluation of Prevention, Fourth Report on the Cost-Effectiveness of Preventive Interventions*; National Institure for Public Health and the Environment: Amsterdam, The Netherlans, 2008.
21. Russo, B.; Velasco, M.; Monjo, R.; Martínez-Gomariz, E.; Domínguez-García, J.L.; Sánchez-Muñoz, D.; Gabàs, A.; Gonzalez, A. Assessment of the resilience of Barcelona urban services in case of flooding. The RESCCUE project. *Ing. del Agua* **2020**, *24*, 2.

22. Locatelli, L.; Russo, B.; Martinez, M. Evaluating health hazard of bathing waters affected by combined sewer overflows. *Nat. Hazards Earth Syst. Sci. Discuss.* **2019**. [CrossRef]

23. Verlaan, J.; Vos, R.; Matthijs, M. The use of equivalent annual cost for cost-benefit analyses in flood risk reduction strategies. *E3S Web Conf.* **2016**, *7*. [CrossRef]

24. Chiabai, A.A.; Galarraga, I.; Markanday, A.; de Murieta, E.S. *Determining Discount Rates: An Application of the Equivalency Principle*; EconAdapt Project: Bath, UK, 2015.

25. Stern, N. The economics of climate change. *Econ. Clim. Chang.* **2004**, *7*, 1–297.

26. Dasgupta, P. Commentary: The Stern Review's Economics of Climate Change. *Natl. Inst. Econ. Rev.* **2007**, *199*, 4–7. [CrossRef]

27. Blue, J.; Hiremath, N.; Gillette, C.; Julius, S. *Evaluating Urban Resilience to Climate Change: A Multi-Sector Approach*; U.S. Environmental Protection Agency: Washington, DC, USA, 2017.

28. Velasco, M.; Cabello, À.; Russo, B. Flood damage assessment in urban areas. Application to the Raval district of Barcelona using synthetic depth damage curves. *Urban Water J.* **2016**, *13*, 426–440. [CrossRef]

29. Russo, B.; Velasco, M.; Martínez-Gomariz, E.; Domínguez, J.-L.; Sánchez, D.; Gabàs, A.; Gonzalez, A. Evaluación de la resiliencia de los servicios urbanos frente a episodios de inundación en Barcelona. El Proyecto RESCCUE. *Ing. del Agua* **2020**, *24*, 101–118. [CrossRef]

30. Martínez-Gomariz, E.; Gómez, M.; Russo, B.; Sánchez, P.; Montes, J.-A. Methodology for the damage assessment of vehicles exposed to flooding in urban areas. *J. Flood Risk Manag.* **2019**, *12*, e12475. [CrossRef]

31. Gulf Engineers & Consultants (GEC). *Depth-Damage Relationships for Structures, Contents, and Vehicles and Content-to-Structure Value Ratios (CSVR) in Support of the Donaldsonville to the Gulf, Luisiana, Feasibility Study*; Gulf Engineers & Consultants (GEC): New Orleans, LA, USA, 2006.

32. Martínez-Gomariz, E.; Guerrero-Hidalga, M.; Russo, B.; Yubero, D.; Gómez, M.; Castán, S. Desarrollo y aplicación de curvas de daño y estanqueidad para la estimación del impacto económico de las inundaciones en zonas urbanas españolas. *Ing. Agua* **2019**, *23*, 229–245.

33. Martínez-Gomariz, E.; Forero-Ortiz, E.; Guerrero-Hidalga, M.; Castán, S.; Gómez, M. Flood Depth-Damage Curves for Spanish Urban Areas. *Sustainability* **2020**, *12*, 2666. [CrossRef]

34. Martínez-Gomariz, E.; Locatelli, L.; Guerrero, M.; Russo, B.; Martínez, M. Socio-Economic Potential Impacts Due to Urban Pluvial Floods in Badalona (Spain) in a Context of Climate Change. *Water* **2019**, *11*, 2658.

35. Turner, B.L.; Kasperson, R.E.; Matsone, P.A.; McCarthy, J.J.; Corell, R.W.; Christensene, L.; Eckley, N.; Kasperson, J.X.; Luers, A.; Martello, M.L.; et al. A framework for vulnerability analysis in sustainability science. *Proc. Natl. Acad. Sci. USA* **2003**, *100*, 8074–8079. [CrossRef]

36. Velasco, M.; Russo, B.; Cabello, À.; Termes, M.; Sunyer, D.; Malgrat, P. Assessment of the effectiveness of structural and nonstructural measures to cope with global change impacts in Barcelona. *J. Flood Risk Manag.* **2018**, *11*, S55–S68. [CrossRef]

37. Martínez-Gomariz, E.; Gómez, M.; Russo, B. Experimental study of the stability of pedestrians exposed to urban pluvial flooding. *Nat. Hazards* **2016**, *82*, 1259–1278.

38. Martínez-Gomariz, E.; Gómez, M.; Russo, B. Experimental study of the stability of pedestrians exposed to urban pluvial flooding. *Nat. Hazards* **2016**, *82*, 1259–1278. [CrossRef]

39. Russo, B.; Gómez, M.; Macchione, F. Pedestrian hazard criteria for flooded urban areas. *Nat. Hazards* **2013**, *69*, 251–265. [CrossRef]

40. Martínez-Gomariz, E.; Gómez, M.; Russo, B.; Djordjević, S. A new experiments-based methodology to define the stability threshold for any vehicle exposed to flooding. *Urban Water J.* **2017**, *14*, 930–939. [CrossRef]

41. Locatelli, L.; Russo, B.; Acero Oliete, A.; Sánchez Catalán, J.C.; Martínez-Gomariz, E.; Martínez, M. Modeling of E. coli distribution for hazard assessment of bathing waters affected by combined sewer overflows. *Nat. Hazards Earth Syst. Sci.* **2020**, *20*, 1219–1232. [CrossRef]

42. Evans, B.; Chen, A.; Djordjevic, S.; Webber, J.; Almeida, M.C.; Morais, M.; Telhado, M.J.; Silva, I.; Duarte, N.; Martínez-Gomariz, E.; et al. *Impact Assessments of Multiple Hazards in Case Study Areas with Adaptation Strategies*; RESCCUE Project: Barcelona, Spain, 2020.

43. Chen, A.S.; Hammond, M.J.; Djordjević, S.; Butler, D.; Khan, D.M.; Veerbeek, W. From hazard to impact: Flood damage assessment tools for mega cities. *Nat. Hazards* **2016**, *82*, 857–890. [CrossRef]

44. National Receptor Dataset (NRD). Risk of Flooding from Rivers and Sea. Available online: https://data.gov.uk/dataset/50545819-8149-4999-9d9f-c082e7234257/risk-of-flooding-from-rivers-and-sea-key-summary-information (accessed on 3 March 2020).

45. Penning-Rowsell, E.; Viavattene, C.; Pardoe, J.; Chatterdon, J.; Parker, D.; Morris, J. *The Benefits of Flood and Coastal Risk Management: A Handbook of Assessment Techniques*; Flood Hazard Research Centre: London, UK, 2010.

46. Bowker, P. *Flood Resistance and Resilience Solutions: An R&D Scoping Study*; R&D Technical Report; Joint Defra/EA Flood and Coastal Erosion Risk Management R&D Programme: London, UK, 2007.

47. Alvarez Lopez, P.; Behrisch, M.; Bieker-Walz, L.; Erdmann, J.; Flötteröd, Y.-P.; Hilbrich, R.; Lücken, L.; Rummel, J.; Wagner, P.; Wießner, E. Microscopic Traffic Simulation using SUMO. In Proceedings of the 21st International Conference on Intelligent Transportation Systems (ITSC), Maui, HI, USA, 4–7 November 2018; pp. 2575–2582. [CrossRef]

48. Pyatkova, K.; Chen, A.S.; Djordjević, S.; Butler, D.; Vojinović, Z.; Abebe, Y.A.; Hammond, M. *Flood Impacts on Road Transportation Using Microscopic Traffic Modelling Techniques BT—Simulating Urban Traffic Scenarios*; Behrisch, M., Weber, M., Eds.; Springer International Publishing: New York, NY, USA, 2015.

49. Sánchez-Muñoz, D.; Domínguez-García, J.L.; Martínez-Gomariz, E.; Russo, B.; Stevens, J.; Pardo, M. Electrical Grid Risk Assessment Against Flooding in Barcelona and Bristol Cities. *Sustainability* **2020**, *12*, 1527. [CrossRef]

50. Brouwer, R.; Barton, D.; Bateman, I.J.; Brander, L.; Georgiou, S.; Martin-Ortega, J.; Navrud, S.; Pulido-Velazquez, M.; Schaafsma, M.; Wagtendonk, A. *Economic Valuation of Environmental and Resource Costs and Benefits in the Water Framework Directive: Technical Guidelines for Practitioners*; Project AquaMoney: Amsterdam, The Netherlands, 2009.

51. TEEB. Available online: http://www.teebweb.org/resources/glossary-of-terms/ (accessed on 25 February 2020).

52. Vojinovic, Z.; Keerakamolchai, W.; Weesakul, S.; Pudar, R.S. Combining Ecosystem Services with Cost-Benefit Analysis for Selection of Green and Grey Infrastructure for Flood Protection in a Cultural Setting. *Environments* **2017**, *4*, 3. [CrossRef]

53. Botzat, A.; Fischer, L.K.; Kowarik, I. Unexploited opportunities in understanding liveable and biodiverse cities. A review on urban biodiversity perception and valuation. *Glob. Environ. Chang.* **2016**, *39*, 220–233. [CrossRef]

54. Cooper, W.; Garcia, F.; Pape, D.; Ryder, D.; Witherell, B. Climate Change Adaptation Case Study: Benefit-Cost Analysis of Coastal Flooding Hazard Mitigation. *J. Ocean Coast. Econ.* **2016**, *3*. [CrossRef]

55. Li, J.; Mullan, M.; Helgeson, J. Improving the practice of economic analysis of climate change adaptation. *J. Benefit Cost Anal.* **2014**, *5*, 445–467. [CrossRef]

56. Fairbrass, A.; Jones, K.; McIntosh, A.; Yao, Z.; Malki-Epshtein, L.; Bell, S. *Green Infrastructure for London: A Review of the Evidence*; A Report by the Engineering Exchange for Just Space and the London Sustainability Exchange; Engineering Exchange: London, UK, 2018.

57. Atkinson, G.; Groom, B.; Hanley, N.; Mourato, S. Environmental Valuation and Benefit-Cost Analysis in U.K. Policy. *J. Benefit Cost Anal.* **2018**, *9*, 97–119. [CrossRef]

58. Feng, H.; Hewage, K.N. *Nature Based Strategies for Urban and Building Sustainability*; Elsevier Inc.: Oxford, UK, 2018; pp. 307–318. ISBN 9780128123249.

59. Bianchini, F.; Hewage, K. Probabilistic social cost-benefit analysis for green roofs: A lifecycle approach. *Build. Environ.* **2012**, *58*, 152–162. [CrossRef]

60. Met Office UKCP UK Climate Projections (UKCP18). Available online: https://www.metoffice.gov.uk/research/approach/collaboration/ukcp/index (accessed on 26 February 2020).

61. Evans, B.; Chen, A.; Djordjević, S.; Webber, J.; González Gómez, A.; Stevens, J. Investigating the Effects of Pluvial Flooding and Climate Change on Traffic Flows in Barcelona and Bristol. *Sustainability* **2020**, *12*, 2330. [CrossRef]

62. Ürge-Vorsatz, D.; Herrero, S.T.; Dubash, N.K.; Lecocq, F. Measuring the Co-Benefits of Climate Change Mitigation. *Annu. Rev. Environ. Resour.* **2014**, *39*, 549–582. [CrossRef]

63. Dai, L.; Wörner, R.; van Rijswick, H.F.M.W. Rainproof cities in the Netherlands: Approaches in Dutch water governance to climate-adaptive urban planning. *Int. J. Water Resour. Dev.* **2018**, *34*, 652–674. [CrossRef]

 *sustainability*

Article

RAF Resilience Assessment Framework—A Tool to Support Cities' Action Planning

Maria Adriana Cardoso [1,*]**, Rita Salgado Brito** [1]**, Cristina Pereira** [1]**, Andoni Gonzalez** [2]**, John Stevens** [3] **and Maria João Telhado** [4]

[1] Urban Water Unit, National Civil Engineering Laboratory, LNEC, Av. Brasil 101, 1700–066 Lisbon, Portugal; rsbrito@lnec.pt (R.S.B.); clpereira@lnec.pt (C.P.)

[2] Barcelona City Council, Ajuntament de Barcelona, Barcelona Torrent de l'Olla 218–220, 4a planta, 08012 Barcelona, Spain; agonzalezgom@bcn.cat

[3] Bristol City Council, 100 Temple Street, Bristol P.O. Box 3176, UK; john.stevens@bristol.gov.uk

[4] Lisbon City Council, Câmara Municipal de Lisboa, CML, Praça José Queirós, n.°1–3° piso–Fração 5, 1800–237 Lisbon, Portugal; joao.telhado@cm-lisboa.pt

* Correspondence: macardoso@lnec.pt

Received: 31 January 2020; Accepted: 13 March 2020; Published: 17 March 2020

Abstract: Urban areas are dynamic, facing evolving hazards, having interacting strategic services and assets. Their management involves multiple stakeholders bringing additional complexity. Potential impacts of climate dynamics may aggravate current conditions and the appearance of new hazards. These challenges require an integrated and forward-looking approach to resilient and sustainable urban development, being essential to identify the real needs for its achievement. Several frameworks for assessing resilience have been developed in different fields. However, considering the focus on climate change and urban services, specific needs were identified, particularly in assessing strategic urban sectors and their interactions with others and with the wider urban system. A resilience assessment framework was developed directing and facilitating an objective-driven resilience diagnosis of urban cities and services. This supports the decision on selection of resilience measures and the development of strategies to enhance resilience, outlining a path to co-build resilience action plans, and to track resilience progress in the city or service over time. This paper presents the framework and the main results of its application to three cities having diverse contexts. It was demonstrated that the framework highlights where cities and urban services stand, regarding resilience to climate change, and identifies the most critical aspects to improve, including expected future impacts.

Keywords: resilience assessment; urban resilience; climate change; urban services; cities

---

1. Introduction

Urban areas are complex, vulnerable and continuously evolving systems. In these dynamic areas, the existence of interacting strategic services and of interdependent services and assets, as well as the involvement of a multiplicity of stakeholders, adds complexity to their management. Besides, the significant impacts of climate dynamics (such as intense precipitation events, tidal effects, droughts or heat waves) in the urban strategic services, people, natural environment and economy, as well as the aggravation of current conditions and the emergence of new hazards, also need to be considered in their management [1,2].

As referred to in [3], following the World Economic Forum 2014, by 2050, exposure of city dwellers to various hazards, including earthquakes, tsunamis, urban floods, cyclones and storm surges, is expected to double. These challenges require an integrated and forward-looking approach to resilient and sustainable urban development, incorporating the interdependencies between systems as well as including stakeholders and citizens perceptions and needs. In order to achieve this, several long-term agendas have been adopted as parts of the United Nations Agenda 2030 for Sustainable Development, such as the Sendai Framework for Disaster Risk Reduction 2015–2030, the Sustainable Development Goals, the New Urban Agenda and the Paris Agreement [3]. A relevant consideration in all of these agendas is the incorporation of assessment steps for tracking their implementation [4].

The resilience concept has evolved over time and among disciplines [5,6]. Herein, urban resilience refers to the ability of human settlements to withstand, recover quickly and adapt from any plausible hazards. Resilience to disruptive events not only refers to reducing risks and damage from disasters, but also the ability to quickly bounce back to a stable state. Besides addressing disaster risk reduction, resilience includes changes in circumstances [7–10].

In order to identify the real needs for enhancing urban resilience, as well as the efficiency and effectiveness of planned or implemented measures, a resilience assessment is essential. Therefore, assessing the current and expected future status of resilience is a basis for cities to know where they are, helping to identify strengths and weaknesses, thus supporting the decision on strategies, actions and measures to be taken, planning for the long-, medium- and short-terms and assessing the progress.

Since the cities are dynamic systems with evolving hazards, it is essential to regularly carry out the assessment of their resilience, considering the principle of continuous improvement [11], and to have tools to support this. Several tools and frameworks for assessing resilience have been developed in different fields of study by a wide variety of stakeholders, such as those created by Local Governments for Sustainability (ICLEI) 2010, UN-Habitat City Resilience Profiling Tool (UN-Habitat CRPT) 2013, Rockefeller and Arup 2014, World Bank 2015, United Nations Office for Disaster Risk Reduction (UNDRR, former UNISDR) 2017, U.S. Environmental Protection Agency (EPA) 2017, among others [5,7–9,12–16]. Within the scope of the current work, i.e., climate change with a focus on water, relevant resilience assessment frameworks are presented in Table 1. It synthetizes the themes, urban sectors and metrics considered in each framework [5,7,13,16,17].

Table 1. Synthesis of resilience assessment frameworks for climate change.

Framework	Themes Addressed						Sectors Addressed							No. of Metrics	Reference
	Governance	Social	Spatial	Built environment	Economy	Natural Environment	Water	Wastewater	Stormwater	Waste	Energy	Mobility	Other(s) *		
EPA conceptual framework	✓	✓	✓	✓	✓	✓	✓	✓	✓		✓	✓	✓	163	[15]
City Resilience Framework	✓	✓	✓										✓	156	[13]
UNDRR Disaster Resilience Scorecard for cities	✓	✓	✓	✓	✓		✓	✓		✓	✓	✓	✓	47 preliminaries 117 detailed	[8,9]
City Resilience Index to Sea Level Rise	✓	✓		✓	✓	✓	✓					✓		13	[18]
Climate Disaster Resilience Index	✓	✓		✓	✓	✓					✓	✓		120	[19]
Climate Disaster Resilience Index	✓	✓		✓	✓	✓					✓	✓		82	[20]
Climate Resilience Screening Index	✓	✓	✓	✓	✓	✓	✓	✓			✓	✓	✓	117	[16]
Flood Resilience Index	✓	✓		✓	✓	✓	✓	✓		✓	✓	✓	✓	91	[21]
Resilience Factor Index	✓	✓		✓	✓							✓		17	[22]
Community disaster resilience	✓	✓		✓	✓	✓						✓		26	[23]

Table 1. *Cont.*

Framework	Themes Addressed						Sectors Addressed							No. of Metrics	Reference
	Governance	Social	Spatial	Built environment	Economy	Natural Environment	Water	Wastewater	Stormwater	Waste	Energy	Mobility	Other(s) *		
NIST (National Institute of Standards and Technology) Community Resilience Assess. Methodology	✓	✓		✓	✓	✓	✓	✓			✓	✓	✓	-	[24]
UKWIR (UK Water Industry Research)						✓	✓	✓	✓					73	[25]
UN-Habitat CRPT	✓	✓	✓	✓	✓	✓	✓	✓	✓	✓	✓	✓	✓	148	[7]

* e.g., Telecommunications, healthcare, education, population.

Taking into account the mentioned scope, the need of a framework (Table 1) that is freely available to be usable by cities and urban services managers was identified, allowing, on the one hand, a structured and objective-driven assessment of their city's resilience considering the integration of all themes and sectors simultaneously and, on the other hand, an assessment of resilience of a single sector considering its interdependencies with other sectors and its contribution to the city resilience.

Grounded in the analysis of these existing frameworks, and in order to bridge the additional gaps and needs identified, particularly in the assessment of strategic urban sectors and their interactions with both other sectors and in the wider urban system, the Resilience Assessment Framework (RAF) was developed—a resilience assessment framework with focus on climate change and the water cycle, herein described.

2. Materials and Methods

2.1. RAF—Resilience Assessment Framework Aims, Assumptions and Development Approach

Considering the challenges of urban areas related to the potential effects of climate dynamics, enhancing urban resilience requires: (i) identification of the real needs, (ii) sustainable action planning and (iii) assessing progress. In order to support the mentioned requirements, bridging the gaps and the abovementioned needs identified, a Resilience Assessment Framework (RAF) was developed with the main purpose of contributing to the referred requirements, namely:

(i) Directing and facilitating a structured resilience diagnosis of the cities and of the strategic urban sectors, following an objective-driven approach [11] with defined assessment criteria and identifying data gaps, opportunities, threats, strengths and weaknesses, highlighting the areas for improvement.

(ii) Outlining a path for the development of cities' resilience action plans by supporting decision-making in the selection of resilience measures and the development of strategies to enhance resilience.

(iii) Monitoring the resilience progress of a city or service over time, by applying it periodically, and facilitating communication among stakeholders.

The RAF described herein considers the following assumptions:

- The scope is urban resilience to climate change (CC), with a focus on the water cycle, meaning that other diverse resilience drivers such as earthquakes, economic crises and cyberattacks, are not taken into account.
- The emphasis is on the city, services and infrastructure resilience, meaning that resilience aspects such as social and political are not developed for diagnosis, but they are incorporated whenever significant for city, services' and infrastructures' resilience.

- The services within the RAF scope are those comprised in the urban water cycle, water supply, wastewater and storm water and those having interconnections and interdependencies, closely related with the water services: waste management, electrical energy supply and mobility.
- The external context of the city and services is considered by a standard characterisation profile of the city and of the services, since it is fundamental to identify the main threats and to support the assessment, particularly the interpretation of results.
- The city and services multi-scale, multi-sectoral, multi-hazards and interdependencies are addressed, meaning that the RAF incorporates: different scales—city, services and infrastructure, the diverse sectors presented above, assessment of several hazards and of aspects related to interdependencies between different services and infrastructures.
- The continuous improvement principle [11] is followed and, since cities are dynamic, it addresses the progress of the strategies' implementation and considers their effect, before, during and after an event and changes in circumstances.
- The long-, medium- and short-terms are incorporated considering three different and aligned assessment levels for the city, services and infrastructures (strategic—overlooking a long-term planning horizon and requiring the involvement of the entire organisation, addressing the overall city and considering its vision; tactical—overlooking a medium-term planning horizon and addressing departmental or sectoral activities in the city, services and infrastructure; and operational—referring to short-term horizon, addresses the actions to be taken in the effective implementation of measures in the city, services and infrastructure) while, as an integrated assessment, addresses the two first.
- A flexible structure is used, based on assessment metrics, allowing it to be expanded to other resilience drivers, dimensions or services.

The development and implementation of the assessment process, in collaboration with different stakeholders, promotes their empowerment and enhance their role in the decision-making process [26], as well as in the implementation of improvement solutions. To consider this, the RAF development was carried out in a stepwise process (Figure 1), comprising the analysis of existing assessment frameworks and related recommendations, and the definition of a preliminary proposal, which was validated to produce the final version.

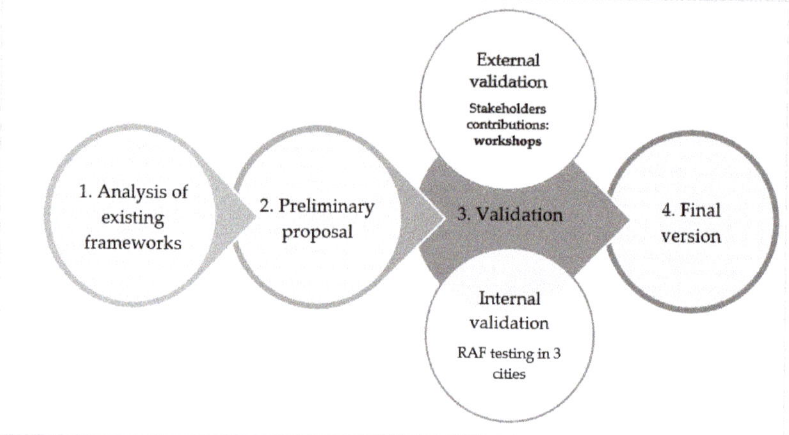

Figure 1. Resilience Assessment Framework (RAF) development process flow chart.

The validation process included an external and an internal validation [26]. The external validation involved different stakeholders, representatives of research organisations, city departments and urban service utilities, allowing for incorporating their concerns as well as their own context and reality through collaborative workshops. Three workshops were implemented in each city, Barcelona (Spain), Lisbon (Portugal) and Bristol (UK), to obtain the stakeholders' opinion on the RAF relevance, structure and applicability, as well as their concerns, own context and reality. Overall, 24 to 38 stakeholders attended each of the sessions, from 13 to 24 different organisations, answering individually and by sector to several surveys.

To ensure coherence, feasibility and effectiveness of the approach, the internal validation was carried out in the abovementioned cities, having different characteristics and contexts, which applied this framework involving the respective stakeholders. Each city and respective services provided their own data and answers to all applicable metrics. From the external and internal validation analysis, it was possible to identify the RAF components that benefited from additional improvements and those that less fitted the cities' available information, thus supporting the development of the final framework herein presented. It is important to take into account that cities are multi-dimension entities and, therefore, urban resilience needs to consider multidisciplinary insights. Additionally, resilience of a city is determined by diverse interacting systems and their relationships. For this reason, resilience also depends on the overall performance, interactions and capacity of its systems in their everyday operation, not solely on its ability to cope with specific natural hazards or to adapt targeted areas to the impacts of climate change [27]. Thus, it is essential to address interdependencies and cascading effects [28]. Another relevant aspect is that it needs to include both sudden crises as well as interacting long-term stressors, address multiple hazards, characterise the specific geographic extent, consider physical dimensions, involve community members and be adaptable and scalable to different communities and changing circumstances [24]. These requirements were considered in the RAF development.

2.2. RAF—Resilience Assessment Framework Description

RAF sought alignment with international frameworks for resilience assessment, particularly with UNDRR Disaster Resilience Scorecard, both preliminary and detailed levels [6,7], and UN-Habitat, and made significant developments with regard to its scope and focus on urban services. The RAF considers the UN-Habitat resilience dimensions [29]: organisational (integrates top-down governance relations and urban population involvement, at the city level), spatial (referring to urban space and environment), functional (resilience of strategic services) and physical (resilience of services infrastructure). Time dimension is implicitly integrated as part of the analysis. The RAF (Table 2) has a hierarchical tree structure (Figure 2) meaning that, for each dimension, resilience objectives are defined, representing the ambitions to be achieved in the medium–long term by the city and services. For those dimensions related to the urban services, they firstly unfold into sub-dimensions, where each sub-dimension represents one service to be assessed. Each objective is described by a set of criteria that translate the different points of view associated with it. Each criterion assembles the respective assessment metrics, through which it is possible to classify the resilience development level by comparison with reference values. Metrics are then defined consisting in questions, parameters or functions used to assess the criteria. Some of the RAF metrics correspond to or were adapted from existing frameworks, mainly from UNDRR framework (former UNISDR)—found to be highly relevant for the scope of the RAF, and others were newly developed. In Appendix A, the complete structure is presented. As an example, Table 3 illustrates the metrics definition to assess, within the spatial dimension, the objective of spatial risk management from the perspective given by the criterion impacts of climate-related events, showing the hierarchical tree structure mentioned above.

Table 2. Overview of the RAF dimensions.

ORGANISATIONAL			SPATIAL		
OBJECTIVE Criterion	No. total metrics	No. essential metrics	**OBJECTIVE** Criterion	No. total metrics	No. essential metrics
COLLECTIVE ENGAGEMENT AND AWARENESS			**SPATIAL RISK MANAGEMENT**		
Citizens and communities' engagement	5	3	General hazard and exposure mapping	5	5
Citizens and communities' awareness and training	5	3	Hazard and exposure for CC	3	3
LEADERSHIP AND MANAGEMENT			Resilient urban development	7	4
Government decision-making and finance	4	3	Impacts of climate-related event	2	2
Coordination and communication with stakeholders	4	2	**PROVISION OF PROTECTIVE INFRASTRUCTURES AND ECOSYSTEMS**		
Resilience engaged city	19	13	Protective infrastructures and ecosystems services	9	6
CITY PREPAREDNESS			Dependence and autonomy regarding other services considering CC	3	2
City preparedness for disaster response	13	8	**TOTAL**	29	22
City preparedness for CC	7	6			
City preparedness for recovery and build back	7	5			
Availability and access to basic services	10	7			
TOTAL	74	50			

FUNCTIONAL			PHYSICAL		
OBJECTIVE Criterion	No. total metrics	No. essential metrics	**OBJECTIVE** Criterion	No. total metrics	No. essential metrics
SERVICE PLANNING AND RISK MANAGEMENT			**SAFE INFRASTRUCTURE**		
Strategic planning	5	5	Infrastructure assets criticality and protection	5	5
Resilience engaged service	5–6	4–5	Infrastructure assets robustness	10–14	4–6
Risk management	7–12	2–7	**AUTONOMOUS AND FLEXIBLE INFRASTRUCTURE**		
Reliable service	6–11	1–5	Infrastructure assets importance to and dependency on other services	3–4	3
Flexible service	4–6	1–4	Infrastructure assets autonomy	1–6	0–4
AUTONOMOUS SERVICE			Infrastructure assets redundancy	1–3	0–3
Service importance to the city	2	1	**INFRASTRUCTURE PREPAREDNESS**		
Service inter-dependency with other services considering CC	2	0	Contribution to city resilience	3–4	2–3
SERVICE PREPAREDNESS			Infrastructure assets exposure to CC	3	0–3
Service preparedness for disaster response	0–4	0–4	Preparedness for CC	2	1
Service preparedness for CC	6–8	4	Preparedness for recovery and build back	7–9	2–4
Service preparedness for recovery and build back	0–15	0–8	**TOTAL**	35–50	17–32
TOTAL	37–71	18–43			

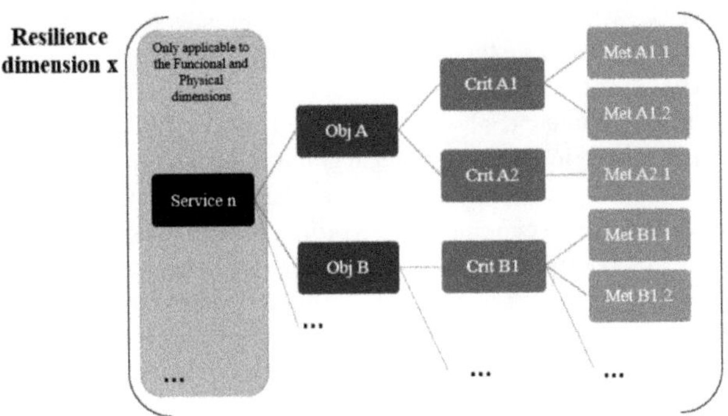

Figure 2. RAF tree structure.

Table 3. Metrics definition—example for spatial dimension, objective spatial risk management, criterion impacts of climate-related event.

DIMENSION: SPATIAL		
OBJECTIVE: SPATIAL RISK MANAGEMENT		
Criterion: Impacts of Climate-Related Event		**Unit**
	Human loss in the last events	
	Human impact of the last climate-related event, with similar or harsher climate variables than the most probable scenario	
Metric: S16		(-)
Definition	Spatial	
	Essential	
Dimension Importance	Open value	
Metric type		
Please answer with an estimated figure [inhab.], disaggregating according to (a) number of casualties, (b)		*Develop. Level*
missing persons and (c) people affected—including severe injuries and displaced. This metric allows to		
answer with a value.		*3 if a, b and c = 0*
Development level: assessment rule		*2 if a and b = 0 and c ≤ 50*
		1 if a = 0, b ≤ 5 and c ≤ 50
		0 if any other answer
- *(a) number of casualties*		
- *(b) missing persons*		
- *(c) people affected—including severe injuries and displaced*		
Metric: S17		
Definition		
	Damages in urban footprint in the last events	(%)
Dimension Importance	Impact on urban footprint of the last	
Metric type	climate-related event, with similar or harsher climate variables than the most probable scenario	
	Spatial	
	Essential	
	Single choice	
Consider urban footprint as a spatial extent of urbanised areas on a regional scale.		
Development level: assessment rule		
- *0%*		
- *Less or equal to 0.5%*		*Develop. level*
- *Between 0.5% and 2.5%*		
- *More or equal to 2.5%*		*3*
		2
		1
		0

The framework considers past, existing and future conditions in the assessment. To incorporate the uncertainties associated to expected variations in climate-related variables, some metrics are specific to CC assessment scenarios, namely those that address preparedness for CC, and that anticipate the city and services' exposure or vulnerability to future scenarios. Besides, the consideration of reference values allows to generally address uncertainties in the assessment.

A relevance degree is assigned to each metric, namely: essential, corresponding to all metrics with higher relevance, required to integrate the resilience assessment of any city or service, complementary, additional metrics to be considered whenever integration of city or service specific aspects is sought, corresponding to a more detailed resilience assessment and comprehensive, additional metrics recommended whenever a more in-depth assessment is aimed, for a city or service with higher maturity in its resilience path. Accordingly, depending on the resilience maturity, the city or service aiming to apply the RAF may select a given set of metrics, according to their relevance.

Additionally, every city or urban service needs to operate in its own specific political, economic, geographical, climatic and cultural context. Considering the context information is fundamental in interpreting any assessment. Following this, city and services' characterisation profiles were developed to integrate the RAF framework, regarding its scope and focus. These profiles require information on geographical characteristics, climate, population, economy and governance, built environment and infrastructures, for the city. Regarding each service, it considers information on context characterisation, climate and infrastructure assets.

2.3. Research Sites

2.3.1. General

In order to test and validate the RAF to assess the cities' resilience to climate change with a focus on the water cycle, it was applied to Bristol (UK), Barcelona (Spain) and Lisbon (Portugal) by the respective cities and strategic services managers. These three cities represent diverse context characteristics as well as different climate change-related concerns. The application was undertaken using the RAF App, a web-based application tool reproducing the RAF structure that allows selection of applicable dimensions and services to assess and allows private submission of answers to the metrics. The results may be visualised graphically (Figures 3–5) and reports are also provided [30].

2.3.2. Bristol

Located in the south-west England, predominantly on a limestone area, Bristol is one of the most densely populated parts of the UK and, after London, the second largest city in the southern region. Most of the urban extent of Bristol is based around the watercourses and river network, with two major rivers flowing through the city (Avon and Frome rivers), resulting in a characteristically hilly landscape. It is one of the warmest cities in the UK and there is a relatively even distribution of rainfall throughout the year, although the autumn and winter seasons tend to be the wettest. Within this context, Bristol has been investing in plans to create and improve resilient systems to tackle its various urban challenges. Based on the analyses conducted by local and international actors working on resilience, the main urban challenges in Bristol can be profiled firstly in terms of natural and environmental hazards and secondly with regards to broader socio-economic issues. Bristol has suffered from significant flooding in the past, with the floating harbour and low-lying city centre being identified as key areas vulnerable to tidal, fluvial and groundwater flooding. The flood of 1968 was one of the most significant and damaging flooding events in the city, caused by both surface water and fluvial flooding that resulted in high damages and impacts to the city and its inhabitants. The construction of large interceptor tunnels in response to this, to divert exceedance flows higher up in the catchment, reduced fluvial flood risk in the city. In 2012, significant flooding occurred across most of the UK due to some of the highest rainfall events since record collection began. During this time, the most notable single flood event lasted two days, with 30 houses internally flooded and many more suffering flooding of gardens, garages and driveways. In order to better manage flood risks in Bristol area, a 'Local Flood Risk Management Strategy' was produced and released in early 2018. The Strategy sets out the Bristol City Council vision for managing flood risk in the city, together with other organisations that have a role in flood-risk management [29]. Bristol City Council has already developed an intensive work towards resilience, and it is proactively committed to increase Bristol's resilience: from social cohesion to economic stresses and by enhancing resilience to all sources of flooding. The resilience of the city to climate change (CC) can be highly related to its urban services' resilience, their interdependencies and cascade effects. For Bristol, the resilience assessment was undertaken for the flooding hazard related to rainfall and sea level variables, by its importance regarding Bristol resilience to CC.

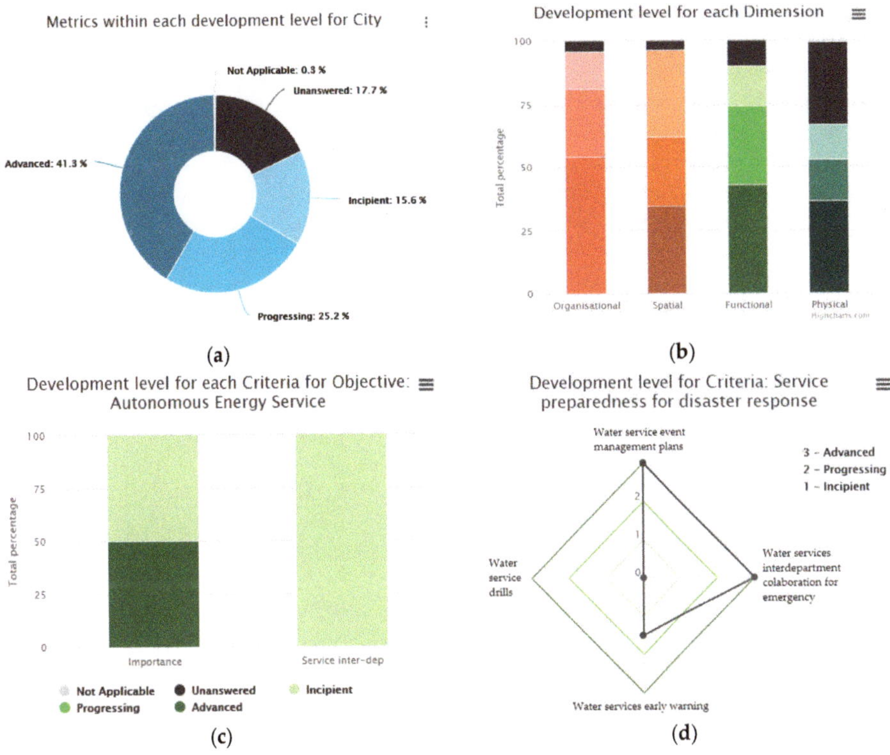

Figure 3. Bristol resilience assessment results for flooding: (**a**) Overall assessment, (**b**) overall assessment per dimension, (**c**) assessment of the objective autonomous electrical energy service, (**d**) assessment of the criterion water service preparedness for disaster response.

2.3.3. Barcelona

Located on the northeast coast of the Iberian Peninsula facing the Mediterranean Sea, Barcelona is the capital city of the autonomous community of Catalonia, Spain. The city is situated on a plain spanning and is bordered by the mountain range of Collserola, the Llobregat river in the southwest and the Besòs river in the north. Barcelona is the second most populous municipality within Spain. However, the population increased slowly but steadily until the 1970's, when the city reached its maximum population, thereafter, it stabilized and even decreased at the beginning of the 21st century, reaching the average population of 1.6 million inhabitants. Barcelona's physical expansion has been limited by the mountains and the sea, resulting in a relatively high population density, among the highest in Europe. Within this context, Barcelona's major vulnerabilities are mainly attributable to the natural and environmental threats faced by the wider Catalonia region. Barcelona's past and recent history has been punctuated with recurrent water crises but also with rainfall events with very strong intensity over short time frames. The most severe and recent disruptive event hitting the urban area was between 2004 and 2008. During that period, four years of scarce precipitation in the Llobregat and Ter rivers' headwaters, coupled with an increased evaporation rate due to high temperatures, culminated in the Spring 2008 water crisis affecting over 5.5 million people in the broader Catalonia. In that context, the Regional Government had to adopt exceptional procedures to minimise water waste, while the City of Barcelona was simultaneously forced to introduce restrictive measures over water use. Since then, several structural measures to ensure water supply have been implemented [29]. In January 2018, the city declared the pre-alert level of the Drought protocol after three consecutive years of low rainfall. The city is affected every year by an average of three intense rainfall events and one extreme flooding event every five years, although these frequencies have been increasing in the last years. Barcelona also has records of one heat wave every four years, a trend that has been increasing notably in the latest years. In 2003, a heatwave that lasted 13 days increased in more than 40% the average mortality. The last heat wave event was in summer 2018, it was 7 days long and caused up to 10 direct deaths. The resilience of the city to climate change can be highly related to its urban services' resilience, their interdependencies and cascade effects. The Barcelona Municipality has already developed an intensive work towards resilience, and it is proactively committed to increase Barcelona's resilience: from social exclusion to economic stresses, flooding, drought and heat waves. For Barcelona, the resilience assessment was carried-out for flooding, combined sewer overflows, drought and heat waves, considering the variables related to rainfall, sea level and temperature.

2.3.4. Lisbon

Located on the northern bank of the Tagus River's estuary, one of the 18 municipalities of the biggest Portuguese metropolitan area, Lisbon is the capital of Portugal and the second largest European port on the Atlantic Ocean. The city has a Mediterranean Climate (Csa), characterised by dry and hot summers and wet and fresh winter periods with a relatively low precipitation rate compared to other Portuguese cities. Lisbon Metropolitan Area, with a population of 2.8 million inhabitants, stretches on both sides of the Tagus River, contributing to 37% of the national economic output. Today, Lisbon is a complex system with more than 1.0 million citizens who live, work, study, circulate and visit the city, Portuguese in the majority, with different ages, cultures, religions, ethnicities, education levels, knowledge and languages. Based on the analyses conducted by both local public stakeholders and international actors working on resilience in Lisbon, one of the urban challenges is related to a combination of contextual environmental, emergency, civil protection and urban planning threats with the contingent impacts of climate change crisis [29]. Since 1950, about 43 relevant events of extreme weather occurred in Lisbon. From these, nine events were related to hot weather, including heat waves, with a maximum temperature of 42 °C recorded in August of 2003, 13 events related to cold weather, including cold waves, with a minimum temperature of −1.2 °C recorded in February 1956, two strong wind and gusts events, with a maximum wind velocity of 108.4 km/h, recorded in January 2014 and 10 rainfall-induced flood events, with a maximum return period of 500 years, recorded in November 1983. The resilience of the city to climate change can be highly related to its urban services' resilience, their interdependencies and cascade effects. Lisbon Municipality has already developed an intensive work towards resilience, and it is proactively committed to increase the resilience of the city: from social exclusion to economic stresses and from seismic shocks to flooding, combined with 17 Sustainable Development Goals' achievement. For Lisbon, the resilience assessment was undertaken for the flooding hazard, related to rainfall and sea level variables.

3. Results

3.1. Bristol

The RAF was applied in Bristol in order to assess the current level of city resilience to flooding. Some results are presented in Figure 3. This could then subsequently identify where the gaps lie and what particular aspects are lacking to help formulate plans to improve or enhance upon the existing status, based on this resilience diagnosis. It went into a great level of detail investigating many aspects of city resilience quite thoroughly. The overall resilience development in the city was deemed as advanced in nearly half of the aspects assessed (Figure 3a). In this same respect, around a quarter were shown as progressing and the remainder incipient, unanswerable or not applicable. Various city services were given consideration including storm water, wastewater, energy, mobility and solid waste management operations.

The analysis highlighted the advancement in organisational areas more so over physical areas (Figure 3b), which were deemed more absent. Infrastructure resilience to climate change is therefore the main concern on reflection of this. In their own respect, the individual services seem resilient to a point, due to a focus on building resilience to historical events in the city and in response to national flood-risk issues. There is, however, susceptibility in the realms of reliance upon inter-related services and a lack of understanding of the cascading impacts and interdependencies between them.

The results from the analysis highlight the coordination between governmental organisations that is not always experienced to the same level externally with all privately run organisations. Engagement with communities is also a dynamic that is not completely to its maximum sufficiency. Availability of service resources is good, since diverse energy sources are used in the city, but the reliance on electricity without alternative provisions is a notable limitation (Figure 3c). Resilience standards to adhere to as well as the position of a Chief Resilience Officer being eliminated make for more areas lacking in Bristol. Learning from past events is a commendable action performed well in Bristol, but the running of emergency scenarios and drills does not appear to be simulated enough to gain its full benefit (Figure 3d). The known threats of a significant proportion from sea level rise and increased rainfall present an extreme level of vulnerability to the city and its inhabitants. There are, however, also opportunities presented, though through the declaration of a climate emergency in Bristol, they require drastic action implemented via a climate strategy. The chance for properly applying climate adaptation measures utilising the knowledge developed of high-risk areas in the city therefore has greater prospect for recognition and the enablement for realisation.

3.2. Barcelona

The RAF enables to highlight where Barcelona and its urban services stand today regarding resilience to climate change, and to identify the most critical aspects to be improved, taking into account both the reference situation and the expected impacts of future climate change scenarios. The diagnosis allowed for understanding those aspects that are being tackled properly from the city and was also to determine gaps and areas of improvement thanks to the great level of detail of the different dimensions that make up the assessment. Some results are presented in Figure 4. The exhaustive analysis led the city to an intense and deep level of self-knowledge about its level of resilience in different ways of approach (Figure 4a). In this sense, the organisational and spatial dimensions yielded good results about the level of response to the metrics considered, reaching a response level of almost 100% (Figure 4b). Regarding the physical and functional dimensions, several services of the city were assessed, namely water, wastewater, storm water, energy, waste management and mobility. The assessment showed those services that are well managed and monitored as waste or water services, but it also highlighted the need of improvement in the energy sector, storm and wastewater and mobility services (Figure 4c,d). For Barcelona, most data gaps can be blamed on the definition of the metrics to be applied and the differences in the way how these metrics are calculated. Most of the time, the indicators did not fit with the ones the city already determines and it would entail a noteworthy effort to address the asked specifications. Without assuming harm, this identification of gaps means an opportunity to improve a new approach to measuring the different aspects of resilience in the city.

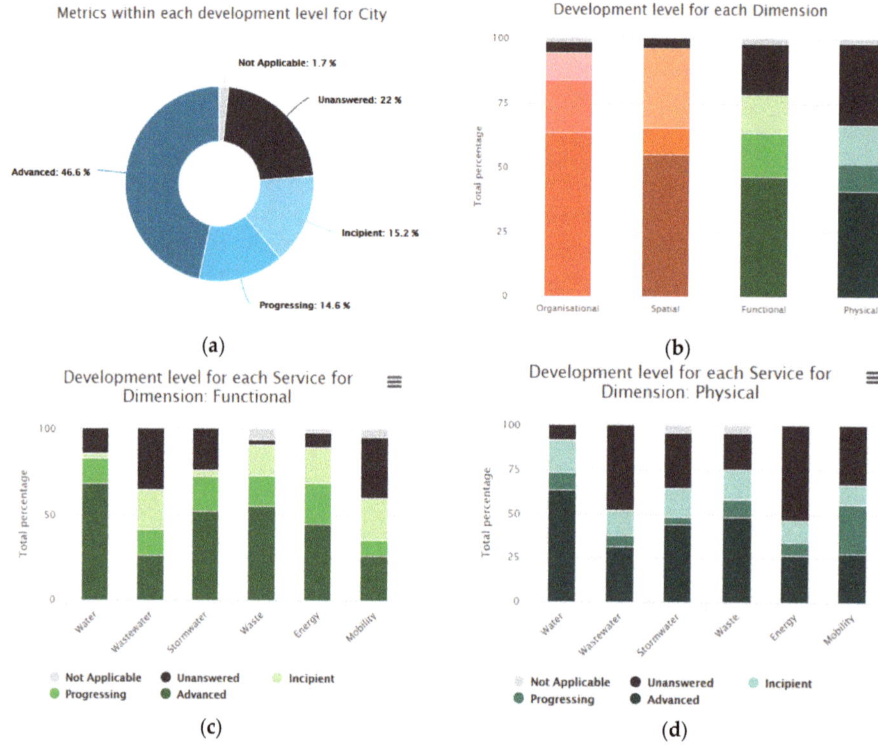

Figure 4. Barcelona resilience assessment results for flooding: (**a**) Overall assessment, (**b**) overall assessment per dimension, (**c**) functional overall assessment per service, (**d**) physical overall assessment per service infrastructure.

The RAF enabled the ability to be realistic with the resilience level of city services. It shed light on the state-of-the-art of urban resilience in Barcelona, highlighting those areas where the city works properly and progresses positively to a high degree of preparedness. At the same time, it has helped to determine those aspects where there is still room for improvement and has also given the chance of applying a methodology capable to reach the deepest areas that make up the operation of a city.

3.3. Lisbon

The RAF was applied in Lisbon in order to assess the current level of city resilience to flooding. The application of a structured resilience assessment framework enables the identification of the resilience criteria, objectives, services and city dimensions with major accomplishments, setbacks or opportunities for improvement. Therefore, it supports identification of resilience measures and development of strategies. Some results are presented in Figure 5. The overall resilience development in the city is advanced in nearly one third of the aspects (Figure 5a). Globally, around a quarter shows progress, meaning that significant steps were already taken, and the city and services are still developing specific aspects. The remainder correspond to incipient, unanswerable or not applicable metrics. Various city services were assessed with more detail, including stormwater, wastewater, energy, mobility and solid waste management.

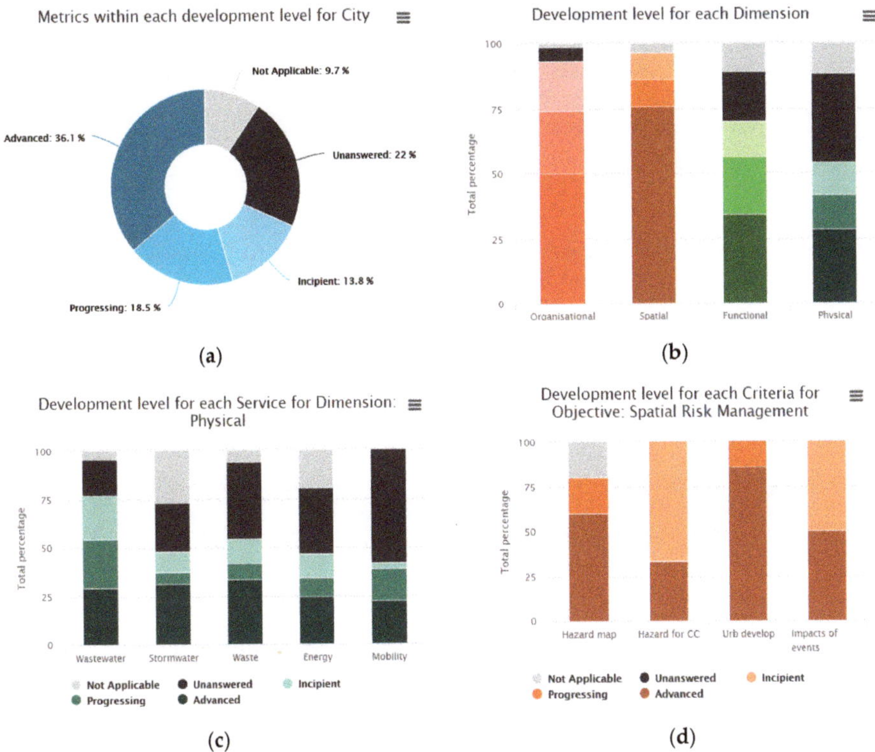

Figure 5. Lisbon resilience assessment results for flooding: (**a**) Overall assessment, (**b**) overall assessment per dimension, (**c**) physical overall assessment per service infrastructure, (**d**) assessment of the objective spatial risk management.

The analysis highlighted a significant advancement in spatial areas more so over physical areas, which were deemed more absent (Figure 5b). The organisational dimension as well as all the services and infrastructures present aspects already having an advanced development level, while still having significant opportunities for improvement. In the mobility service, considering the significant percentage of metrics that were not answered, data may be not be easily applicable to the metrics provided or some lack of information may exist. This is also applicable to the infrastructure assessment of the stormwater, waste and energy services (Figure 5c). Infrastructure resilience to climate change is therefore the main concern on reflection of this. For all services, the contribution of infrastructure to city resilience needs to be more exploited.

The results from the organisational analysis highlight that citizens and communities' awareness and training is one of the aspects that needs further development, followed by the city preparedness for disaster response and for recovery and build back. Engagement with communities is also a dynamic that is not completely to its maximum sufficiency as well as the coordination of financial plans and budgets for resilience.

Concerning the spatial analysis, the provision of protective infrastructures and ecosystems is well developed, while the knowledge on climate change hazard and exposure as well as impacts are highlighted as opportunities to be further developed (Figure 5d).

Generally, there is strong development of strategic planning and there is limited preparedness in the wastewater service for climate change, as well as limited autonomy for the majority of the services, with the exception of the stormwater service. There are, however, some susceptibilities in the realms of reliance upon inter-related services and a lack of understanding of the cascading impacts and interdependencies between those for climate change.

This diagnosis of the main strengths and weaknesses supports the identification of the adequate measures for resilience enhancement to climate change. This assessment is a step up in Lisbon's Climate Change Resilience Process and one diagnosis to be integrated in the ongoing Climate Action Plan of the city.

4. Discussion

By applying the RAF (Sections 2.1 and 2.2) to Bristol, Barcelona and Lisbon (Section 2.3), from the results obtained (Sections 3.1–3.3), it was possible to validate that it provides information on the assessment of the current level of the cities' resilience to climate change with a focus on the water cycle. The framework delivers a structured assessment clearly identifying the work already carried out, translating the strengths of the cities' resilience and which dimensions of resilience they fit into most. This is illustrated by the advanced or progressing values in Figures 3a,b, 4a,b and 5a,b. Besides the assessment of the organisational and spatial dimensions of the city, one particular aspect to emphasize is the identification of the contribution of the urban services to cities' resilience, as evident in Figures 3b, 4b–d and 5b,c. At the same time, the framework highlights the gaps, including limitations on data related to unanswered metrics. It also indicates particular aspects that are lacking, as can be seen by incipient values in Figures 3c,d and 5d, as well as those in more need of further development, given by progressing values in the same figures.

It is evident that the RAF enables to highlight where the cities and respective urban services stand today regarding resilience to climate change, and to identify the most critical aspects to be improved. It should, however, be noted that results of unanswered metrics, corresponding to limitations on data, may be due both to a lack of information or to the alignment in the way existing information is processed in the city with the way the metrics are calculated, as in the Barcelona case (Section 3.2). This last case is likely to occur in cities already using other assessment frameworks. Whenever the framework in use allows to assess the same concerns, i.e., the resilience objectives and criteria corresponding to those of the RAF, they may be used instead. Nevertheless, this provides the challenge to align the RAF with other existing frameworks in this scope. In these circumstances, it is fundamental to clearly identify actual data gaps in the cities and services that need to be filled.

Considering the assignment of a relevance degree described in Section 2.2, it is possible to undertake a stepwise process going into a gradually deeper assessment, depending on the resilience maturity of a city, allowing replicability of the methodology to other cities and services. The framework allows to go into a considerable level of detail investigating many aspects of city resilience quite thoroughly. The whole assessment provides a resilience diagnosis that helps with formulating plans to improve or enhance upon the existing status.

It is feasible to use the RAF to assess diverse hazards such as flooding, combined sewer overflows, drought and heat waves, as it was in the case of Barcelona (Section 2.3.2). The framework may be applicable to provide an overall response regarding the cities' resilience assessment or it may be applied to assess a certain urban service within its scope (Section 2.1).

5. Conclusions

The resilience assessment framework (RAF) herein presented enables to highlight where the cities and respective urban services stand today regarding resilience to climate change, and to identify the most critical aspects to be improved, taking into account both the reference situation and the expected impacts of future climate change scenarios. The diagnosis allows for understanding those aspects that are being tackled properly and also to determine gaps and areas of improvement thanks to the great level of detail of the different dimensions that make up the assessment. It also provides a means to assess resilience progress, therefore contributing to an integrated and forward-looking approach to resilient and sustainable urban development. Additionally, it may facilitate communication among different stakeholders and between different decision levels.

The application of this framework to Bristol, Barcelona and Lisbon cities have demonstrated that the RAF is a tool that provides support to a structured assessment of urban resilience to climate change with a focus on water. Even though it was developed within the scope of climate change and with a focus on the water cycle, replication to other hazards and services is considered on its foundation. Given its different assessment levels, it may be used by any city, service or organisation that intends to undertake a resilience assessment with this scope and focus, regardless of their resilience maturity. The RAF allows to align with the resilience path and integrate the work already in place in the cities and services, as well as to consider the information provided by diverse analysis approaches and tools, already in use or to be used by the city and service managers. Given the adopted structure, an effective and robust implementation requires the involvement of multiple parties, in a collaborative process allowing incorporation of the best available information.

The RAF is a flexible framework allowing further inclusion of additional dimensions, such as social or economic, and of other objectives, criteria and metrics, for the services already addressed. Moreover, it may be strengthened with the incorporation of other services, such as telecommunication, education or health. Other development opportunities are the consideration of other hazards, such as earthquakes, or of other risks.

Author Contributions: M.A.C. supervised this entire study, co-developed the methodology, the framework and the validation, co-analysed the results, drafted manuscript and finalised it. R.S.B. co-developed the methodology, the framework and the validation, co-analysed the results and provided suggestions on the draft manuscript. C.P. co-developed the framework and the validation, co-analysed the results, contributed to the draft manuscript. A.G. coordinated the framework application in Barcelona and contributed to the draft manuscript. J.S. coordinated the framework application in Bristol and contributed to the draft manuscript. M.J.T. coordinated the framework application in Lisbon, provided suggestions on the framework and contributed to the draft manuscript. All authors have read and agreed to the published version of the manuscript.

Funding: This research was funded by EUROPEAN UNION'S HORIZON 2020 RESEARCH AND INNOVATION PROGRAM, under the Grant Agreement number 700174.

Acknowledgments: The work presented was developed within the EU H2020 RESCCUE project—Resilience to Cope with Climate change in Urban areas. Acknowledgment is due to all RESCCUE partners, particularly from UN-Habitat and Luís Mesquita David and Maria do Céu Almeida from LNEC regarding contributions to the framework development, as well as to all participants of the Bristol, Barcelona and Lisbon workshops, particularly the external contributors, the organisers and facilitators fundamental for the validation.

Conflicts of Interest: The authors declare no conflict of interest.

Appendix A

Resilience Assessment Framework Including Metrics Overview

Table A1. Organisational dimension.

OBJECTIVE Criterion PI		PI Unit
COLLECTIVE ENGAGEMENT AND AWARENESS		
Citizens and communities' engagement		
O01	Community or "grassroots" organisations, networks and training	(-)
	Are grassroots or community organisations participating in pre-event planning and post-event response for each neighbourhood in the city? (UNISDR Scorecard P7.1)	
O02	Civil society links	(-)
	Are civil society organisations engaged? (UNISDR Scorecard D4.1.4 (adapted))	
O03	Engagement of vulnerable groups of the population	(-)
	There is evidence of disaster resilience planning with or for the relevant groups of vulnerable population, and there is a confirmation from those groups of effective engagement. (UNISDR Scorecard D7.2.2 (adapted))	
O04	Citizen engagement techniques	(-)
	How effective is the city at citizen engagement and communications in relation to disaster risk reduction (DRR)? (UNISDR Scorecard P7.4)	
O05	Use of mobile and e-mail "systems of engagement" to enable citizens to receive and give updates before and after a disaster	(-)
	Use of mobile and social computing-enabled systems of engagement. All information before, during and after an event is supported by email, available on mobile devices, supported by alerts on social media, used to enable an in-bound "citizen to government" flow allowing crowd sourcing of data on events and issues. (UNISDR Scorecard D7.4.2 (adapted))	
Citizens and communities' awareness and training		
O06	Public education and awareness	(-)
	Existence and reach of a co-ordinated public relations and education campaign, with structured messaging and channels to ensure hazard, risk and disaster information is disseminated to the public. (UNISDR Scorecard P6.2)	
O07	Training delivery	(-)
	Existence and reach (to all sectors) of training courses covering risk and resilience issues. (UNISDR Scorecard P6.4)	
O08	Drills	(-)
	Do practices and drills involve both the public and professionals? (UNISDR Scorecard P9.7)	
O09	Social networks	(-)
	Are there regular training programmes provided to the most vulnerable and at need populations in the city?	
O10	Validation of effectiveness of education	(-)
	Knowledge of "most probable" risk scenario and knowledge of key response and preparation steps is widespread throughout city. Tested by sample survey. (UNISDR Scorecard D7.4.3 (adapted))	
O11	Consultative planning process	(-)
	Existence and characteristics of formal planning consultative process?	
O12	Planning approval process	(-)
	Characteristics of the planning approval process?	
O13	Public finances	(-)
	Are the objectives of the city Strategy and/or Planning portfolio matched by adequate public finances?	
O14	Financial plan and budget for resilience, including contingency funds	(-)
	Does the city have in place a specific 'ring fenced' (protected) budget, the necessary resources and contingency fund arrangements for local disaster risk reduction (DRR) (mitigation, prevention, response and recovery)? (UNISDR Scorecard P3.2)	

Table A1. *Cont.*

OBJECTIVE Criterion PI		PI Unit
Coordination and communication with stakeholders		
O15	**Co-ordination with other government bodies** Does the city have a formal mechanism (e.g., Office, Committee, National/Regional Platform) to coordinate actions between city and other international, national, regional or local governments, which ensures integrated and flexible communication and collaboration between them?	(-)
O16	**Multi-stakeholder collaboration** Does the city have a formal stakeholder engagement programme (including the most socially vulnerable and at need populations)?	(-)
O17	**Access and use of digital services** In its stakeholder engagement programme, does the city encourage access and use of digital services?	(-)
O18	**Collaboration mechanisms** In its stakeholder engagement programme, does the city have mechanisms to ensure: a) regular, proactive and inclusive multi-stakeholder collaboration (including the most socially vulnerable and at need populations) (…)	(-)
Resilience-engaged city		
O19	**City Master Plan making and implementation** Does the city master plan (or relevant strategy/plan) include and localise and/or implement objectives of Agenda 2030?	(-)
O20	**City Master Plan monitoring and review** Is the City Master Plan periodically monitored and reviewed, ensuring it remains relevant and is properly operational?	(-)
O21	**Hazard Assessment** Existence of hazard assessment(s) (knowledge of key hazards that the city faces, including likelihood of occurrence)? (UNISDR Scorecard P2.1 (adapted))	(-)
O22	**Damage and loss estimation** Does risk assessment include estimations of damage and loss from potential disasters, based on current development and future urban and population growth? (UNISDR Scorecard D2.2.2 (adapted))	(-)
O23	**Shared understanding of infrastructure risk** Is there a shared understanding of risks between the city and various utility providers and other regional and national agencies that have a role in managing infrastructure such as power, water, roads and trains, of the points of stress on the system and city scale risks? (UNISDR Scorecard P2.2)	(-)
O24	**Plan for resilience** Does the city have a municipally approved resilience plan (strategy or action plan)? And what is its timeframe?	(-)
O25	**Plan for resilience and Climate Change** Does the resilience plan consider climate change (projection, scenarios, impacts, etc.)?	(-)
O26	**Plan integration in the City Master Plan** Is the resilience plan integrated with the City Master Plan?	(-)
O27	**External support for the resilience plan** Is the document being developed by the city alone or with support from INGOs/UN bodies working on the subject?	(-)
O28	**Robustness of resilience plan** How robust is the resilience plan?	(-)
O29	**Resilience Plan monitoring and review** Is the resilience plan periodically monitored and reviewed, ensuring it remains relevant and operational?	(-)
O30	**Knowledge of resilience scenarios** Are there agreed scenarios for resilience (with relevant background information and supporting notes, updated at agreed intervals), setting out city-wide exposure and vulnerability from each hazard, or groups of hazards? (UNISDR Scorecard P2.3 (adapted))	(-)

Table A1. *Cont.*

OBJECTIVE Criterion PI		PI Unit
Resilience-engaged city		
O31	Data sharing Extent to which data on the city's resilience context is shared with other organisations involved with the city's resilience. (UNISDR Scorecard P6.3)	(-)
O32	Integration Is resilience properly integrated with other key city functions/portfolios? (UNISDR Scorecard P1.3)	(-)
O33	Organisation, coordination and participation Is there a multi-agency/sectoral mechanism with appropriate authority and resources to address resilience?	(-)
O34	Critical infrastructure as a priority Is critical infrastructure resilience a city priority? (UNISDR Scorecard P8.1 (adapted))	(-)
O35	Critical infrastructure plan overview Does the city own and implement a critical infrastructure plan or strategy? (UNISDR Scorecard P8.1 (adapted))	(-)
O36	Cascading impacts Is there a collective understanding of potentially cascading failures between different city and infrastructure systems, under different scenarios, and a mapping of such cascading effects is available? (UNISDR Scorecard P2.4 (adapted))	(-)
O37	Learning from others Is the city proactively seeking to exchange knowledge and learn from other cities facing similar challenges? (UNISDR Scorecard P6.6 (adapted))	(-)
CITY PREPAREDNESS		
City preparedness for disaster response		
O38	Early warning Existence of Early Warning System for monitoring, forecasting and doing predictions on hazards (including climate change-related events) (UNISDR Scorecard P9.1 (adapted))	(-)
O39	Reach of warning Percentage of population reachable by early warning systems (UNISDR Scorecard P9.1.1.1 (adapted))	(-)
O40	Communications Would a significant loss of service be expected for a significant proportion of the city in the 'worst case' scenario event? (UNISDR Scorecard P8.6)	(-)
O41	Event management plans Is there a disaster management/preparedness/emergency response plan outlining city mitigation, preparedness and response to local emergencies? (UNISDR Scorecard P9.2)	(-)
O42	Staffing/responder needs Does the responsible disaster management authority have sufficient staffing capacity to support first responder duties in surge event scenario? (UNISDR Scorecard P9.3)	(-)
O43	Equipment and relief supply needs Are equipment and supply needs, as well as the availability of equipment, clearly defined? (UNISDR Scorecard P9.4)	(-)
O44	Definition of human resources, equipment and supply needs, and availability of equipment Has an estimated shortfall in human resources and equipment been identified?	(-)
O45	Existence of agreements If yes, have MOUs - or several ones - been signed, regarding mutual agreements with other cities or private sector resources, in order to cover the detected shortfall?	(-)
O46	Health care Would there be sufficient acute healthcare capabilities to deal with expected major injuries in 'worst case' scenario? (UNISDR Scorecard P8.7)	(-)
O47	Food, shelter, staple goods and fuel supply Would the city be able to continue to feed and shelter its population post-event? (UNISDR Scorecard P9.5)	(-)
O48	Interoperability and interagency working Is there an emergency operations' centre, with participation from all agencies, automating standard operating procedures specifically designed to deal with "most probable" and "most severe" scenarios? (UNISDR Scorecard P9.6)	(-)

Table A1. *Cont.*

OBJECTIVE Criterion PI		PI Unit
CITY PREPAREDNESS		
City preparedness for disaster response		
O49	Existence of civil society focal points for citizens Existence of volunteers and civil society organisations acting as focal points for citizens after an event, and regularly thereafter, to confirm safety issues, needs etc.	(-)
O50	Social connectedness and neighbourhood cohesion What is the estimated percentage of population that would be contacted by volunteers, within the 12 hours following an event and regularly thereafter? (UNISDR Scorecard D7.2.1 (adapted))	(%)
City preparedness for climate change		
O51	Management plans for climate-related events Does the city have a plan addressing climate-related events, either consisting of a specific document or integrated into the city's planning portfolio?	(-)
O52	Implementation of management plans for climate-related events If existing, is this document being implemented through defined standard operational procedures?	(-)
O53	Management plans for climate-related events monitoring and review If existing, is this document being monitored and reviewed in less than a 5-year interval?	(-)
O54	Knowledge of exposure and vulnerability for climate change scenarios Are there agreed climate change scenarios setting out city-wide exposure and vulnerability from each hazard, or groups of hazards? (UNISDR Scorecard P2.3 (adapted))	(-)
O55	City status when addressing contribution to climate change Comparing to the mean GHG emission per inhabitant that was considered to elaborate the official RCP scenarios, what are the current city's emissions?	(-)
O56	City commitment with mitigation of climate change effects Has the city signed any formal agreement in order to reach an established mitigation target for GHG reduction by 2050, when comparing to 1990 values?	(%)
O57	Planning for mitigation of climate change effects Are the mitigation targets for GHG (emission reduction by 2050) being considered in the city plans and being enforced in new projects?	(-)
City preparedness for recovery and build back		
O58	Post event recovery planning—pre event Is there a strategy or process in place for post-event recovery and reconstruction, including economic reboot, societal aspects etc.? (UNISDR Scorecard P10.1)	(-)
O59	Coordination of post event recovery Is the coordinating body for all post-disaster processes identified and structured, including the distribution of roles and responsibilities between relevant organisations? (UNISDR Scorecard D9.6.3 (adapted))	(-)
O60	Lessons learnt Do post-event assessment processes include failure analysis?	(-)
O61	Learning loops If yes, does this process allow to capture lessons learned, which then feed into design and delivery of rebuilding projects? (UNISDR Scorecard P10.2 (adapted))	(-)
O62	Insurance What level of insurance cover exists in the city, across all sectors - business and community? (UNISDR Scorecard P3.3)	(-)
O63	Damage and loss post-event assessment Does the city have a system in place to provide Post-Disaster Needs Assessment?	(-)
O64	Current post-event assessment system If yes, has such system been defined, implemented, tested and historic data is registered?	(-)

Table A1. *Cont.*

OBJECTIVE Criterion PI		PI Unit
Availability and access to basic services		
O65	Water supply	(%)
	Percentage of households with access to safe drinking water distribution.	
O66	Wastewater collection	(%)
	Percentage of households served by wastewater collection.	
O67	Wastewater treatment	(-)
	Provision of adequate treatment to wastewater through wastewater treatment plant.	
O68	Urban waste collection	(%)
	Percentage of population served by regular solid waste collection (having waste picked up within 200 m from households, by a legally established entity, on at least a weekly basis).	
O69	Urban waste treatment	(-)
	Provision of adequate treatment to solid waste through recovery methods or disposal in landfill?	
O70	Urban electrical energy network	(%)
	Percentage of households with regular connection to the electricity network.	
O71	Urban electrical energy alternative source	(%)
	Estimated percentage of households connected to alternative sources of electricity.	
O72	Urban gas energy network	(%)
	Percentage of households with regular access to the gas distribution network.	
O73	Urban mobility accessing collective transportation	(%)
	Percentage of population living less than 500 m. from any type of public stop, including trains, subway, tram, bus transportation.	
O74	Urban cycling mobility	(-)
	Is there a public plan/strategy to develop cycling paths in the city or expend the existing network?	

Table A2. Spatial dimension.

OBJECTIVE Criterion PI		PI Unit
SPATIAL RISK MANAGEMENT		
General hazard and exposure mapping		
S01	Presentation process for risk information	(-)
	Do clear hazard maps and data on risk exist? (UNISDR Scorecard P2.5 (adapted))	
S02	Update process for risk information	(-)
	If yes, are these maps regularly updated? (UNISDR Scorecard P2.5 (adapted))	
S03	Knowledge of exposure and vulnerability	(-)
	Existence of scenarios setting out city-wide exposure and vulnerability from each hazard level. (UNISDR Scorecard D2.2.1)	
S04	Scenarios and update process for risk information	(-)
	Risk scenarios are updated at least every three years for the following. (UNISDR Scorecard D2.5.1 (adapted))	
S05	Damage and loss estimation	(-)
	Damage and loss aspects taken into account by risk assessments for key identified scenarios. (UNISDR Scorecard D2.2.2)	
Hazard and exposure for climate change		
S06	Potential population at risk of displacement for climate change scenarios	(-)
	Percentage of population at risk of displacement for three months or longer according to climate change scenarios. (UNISDR Scorecard D4.1.1 (adapted))	
S07	Urban footprint at risk for climate change scenarios	(-)
	Percentage of urban footprint at risk, according to climate change scenarios.	
S08	Economic activity at risk for climate change scenarios	(-)
	Percentage of economic activity at risk from climate change scenarios. (UNISDR Scorecard D4.1.2.1 (adapted))	

Table A2. *Cont.*

OBJECTIVE Criterion PI		PI Unit
Resilient urban development		
S09	**Land use zoning and planning** Is the land use plan - including zoning - informed by risk scenarios?	(-)
S10	**Land use plan monitoring and review** Is this plan regularly monitored and reviewed? (UNISDR Scorecard P4.1 (adapted))	(-)
S11	**Land use zoning implementation** Extent to which land use zoning is implemented in the city and complied with. (UNISDR Scorecard D4.4.1 (adapted))	(-)
S12	**New urban development** Is there a policy promoting physical measures in new development that enhance resilience to one or multiple hazards? (UNISDR Scorecard P4.2 (adapted))	(-)
S13	**Urban design solutions that increase resilience** Does the city implement urban design solutions tasked to improve resilience? (UNISDR Scorecard D4.2.1 (adapted))	(-)
S14	**Building codes and standards** Do building codes or standards exist, and do they address specific known hazards and risks for the city? Are these standards regularly updated? (UNISDR Scorecard P4.3)	(-)
S15	**Application of building codes** Implementation of building codes on relevant structures, certified as such by a 3rd party. (UNISDR Scorecard D4.4.2)	(-)
Impacts of climate-related event		
S16	**Human loss in the last events** Human impact of the last climate-related event, with similar or harsher climate variables than the most probable scenario.	(-)
S17	**Damages in urban footprint in the last events** Impact on urban footprint of the last climate-related event, with similar or harsher climate variables than the most probable scenario.	(%)
PROVISION OF PROTECTIVE INFRASTRUCTURES AND ECOSYSTEMS		
Protective infrastructures and ecosystems services		
S18	**Existing protective infrastructure** Is existing protective infrastructure designed and built according to risk information? (UNISDR Scorecard P8.2 (adapted))	(-)
S19	**New protective infrastructure** Is new protective infrastructure (in design or construction process) under development and consistent with best practice (for asset design, building and management, based on relevant risk information)?	(-)
S20	**Maintenance of protective infrastructure** Is protective infrastructure regularly maintained?	(-)
S21	**Awareness and understanding of ecosystem services/functions** Beyond just an awareness of the natural assets, does the city understand the functions that this natural capital provides for the city? (UNISDR Scorecard P5.1)	(-)
S22	**Awareness of the role that assets that provide ecosystem services play in the city's resilience** Assets that provide ecosystem services are specifically identified and managed as critical assets?	(-)
S23	**Trends in ecosystem services health** Change in health, extent or benefit of each ecosystem service in last 5 years. (UNISDR Scorecard D5.1.2)	(-)

Table A2. *Cont.*

OBJECTIVE Criterion PI		PI Unit
PROVISION OF PROTECTIVE INFRASTRUCTURES AND ECOSYSTEMS		
Protective infrastructures and ecosystems services		
S24	**Maintenance of ecosystem services** Are ecosystem services specifically maintained and annually monitored on a defined set of key health/performance indicators?	(-)
S25	**Availability of green and blue infrastructures** Estimated green and blue area per inhabitant.	$(m^2$/inhabitant)
S26	**Integration of green and blue infrastructure into city policy and projects** Is green and blue infrastructure being promoted on major urban development and infrastructure projects through policy?	(-)
Dependence and autonomy regarding other services considering climate change		
S27	**Critical services dependence of protective infrastructures and ecosystems under climate change scenarios** Critical services (CS -RESCCUE services) dependence of protective infrastructures and ecosystems under climate change scenarios.	(-)
S28	**Autonomy from other services under climate change scenarios** Protective infrastructure and ecosystems autonomy regarding critical services (CS -RESCCUE services) loss under climate change scenarios.	(-)
S29	**Transboundary environmental issues** Is the city aware of ecosystem services being provided to the city from natural capital beyond its administrative borders? Are agreements in place with neighbouring administrations to support the protection and management of these assets? (UNISDR Scorecard P5.3)	(-)

Table A3. Functional dimension for the Water Service.

OBJECTIVE Criterion PI		PI Unit
WATER SERVICE PLANNING AND RISK MANAGEMENT		
Strategic planning		
FWts01	**Water service strategic plan making and implementation** Does the service have a strategic plan and is it implemented? (UNISDR Scorecard P1.1 (adapted))	(-)
FWts02	**Plan alignment with the City Master Plan** If yes, is the plan aligned with the city main planning document?	(-)
FWts03	**Service plan monitoring and review** If existing, is the plan periodically monitored and reviewed, ensuring it remains relevant and operational?	(-)
FWts04	**Exchange of information to the city** Is there regular exchange of data and information between service and the city concerning the review of planning documents?	(-)
FWts05	**Land use zoning compliance** Do the service-specific plans comply with up-to-date land use and zoning regulations?	(-)
Resilience engaged service		
FWts06	**Resilience in water service strategy and alignment with City Master Plan** Does the service have a resilience plan (either as an autonomous action plan or as a strategy included in the service's strategic plan) and what is its timeframe?	(-)
FWts07	**Service strategic plan for resilience and CC** Does the resilience plan consider climate change (projection, scenarios, impacts, etc.)?	(-)
FWts08	**Service financial plan and budget for resilience** Do the service financial plans have dedicated allocations for resilience-building actions including disaster risk reduction (DRR))?	(-)
FWts09	**Water service business continuity** Do business continuity plans exist?	(-)
FWts10	**Co-ordination with other water services in the city** Is there any coordination mechanism in place with other water services/entities either at municipal or metropolitan level?	(-)
FWts11	**Learning from other water services** Is there any knowledge exchange with other services?	(-)

Table A3. *Cont.*

OBJECTIVE Criterion PI		PI Unit
Risk management		
FWts12	Risk information related to the water service Do specific service plans include risk information (such as exposure and vulnerability, damage and loss quantification, etc.) related to the service and are regularly updated?	(-)
FWts13	Damage and loss estimation Does risk assessment include estimations of damage and loss for agreed climate change scenarios, based on current development and future urban and population growth?	(-)
FWts14	Expected water supply interruptions, not caused by water quality problems, in the city area according to CC scenarios Percentage of the city area expected to be affected by water supply interruptions exceeding 6 h, not caused by water quality problems, according to climate change scenarios.	(% city area)
FWts15	Expected water supply interruptions caused by water quality problems, in the city area according to CC scenarios Percentage of the city area expected to be affected by interruptions exceeding 6 h, caused by water quality problems, according to climate change scenarios.	(% city area)
FWts16	Expected water supply interruptions, not caused by water quality problems, for sensitive customers according to CC scenarios % of sensitive customers expected to be affected by water supply interruptions exceeding 6 h, not caused by water quality problems, according to climate change scenarios.	(% sensitive customers)
FWts17	Expected water supply interruptions caused by water quality problems, for sensitive customers according to CC scenarios % of sensitive customers expected to be affected by interruptions exceeding 6 h, caused by water quality problems, according to climate change scenarios.	(% sensitive customers)
FWts18	Expected water supply interruptions, not caused by water quality problems, for other services according to CC scenarios % of customers of other services expected to be affected by water supply interruptions exceeding 6 h, not caused by water quality problems, according to climate change scenarios.	(% customers other services)
FWts19	Expected water supply interruptions caused by water quality problems, for other services according to CC scenarios % of customers of other services expected to be affected by interruptions exceeding 6 h, caused by water quality problems, according to climate change scenarios.	(% customers other services)
FWts20	Expected water supply interruptions, not caused by water quality problems, for households according to CC scenarios % of households expected to be affected by water supply interruptions exceeding 6 h, not caused by water quality problems, according to climate change scenarios.	(% households)
FWts21	Expected water supply interruptions caused by water quality problems, for households according to CC scenarios % of households expected to be affected by interruptions exceeding 6 h, caused by water quality problems, according to climate change scenarios.	(% households)
FWts22	Expected total duration of water supply interruption, not caused by water quality problems, according to CC scenarios Total duration (days) of expected water supply interruption, not caused by water quality problems, according to climate change scenarios.	(Days)
FWts23	Expected total duration of water supply interruption, caused by water quality problems, according to CC scenarios Total duration (days) of expected water supply interruption, caused by water quality problems, according to climate change scenarios.	(Days)

Table A3. *Cont.*

OBJECTIVE Criterion PI		PI Unit
Reliable service		
FWts24	Water supply interruptions, not caused by water quality problems, in the city area last year Percentage of the city area affected by water supply interruptions exceeding 6 h, not caused by water quality problems, last year.	(% city area)
FWts25	Water supply interruptions caused by water quality problems, in the city area last year Percentage of the city area affected by water supply interruptions exceeding 6 h, caused by water quality problems, last year.	(% city area)
FWts26	Water supply interruptions, not caused by water quality problems, for sensitive customers last year % of sensitive customers affected by water supply interruptions exceeding 6 h, not caused by water quality problems, last year.	(% sensitive customers)
FWts27	Water supply interruptions caused by water quality problems, for sensitive customers last year % of sensitive customers affected by water supply interruptions exceeding 6 h, caused by water quality problems, last year.	(% sensitive customers)
FWts28	Water supply interruptions, not caused by water quality problems, for other services last year % of customers of other services affected by water supply interruptions exceeding 6 h, not caused by water quality problems, last year.	(% customers other services)
FWts29	Water supply interruptions caused by water quality problems, for other services last year % of customers of other services affected by water supply interruptions exceeding 6 h, caused by water quality problems, last year.	(% customers other services)
FWts30	Water supply interruptions, not caused by water quality problems, for households last year % of households affected by water supply interruptions exceeding 6 h, not caused by water quality problems, last year.	(% households)
FWts31	Water supply interruptions caused by water quality problems, for households last year % of households affected by water supply interruptions exceeding 6 h, caused by water quality problems, last year.	(% households)
FWts32	Total duration of water supply interruption, not caused by water quality problems, last year Total duration (days) of water supply interruption, not caused by water quality problems, last year.	(Days)
FWts33	Total duration of water supply interruption, caused by water quality problems, last year Total duration (days) of water supply interruption, caused by water quality problems, last year.	(Days)
FWts34	Water losses last year Water losses last year (water loss volume in the supply system/(total pipe length × 365))	$(m^3/(km.day))$
Flexible service		
FWts35	Water uses % of drinking water being used for irrigation, street cleaning, firefighting, or other public uses.	(% drinking water)
FWts36	Water sources Which types of water supply sources are being used in the city?	(-)
FWts37	Water sources location Where are the city's water supply sources located?	(-)
FWts38	Service management Services are appropriately managed, i.e., technological tools are used, existing competences are adequate, and a command chain is at place?	(-)

Table A3. *Cont.*

OBJECTIVE Criterion PI		PI Unit
AUTONOMOUS WATER SERVICE		
Service importance to the city		
FWts39	Stakeholders perception Is there a mechanism to provide service score, based on stakeholders' perception and is it applied? If yes quantify the service score from stakeholder perception.	(-)
FWts40	Cascading impacts Is there an understanding of potentially cascading failures between different services, under different scenarios? (UNISDR Scorecard P2.4 (adapted))	(-)
Service inter-dependency with other services considering climate change		
FWts41	Critical services dependence on water service according to CC scenarios To what extent are critical services (CS -RESCCUE services) dependent on the water service, based on climate change scenarios?	(-)
FWts42	Water services autonomy from other critical services according to CC scenarios To what extent is the water service dependent on other critical services (CS -RESCCUE services), based on climate change scenarios?	(-)
WATER SERVICE PREPAREDNESS		
Service preparedness for disaster response		
FWts43	Water service event management plans Is there a disaster management/preparedness/emergency response plan outlining service mitigation, preparedness and response to local emergencies? (UNISDR Scorecard P9.2 (adapted))	(-)
FWts44	Water services interdepartmental collaboration for emergency Is there an emergency operations' centre, automating standard operating procedures specifically designed to deal with "most probable" and "most severe" scenarios? (UNISDR Scorecard P9.6 (adapted))	(-)
FWts45	Water services early warning Does the service have a plan or standard operating procedure to act on early warnings and forecasts? Is the city warned by this system? (UNISDR Scorecard P9.1 (adapted))	(-)
FWts46	Water service drills Are practices and drills carried out internally and periodically?	(-)
Service preparedness for climate change		
FWts47	Service commitment with mitigation of CC effects Is the service committed with an established mitigation target regarding reduction of GHG within its strategic planning?	(% reduction GHG)
FWts48	Existence of agreed CC scenarios and alignment with the city CC scenarios Are there agreed climate change scenarios, setting out service exposure and vulnerability, from each hazard level? Are they aligned with the city-wide climate change scenarios?	(-)
FWts49	Knowledge of exposure and service vulnerability for CC scenarios The analysis of exposure and service vulnerability for climate change scenarios addresses: a) People (…)	(-)
FWts50	Service planning for adaptation to CC Is adaptation to climate change being considered in the service plans and enforced in new projects?	(-)
FWts51	Implemented measures to address CC mitigation and adaptation What type of measures has the service implemented to address climate change mitigation and adaptation?	(-)
FWts52	Planned measures to address CC mitigation and adaptation What type of measures is the service planning to implement to address climate change mitigation and adaptation?	(-)
FWts53	Equipment capacity of the service Has the service adequate equipment capacity, in normal and emergency circumstances?	(-)
FWts54	Staffing capacity of the service Has the service adequate staffing capacity, in normal and emergency circumstances?	(-)

Table A3. *Cont.*

OBJECTIVE Criterion PI		PI Unit
Service preparedness for recovery and build back		
FWts55	Water service CC recovery planning Is there a strategy or process in place for post-event service recovery and reconstruction? (UNISDR Scorecard P10.1)	(-)
FWts56	Water service damage and loss post-event assessment Does the service have a system in place to provide Post-Disaster Needs Assessment?	(-)
FWts57	Current post-event assessment system If yes, has such system been defined, implemented, tested and historic data is registered?	(-)
FWts58	Water supply interruption, not caused by water quality problems, in the city area in the last relevant climate-related event Percentage of the city area affected by water supply interruptions exceeding 6 h, not caused by water quality, in the last climate-related event, with similar or harsher climate variables than the most probable scenario.	(% city area)
FWts59	Water supply interruptions caused by water quality problems, in the city area, in the last relevant climate-related event Percentage of the city area affected by water supply interruptions exceeding 6 h, caused by water quality problems, in the last climate-related event, with similar or harsher climate variables than the most probable scenario.	(% city area)
FWts60	Water supply interruptions, not caused by water quality problems, for sensitive customers in the last relevant climate-related event % of sensitive customers affected by water supply interruptions exceeding 6 h, not caused by water quality problems, in the last climate-related event, with similar or harsher climate variables than the most probable scenario.	(% sensitive customers)
FWts61	Water supply interruptions caused by water quality problems, for sensitive customers in the last relevant climate-related event % of sensitive customers affected by water supply interruptions exceeding 6 h, caused by water quality problems, in the last climate-related event, with similar or harsher climate variables than the most probable scenario.	(% sensitive customers)
FWts62	Water supply interruptions, not caused by water quality problems, for other services in the last relevant climate-related event % of customers of other services affected by water supply interruptions exceeding 6 h, not caused by water quality problems, in the last climate-related event, with similar or harsher climate variables than the most probable scenario.	(% customers other services)
FWts63	Water supply interruptions caused by water quality problems, for other services in the last relevant climate-related event % of customers of other services affected by water supply interruptions exceeding 6 h, caused by water quality problems, in the last climate-related event, with similar or harsher climate variables than the most probable scenario.	(% customers other services)
FWts64	Water supply interruptions, not caused by water quality problems, for households in the last relevant climate-related event % of households affected by water supply interruptions exceeding 6 h, not caused by water quality problems, in the last climate-related event, with similar or harsher climate variables than the most probable scenario.	(% households)
FWts65	Water supply interruptions caused by water quality problems, for households in the last relevant climate-related event % of households affected by water supply interruptions exceeding 6 h, caused by water quality problems, in the last climate-related event, with similar or harsher climate variables than the most probable scenario.	(% households)
FWts66	Total duration of water supply interruption, caused by water quality problems, in the last relevant climate-related event Days of water supply interruption, not caused by water quality problems, in the last climate-related event, with similar or harsher climate variables than the most probable scenario.	(Days)
FWts67	Total duration of water supply interruption, caused by water quality problems in the last relevant climate-related event Days of water supply interruption, caused by water quality problems, in the last climate-related event, with similar or harsher climate variables than the most probable scenario.	(Days)
FWts68	Water service lessons learnt and learning loops Are service-specific processes in place for lessons learnt, including failure analysis? If yes, are service-specific plans informed by them?	(-)
FWts69	Insurance What level of insurance cover exists in the service?	(-)

Table A4. Functional dimension for Wastewater Service.

OBJECTIVE Criterion PI		PI Unit
WASTEWATER SERVICE PLANNING AND RISK MANAGEMENT		
Strategic planning		
FWwt01	Wastewater service strategic plan making and implementation Does the service have a strategic plan and is it implemented? (UNISDR Scorecard P1.1 (adapted))	(-)
FWwt02	Plan alignment with the City Master Plan If yes, is the plan aligned with the city main planning document?	(-)
FWwt03	Service plan monitoring and review If existing, is the plan periodically monitored and reviewed, ensuring it remains relevant and operational?	(-)
FWwt04	Exchange of information to the city Is there regular exchange of data and information between service and the city concerning the review of planning documents?	(-)
FWwt05	Land use zoning compliance Do the service-specific plans comply with up-to-date land use and zoning regulations?	(-)
Resilience engaged service		
FWwt06	Resilience in wastewater service strategy and alignment with City Master Plan Does the service have a resilience plan (either as an autonomous action plan or as a strategy included in the service's strategic plan) and what is its timeframe?	(-)
FWwt07	Service strategic plan for resilience and CC Does the resilience plan consider climate change (projection, scenarios, impacts, etc.)?	(-)
FWwt08	Service financial plan and budget for resilience Do the service financial plans have dedicated allocations for resilience-building actions (including disaster risk reduction (DRR))?	(-)
FWwt09	Wastewater service business continuity Do business continuity plans exist?	(-)
FWwt10	Co-ordination with other wastewater services in the city Is there any coordination mechanism in place with other wastewater services/entities either at municipal or metropolitan level?	(-)
FWwt11	Learning from other wastewater services Is there any knowledge exchange with other services?	(-)
Risk management		
FWwt12	Risk information related to the wastewater service Do specific service plans include risk information (such as exposure and vulnerability, damage and loss quantification, etc.) related to the service and are regularly updated?	(-)
FWwt13	Damage and loss estimation Does risk assessment include estimations of damage and loss for agreed climate change scenarios, based on current development and future urban and population growth?	(-)
FWwt14	Expected wastewater flooding in the city area according to CC scenarios Percentage of the city area expected to be affected by flooding due to wastewater collection interruption, according to climate change scenarios.	(% city area)
FWwt15	Expected wastewater treatment failures in the city area according to CC scenarios Percentage of the city area expected to be affected by wastewater treatment failures, according to climate change scenarios.	(% city area)

Table A4. *Cont.*

OBJECTIVE Criterion PI		PI Unit
Risk management		
FWwt16	Expected wastewater flooding in sensitive customers according to CC scenarios	(% sensitive customers)
	% of sensitive customers expected to be affected by flooding due to wastewater collection interruption, according to climate change scenarios.	
FWwt17	Expected wastewater discharges, due to failure in wastewater service to ecosystem services according to CC scenarios	(-)
	Number of expected wastewater discharges into ecosystems services due to wastewater service interruption, according to climate change scenarios.	
FWwt18	Expected wastewater flooding in other services according to CC scenarios	(% customers other services)
	% of customers of other services expected to be affected by flooding due to wastewater collection interruption, according to climate change scenarios.	
FWwt19	Expected wastewater flooding in households according to CC scenarios	(% households)
	% of households expected to be affected by flooding due to wastewater collection interruption, according to climate change scenarios.	
FWwt20	Expected total duration of wastewater flooding period according to CC scenarios	(Days)
	Total duration (days) of expected wastewater flooding due to wastewater collection interruption, according to climate change scenarios.	
FWwt21	Expected total duration of wastewater treatment failure period according to CC scenarios	(Days)
	Total duration (days) of expected wastewater treatment failures, according to climate change scenarios.	
Reliable service		
FWwt22	Wastewater flooding in the city area last year	(% city area)
	Percentage of the city area affected by flooding due to wastewater collection interruption, last year.	
FWwt23	Wastewater treatment failures in the city area in the city area last year	(% city area)
	Percentage of the city area affected by wastewater treatment failures, last year.	
FWwt24	Wastewater flooding in sensitive customers last year	(% sensitive customers)
	% of sensitive customers affected by flooding due to wastewater collection interruption, last year.	
FWwt25	Wastewater discharges, due to failure in wastewater service, to ecosystem services last year	(-)
	Number of wastewater discharges into ecosystems services due to wastewater service interruption, last year.	
FWwt26	Wastewater flooding in other services last year	(% customers other services)
	% of customers of other services affected by flooding due to wastewater collection interruption, last year.	
FWwt27	Wastewater effective treatment in the city area last year	(%)
	Percentage of wastewater that was collected and safely treated, last year.	
FWwt28	Wastewater flooding in households last year	(% households)
	% of households affected by flooding due to wastewater collection interruption, last year.	
FWwt29	Total duration of wastewater flooding period last year	(Days)
	Total duration (days) of wastewater flooding, last year.	
FWwt30	Total duration of wastewater treatment failure period last year	(Days)
	Total duration (days) of wastewater treatment failure, last year.	
FWwt31	Estimated undue inflows into wastewater system last year	$(m^3/(km.day))$
	Undue inflows (e.g., stormwater, industrial, saline, water supply inflows) into the system last year (undue wastewater inflow volume in the collection system/(total pipe length × 365)).	

Table A4. *Cont.*

OBJECTIVE Criterion PI		PI Unit
Flexible service		
FWwt32	Treated wastewater uses	(% treated wastewater)
	Percentage of treated wastewater being recycled or reused (for e.g., irrigation, urban cleaning, firefighting).	
FWwt33	Wastewater disposal	(-)
	Which solutions for wastewater disposal are used in the city?	
FWwt34	Wastewater disposal location	(-)
	Where are the city's wastewater disposal points located?	
FWwt35	Service management	(-)
	Services are appropriately managed, i.e., technological tools are used, existing competences are adequate, and a command chain is in place?	
AUTONOMOUS WASTEWATER SERVICE		
Service importance to the city		
FWwt36	Stakeholders perception	(-)
	Is there a mechanism to provide service score, based on stakeholders' perception and is it applied? If yes quantify the service score from stakeholder perception.	
FWwt37	Cascading impacts	(-)
	Is there an understanding of potentially cascading failures between different services, under different scenarios? (UNISDR Scorecard P2.4 (adapted))	
Service inter-dependency with other services considering climate change		
FWwt38	Critical services dependence on wastewater service according to CC scenarios	(-)
	To what extent are critical services (CS -RESCCUE services) dependent on the wastewater service, based on climate change scenarios?	
FWwt39	Wastewater services autonomy from other critical services according to CC scenarios	(-)
	To what extent is the wastewater service dependent on other critical services (CS -RESCCUE services), based on climate change scenarios?	
WASTEWATER SERVICE PREPAREDNESS		
Service preparedness for disaster response		
FWwt40	Wastewater service event management plans	(-)
	Is there a disaster management/preparedness/emergency response plan outlining service mitigation, preparedness and response to local emergencies? (UNISDR Scorecard P9.2 (adapted))	
FWwt41	Wastewater services interdepartmental collaboration for emergency	(-)
	Is there an emergency operations' centre, automating standard operating procedures specifically designed to deal with "most probable" and "most severe" scenarios? (UNISDR Scorecard P9.6 (adapted))	
FWwt42	Wastewater services early warning	(-)
	Does the service have a plan or standard operating procedure to act on early warnings and forecasts? Is the city warned by this system? (UNISDR Scorecard P9.1 (adapted))	
FWwt43	Wastewater service drills	(-)
	Are practices and drills carried out internally and periodically?	
Service preparedness for climate change		
FWwt44	Service commitment with mitigation of CC effects	(% reduction GHG)
	Is the service committed with an established mitigation target regarding reduction of GHG within its strategic planning?	
FWwt45	Existence of agreed CC scenarios and alignment with the city CC scenarios	(-)
	Are there agreed climate change scenarios, setting out service exposure and vulnerability, from each hazard level? Are they aligned with the city-wide climate change scenarios?	
FWwt46	Knowledge of exposure and service vulnerability for CC scenarios	(-)
	The analysis of exposure and service vulnerability for climate change scenarios addresses: a) People (…)	
FWwt47	Service planning for adaptation to CC	(-)
	Is adaptation to climate change being considered in the service plans and enforced in new projects?	

Table A4. *Cont.*

OBJECTIVE Criterion PI		PI Unit
Service preparedness for climate change		
FWwt48	Implemented measures to address CC mitigation and adaptation What type of measures has the service implemented to address climate change mitigation and adaptation?	(-)
FWwt49	Planned measures to address CC mitigation and adaptation What type of measures is the service planning to implement to address climate change mitigation and adaptation?	(-)
FWwt50	Equipment capacity of the service Has the service adequate equipment capacity, in normal and emergency circumstances?	(-)
FWwt51	Staffing capacity of the service Has the service adequate staffing capacity, in normal and emergency circumstances?	(-)
Service preparedness for recovery and build back		
FWwt52	Wastewater service CC recovery planning Is there a strategy or process in place for post-event service recovery and reconstruction? (UNISDR Scorecard P10.1)	(-)
FWwt53	Wastewater service damage and loss post-event assessment Does the service have a system in place to provide Post-Disaster Needs Assessment?	(-)
FWwt54	Current post-event assessment system If yes, has such system been defined, implemented, tested and historic data is registered?	(-)
FWwt55	Wastewater flooding in the city area in the last relevant climate-related event Percentage of the city area affected by flooding due to wastewater collection interruption, in the last climate-related event, with similar or harsher climate variables than the most probable scenario.	(% city area)
FWwt56	Wastewater treatment failures in the city area in the last relevant climate-related event Percentage of the city area affected by wastewater treatment failures, in the last climate-related event, with similar or harsher climate variables than the most probable scenario.	(% city area)
FWwt57	Wastewater flooding in sensitive customers in the last relevant climate-related event % of sensitive customers affected by flooding due to wastewater collection interruption, in the last climate-related event, with similar or harsher climate variables than the most probable scenario.	(% sensitive customers)
FWwt58	Wastewater discharges, due to failure in wastewater service, to ecosystem services in the last relevant climate-related event Number of wastewater discharges into ecosystems services due to wastewater collection interruption, in the last climate-related event, with similar or harsher climate variables than the most probable scenario	(-)
FWwt59	Wastewater flooding for other services in the last relevant event % of customers of other services affected by flooding due to wastewater collection interruption, in the last climate-related event, with similar or harsher climate variables than the most probable scenario.	(% customers other services)
FWwt60	Wastewater effective treatment in the city area in the last relevant climate-related event Percentage of wastewater that was collected and safely treated, in the last climate-related event, with similar or harsher climate variables than the most probable scenario.	(%)
FWwt61	Wastewater flooding in households in the last relevant climate-related event % of households affected by flooding due to wastewater collection interruption, in the last climate-related event, with similar or harsher climate variables than the most probable scenario.	(% households)
FWwt62	Total duration of wastewater flooding period in the last relevant climate-related event Days of wastewater flooding, in the last climate-related event, with similar or harsher climate variables than the most probable scenario.	(Days)

Table A4. *Cont.*

OBJECTIVE Criterion PI		PI Unit
Service preparedness for recovery and build back		
FWwt63	Total duration of wastewater treatment failure period in the last relevant climate-related event Days of wastewater treatment failure, in the last climate-related event, with similar or harsher climate variables than the most probable scenario.	(Days)
FWwt64	Wastewater service lessons learnt and learning loops Are service-specific processes in place for lessons learnt, including failure analysis? If yes, are service-specific plans informed by them?	(-)
FWwt65	Insurance What level of insurance cover exists in the service?	(-)

Table A5. Functional resilience assessment framework of the Stormwater Service.

OBJECTIVE Criterion PI		PI Unit
STORMWATER SERVICE PLANNING AND RISK MANAGEMENT		
Strategic planning		
FSwt01	Stormwater service strategic plan making and implementation Does the service have a strategic plan and is it implemented? (UNISDR Scorecard P1.1 (adapted))	(-)
FSwt02	Plan alignment with the City Master Plan If yes, is the plan aligned with the city main planning document?	(-)
FSwt03	Service plan monitoring and review If existing, is the plan periodically monitored and reviewed, ensuring it remains relevant and operational?	(-)
FSwt04	Exchange of information to the city Is there regular exchange of data and information between service and the city concerning the review of planning documents?	(-)
FSwt05	Land use zoning compliance Do the service-specific plans comply with up-to-date land use and zoning regulations?	(-)
Resilience engaged service		
FSwt06	Resilience in stormwater service strategy and alignment with City Master Plan Does the service have a resilience plan (either as an autonomous action plan or as a strategy included in the service's strategic plan) and what is its timeframe?	(-)
FSwt07	Service strategic plan for resilience and CC Does the resilience plan consider climate change (projection, scenarios, impacts, etc.)?	(-)
FSwt08	Service financial plan and budget for resilience Do the service financial plans have dedicated allocations for resilience-building actions (including disaster risk reduction (DRR))?	(-)
FSwt09	Stormwater service business continuity Do business continuity plans exist?	(-)
FSwt10	Co-ordination with other stormwater services in the city Is there any coordination mechanism in place with other stormwater services/entities either at municipal or metropolitan level?	(-)
FSwt11	Learning from other stormwater services	(-)

Table A5. *Cont.*

OBJECTIVE Criterion PI		PI Unit
	Is there any knowledge exchange with other services?	
Risk management		
FSwt12	Risk information related to the stormwater service	(-)
	Do specific service plans include risk information (such as exposure and vulnerability, damage and loss quantification, etc.) related to the service and are regularly updated?	
FSwt13	Damage and loss estimation	(-)
	Does risk assessment include estimations of damage and loss for agreed climate change scenarios, based on current development and future urban and population growth?	
FSwt14	Expected stormwater flooding in the city area according to CC scenarios	(% city area)
	Percentage of the city area expected to be affected by flooding due to stormwater drainage problems, according to climate change scenarios.	
FSwt15	Expected stormwater flooding in sensitive customers according to CC scenarios	(% sensitive customers)
	% of sensitive customers expected to be affected by flooding due to stormwater drainage problems, according to climate change scenarios.	
FSwt16	Expected stormwater flooding in other services according to CC scenarios	(% customers other services)
	% of customers of other services expected to be affected by flooding due to stormwater drainage problems, according to climate change scenarios.	
FSwt17	Expected stormwater flooding in households according to CC scenarios	(% households)
	% of households expected to be affected by flooding due to stormwater drainage problems, according to climate change scenarios.	
FSwt18	Expected total duration of stormwater flooding period according to CC scenarios	(Days)
	Total duration (days) of expected stormwater flooding due to stormwater drainage problems, according to climate change scenarios.	
Reliable service		
FSwt19	Stormwater flooding in the city area last year	(% city area)
	Percentage of the city area affected by flooding due to stormwater drainage problems, last year.	
FSwt20	Stormwater flooding in sensitive customers last year	(% sensitive customers)
	% of sensitive customers affected by flooding due to stormwater drainage problems, last year.	
FSwt21	Stormwater flooding in other services last year	(% customers other services)
	% of customers of other services affected by flooding due to stormwater drainage problems, last year.	
FSwt22	Stormwater flooding in households last year	(% households)
	% of households affected by flooding due to stormwater drainage problems, last year.	
FSwt23	Total duration of stormwater flooding period last year	(Days)
	Total duration (days) of stormwater flooding, due to stormwater drainage problems, last year.	
FSwt24	Estimated undue inflows into stormwater system last year	$(m^3/(km.day))$
	Undue inflows (e.g., wastewater, industrial, saline, water supply inflows) into the system last year (undue wastewater inflow volume in the collection system/(total pipe length × 365)).	

Table A5. *Cont.*

OBJECTIVE Criterion PI		PI Unit
Flexible service		
FSwt25	Treated stormwater uses	(% treated stormwater)
	% of collected stormwater being recycled or reused (for e.g., irrigation, urban cleaning, firefighting).	
FSwt26	Stormwater disposal	(-)
	Which solutions for stormwater disposal are used in the city?	
FSwt27	Stormwater disposal location	(-)
	Where are the city's stormwater disposal points located?	
FSwt28	Service management	(-)
	Services are appropriately managed, i.e., technological tools are used, existing competences are adequate, and a command chain is at place?	
AUTONOMOUS STORMWATER SERVICE		
Service importance to the city		
FSwt29	Stakeholders perception	(-)
	Is there a mechanism to provide service score, based on stakeholders' perception and is it applied? If yes quantify the service score from stakeholder perception.	
FSwt30	Cascading impacts	(-)
	Is there an understanding of potentially cascading failures between different services, under different scenarios? (UNISDR Scorecard P2.4 (adapted))	
Service inter-dependency with other services considering climate change		
FSwt31	Critical services dependence on stormwater service according to CC scenarios	(-)
	To what extent are critical services (CS -RESCCUE services) dependent on the stormwater service, based on climate change scenarios?	
FSwt32	Stormwater services autonomy from other critical services according to CC scenarios	(-)
	To what extent is the stormwater service dependent on other critical services (CS -RESCCUE services), based on climate change scenarios?	
STORMWATER SERVICE PREPAREDNESS		
Service preparedness for disaster response		
FSwt33	Stormwater service event management plans	(-)
	Is there a disaster management/preparedness/emergency response plan outlining service mitigation, preparedness and response to local emergencies? (UNISDR Scorecard P9.2 (adapted))	
FSwt34	Stormwater services interdepartmental collaboration for emergency	(-)
	Is there an emergency operations' centre, automating standard operating procedures specifically designed to deal with "most probable" and "most severe" scenarios? (UNISDR Scorecard P9.6 (adapted))	
FSwt35	Stormwater services early warning	(-)
	Does the service have a plan or standard operating procedure to act on early warnings and forecasts? Is the city warned by this system? (UNISDR Scorecard P9.1 (adapted))	
FSwt36	Stormwater service drills	(-)
	Are practices and drills carried out internally and periodically?	

Table A5. *Cont.*

OBJECTIVE Criterion PI		PI Unit
Service preparedness for climate change		
FSwt37	Service commitment with mitigation of CC effects	(% reduction GHG)
	Is the service committed with an established mitigation target regarding reduction of GHG within its strategic planning?	
FSwt38	Existence of agreed CC scenarios and alignment with the city CC scenarios	(-)
	Are there agreed climate change scenarios, setting out service exposure and vulnerability, from each hazard level? Are they aligned with the city-wide climate change scenarios?	
FSwt39	Knowledge of exposure and service vulnerability for CC scenarios	(-)
	The analysis of exposure and service vulnerability for climate change scenarios addresses: a) People (…)	
FSwt40	Service planning for adaptation to CC	(-)
	Is adaptation to climate change being considered in the service plans and enforced in new projects?	
FSwt41	Implemented measures to address CC mitigation and adaptation	(-)
	What type of measures has the service implemented to address climate change mitigation and adaptation?	
FSwt42	Planned measures to address CC mitigation and adaptation	(-)
	What type of measures is the service planning to implement to address climate change mitigation and adaptation?	
FSwt43	Equipment capacity of the service	(-)
	Has the service adequate equipment capacity, in normal and emergency circumstances?	
FSwt44	Staffing capacity of the service	(-)
	Has the service adequate staffing capacity, in normal and emergency circumstances?	
Service preparedness for recovery and build back		
FSwt45	Stormwater service CC recovery planning	(-)
	Is there a strategy or process in place for post-event service recovery and reconstruction? (UNISDR Scorecard P10.1)	
FSwt46	Stormwater service damage and loss post-event assessment	(-)
	Does the service have a system in place to provide Post-Disaster Needs Assessment?	
FSwt47	Current post-event assessment system	(-)
	If yes, has such system been defined, implemented, tested and historic data is registered?	
FSwt48	Stormwater flooding in the city area in the last relevant climate-related event	(% city area)
	Percentage of the city area affected by flooding due to stormwater drainage problems in the last climate-related event, with similar or harsher climate variables than the most probable scenario.	
FSwt49	Stormwater flooding in sensitive customers in the last relevant climate-related event	(% sensitive customers)
	% of sensitive customers affected by flooding due to stormwater drainage problems in the last climate-related event, with similar or harsher climate variables than the most probable scenario.	
FSwt50	Stormwater flooding in other services in the last relevant climate-related event	(% customers other services)
	% of customers of other services affected by flooding due to stormwater drainage problems in the last climate-related event, with similar or harsher climate variables than the most probable scenario.	
FSwt51	Stormwater flooding in households in the last relevant climate-related event	(% households)
	% of households affected by flooding due to stormwater drainage problems in the last climate-related event, with similar or harsher climate variables than the most probable scenario.	
FSwt52	Total duration of stormwater flooding in the last relevant climate-related event	(Days)
	Days of stormwater flooding due to stormwater drainage problems in the last climate-related event, with similar or harsher climate variables than the most probable scenario.	
FSwt53	Stormwater service lessons learnt and learning loops	(-)
	Are service-specific processes in place for lessons learnt, including failure analysis? If yes, are service-specific plans informed by them?	
FSwt54	Insurance	(-)
	What level of insurance cover exists in the service?	

Table A6. Functional dimension for Waste Service.

OBJECTIVE Criterion PI		PI Unit
WASTE SERVICE PLANNING AND RISK MANAGEMENT		
Strategic planning		
FSlw01	Waste service strategic plan making and implementation Does the service have a strategic plan and is it implemented? (UNISDR Scorecard P1.1 (adapted))	(-)
FSlw02	Plan alignment with the City Master Plan If yes, is the plan aligned with the city main planning document?	(-)
FSlw03	Service plan monitoring and review If existing, is the plan periodically monitored and reviewed, ensuring it remains relevant and operational?	(-)
FSlw04	Exchange of information to the city Is there regular exchange of data and information between service and the city concerning the review of planning documents?	(-)
FSlw05	Land use zoning compliance Do the service-specific plans comply with up-to-date land use and zoning regulations?	(-)
Resilience engaged service		
FSlw06	Resilience in waste service strategy and alignment with City Master Plan Does the service have a resilience plan (either as an autonomous action plan or as a strategy included in the service's strategic plan) and what is its timeframe?	(-)
FSlw07	Service strategic plan for resilience and CC Does the resilience plan consider climate change (projection, scenarios, impacts, etc.)?	(-)
FSlw08	Service financial plan and budget for resilience Do the service financial plans have dedicated allocations for resilience-building actions (including disaster risk reduction (DRR))?	(-)
FSlw09	Waste service business continuity Do business continuity plans exist?	(-)
FSlw10	Co-ordination with other waste services in the city Is there any coordination mechanism in place with other solid waste services/entities either at municipal or metropolitan level?	(-)
FSlw11	Learning from other waste services Is there any knowledge exchange with other services?	(-)
Risk management		
FSlw12	Risk information related to the waste service Do specific service plans include risk information (such as exposure and vulnerability, damage and loss quantification, etc.) related to the service and are regularly updated?	(-)
FSlw13	Damage and loss estimation Does risk assessment include estimations of damage and loss for agreed climate change scenarios, based on current development and future urban and population growth?	(-)
FSlw14	Expected solid waste collection interruption in the city area according to CC scenarios. Percentage of the city area expected to be affected by solid waste collection interruptions exceeding 4 days, according to climate change scenarios.	(% city area)
FSlw15	Expected solid waste treatment failure in the city area according to CC scenarios Percentage of the city area expected to be affected by solid waste treatment problems exceeding 4 days, according to climate change scenarios.	(% city area)

Table A6. *Cont.*

OBJECTIVE Criterion PI		PI Unit
Risk management		
FSlw16	Expected solid waste collection interruption of sensitive customers according to CC scenarios % of sensitive customers expected to be affected by solid waste collection interruption exceeding 4 days, according to climate change scenarios.	(% sensitive customers)
FSlw17	Expected solid waste collection interruption for other services according to CC scenarios % of customers of other services expected to be affected by solid waste collection interruption exceeding 4 days, according to climate change scenarios.	(% customers other services)
FSlw18	Expected solid waste collection interruption in households according to CC scenarios % of households expected to be affected by solid waste collection interruption exceeding 4 days, according to climate change scenarios.	(% households)
FSlw19	Expected total duration of solid waste collection interruption period according to CC scenarios Total duration (days) of expected solid waste collection interruption, according to climate change scenario.	(Days)
FSlw20	Expected total duration of solid waste treatment failure period according to CC scenarios Total duration (days) of expected solid waste treatment failure, according to climate change scenarios.	(Days)
Reliable service		
FSlw21	Solid waste collection interruption in the city area last year Percentage of the city area affected by solid waste collection interruptions exceeding 4 days, last year.	(% city area)
FSlw22	Solid waste effective treatment failure in the city area last year Percentage of the city area affected by solid waste treatment problems exceeding 4 days, last year.	(% city area)
FSlw23	Solid waste collection interruption for sensitive customers last year % of sensitive customers affected by solid waste collection interruption exceeding 4 days, last year.	(% sensitive customers)
FSlw24	Solid waste collection interruption for other services, last year % of customers of other services affected by solid waste collection interruption exceeding 4 days, last year.	(% customers other services)
FSlw25	Solid waste effective treatment in the city area last year Percentage of solid waste that was collected and safely treated, last year.	(% safely treated solid waste)
FSlw26	Solid waste collection interruption in households, last year % of households affected by solid waste collection interruption exceeding 4 days, last year.	(% households)
FSlw27	Total duration of solid waste collection interruption period last year Total duration (days) of solid waste collection interruption, last year.	(Days)
FSlw28	Total duration of solid waste treatment failure period last year Total duration (days) of solid waste treatment failure, last year.	(Days)
FSlw29	Estimated undue wastes into solid waste system last year Types of undue wastes into the solid waste system.	(-)
Flexible service		
FSlw30	Treated solid waste recovered % of treated solid waste being recovered (from recycling and reuse, energy recovery, composting …)	(% treated solid waste being recovered)
FSlw31	Solid waste disposal Which solutions for solid waste disposal are used in the city?	(-)
FSlw32	Solid waste disposal location Where are the city's solid waste disposal points located?	(-)
FSlw33	Service management Services are appropriately managed, i.e., technological tools are used, existing competences are adequate, and a command chain is at place?	(-)

Table A6. *Cont.*

OBJECTIVE Criterion PI		PI Unit
AUTONOMOUS WASTE SERVICE		
Service importance to the city		
FSlw34	Stakeholders perception Is there a mechanism to provide service score, based on stakeholders' perception and is it applied? If yes quantify the service score from stakeholder perception.	(-)
FSlw35	Cascading impacts Is there an understanding of potentially cascading failures between different services, under different scenarios? (UNISDR Scorecard P2.4 (adapted))	(-)
Service inter-dependency with other services considering climate change		
FSlw36	Critical services dependence on solid waste service according to CC scenarios To what extent are critical services (CS -RESCCUE services) dependent on the waste service, based on climate change scenarios?	(-)
FSlw37	Solid waste services autonomy from other critical services according to CC scenarios To what extent is the waste service dependent on other critical services (CS -RESCCUE services), based on climate change scenarios?	(-)
WASTE SERVICE PREPAREDNESS		
Service preparedness for disaster response		
FSlw38	Solid waste service event management plans Is there a disaster management/preparedness/emergency response plan outlining service mitigation, preparedness and response to local emergencies? (UNISDR Scorecard 9.2 (adapted))	(-)
FSlw39	Solid waste services interdepartmental collaboration for emergency Is there an emergency operations' centre, automating standard operating procedures specifically designed to deal with "most probable" and "most severe" scenarios? (UNISDR Scorecard P9.6 (adapted))	(-)
FSlw40	Solid waste services early warning Does the service have a plan or standard operating procedure to act on early warnings and forecasts? Is the city warned by this system? (UNISDR Scorecard P9.1 (adapted))	(-)
FSlw41	Solid waste service drills Are practices and drills carried out internally and periodically?	(-)
Service preparedness for climate change		
FSlw42	Service commitment with mitigation of CC effects Is the service committed with an established mitigation target regarding reduction of GHG within its strategic planning?	(% reduction GHG)
FSlw43	Existence of agreed CC scenarios and alignment with the city CC scenarios Are there agreed climate change scenarios, setting out service exposure and vulnerability, from each hazard level? Are they aligned with the city-wide climate change scenarios?	(-)
FSlw44	Knowledge of exposure and service vulnerability for CC scenarios The analysis of exposure and service vulnerability for climate change scenarios addresses: a) People (…)	(-)
FSlw45	Service planning for adaptation to CC Is adaptation to climate change being considered in the service plans and enforced in new projects?	(-)
FSlw46	Implemented measures to address CC mitigation and adaptation What type of measures has the service implemented to address climate change mitigation and adaptation?	(-)
FSlw47	Planned measures to address CC mitigation and adaptation What type of measures is the service planning to implement to address climate change mitigation and adaptation?	(-)
FSlw48	Equipment capacity of the service Has the service adequate equipment capacity, in normal and emergency circumstances?	(-)
FSlw49	Staffing capacity of the service Has the service adequate staffing capacity, in normal and emergency circumstances?	(-)
Service preparedness for recovery and build back		
FSlw50	Solid waste service CC recovery planning Is there a strategy or process in place for post-event service recovery and reconstruction? (UNISDR Scorecard 10.1)	(-)
FSlw51	Solid waste service damage and loss post-event assessment Does the service have a system in place to provide Post-Disaster Needs Assessment?	(-)

Table A6. *Cont.*

OBJECTIVE Criterion PI		PI Unit
Service preparedness for recovery and build back		
FSlw52	Current post-event assessment system If yes, has such system been defined, implemented, tested and historic data is registered?	(-)
FSlw53	Solid waste collection interruption in the city area in the last relevant climate-related event % of city area with solid waste collection interruption in the last climate-related event, with similar or harsher climate variables than the most probable scenario.	(% city area)
FSlw54	Solid waste effective treatment failure in the city area in the last relevant climate-related event Percentage of the city area affected by solid waste treatment problems, in the last climate-related event, with similar or harsher climate variables than the most probable scenario.	(% city area)
FSlw55	Solid waste collection interruption in sensitive customers in the last relevant climate-related event % of sensitive customers affected by solid waste collection interruption, in the last climate-related event, with similar or harsher climate variables than the most probable scenario.	(% sensitive customers)
FSlw56	Solid waste collection interruption for other services in the last relevant climate-related event % of customers of other services affected by solid waste collection interruption in the last climate-related event, with similar or harsher climate variables than the most probable scenario.	(% customers other services)
FSlw57	Solid waste effective treatment in the city area in the last relevant climate-related event Percentage of solid waste that was collected and safely treated in the last climate-related event, with similar or harsher climate variables than the most probable scenario.	(% solid waste safely treated)
FSlw58	Solid waste collection interruption in households in the last relevant climate-related event % of households affected by solid waste collection interruption in the last climate-related event, with similar or harsher climate variables than the most probable scenario.	(% households)
FSlw59	Total duration of solid waste collection interruption in the last relevant climate-related event Days of solid waste collection interruption, in the last climate-related event, with similar or harsher climate variables than the most probable scenario.	(Days)
FSlw60	Total duration of solid waste treatment failure in the last relevant climate-related event Days of solid waste treatment failure, in the last climate-related event, with similar or harsher climate variables than the most probable scenario.	(Days)
FSlw61	Solid waste service lessons learnt and learning loops Are service-specific processes in place for lessons learnt, including failure analysis? If yes, are service-specific plans informed by them?	(-)
FSlw62	Insurance What level of insurance cover exists in the service?	(-)

Table A7. Functional dimension for the Energy Service.

OBJECTIVE Criterion PI		PI Unit
ENERGY SERVICE PLANNING AND RISK MANAGEMENT		
Strategic planning		
FEne01	**Energy service strategic plan making and implementation** Does the service have a strategic plan and is it implemented? (UNISDR Scorecard P1.1 (adapted))	(-)
FEne02	**Plan alignment with the City Master Plan** If yes, is the plan aligned with the city main planning document?	(-)
FEne03	**Service plan monitoring and review** If existing, is the plan periodically monitored and reviewed, ensuring it remains relevant and operational?	(-)
FEne04	**Exchange of information to the city** Is there regular exchange of data and information between service and the city concerning the review of planning documents?	(-)
FEne05	**Land use zoning compliance** Do the service-specific plans comply with up-to-date land use and zoning regulations?	(-)
Resilience engaged service		
FEne06	**Resilience in energy service strategy and alignment with City Master Plan** Does the service have a resilience plan (either as an autonomous action plan or as a strategy included in the service's strategic plan) and what is its timeframe?	(-)
FEne07	**Service strategic plan for resilience and CC** Does the resilience plan consider climate change (projection, scenarios, impacts, etc.)?	(-)
FEne08	**Service financial plan and budget for resilience** Do the service financial plans have dedicated allocations for resilience-building actions (including disaster risk reduction (DRR))?	(-)
FEne09	**Energy service business continuity** Do business continuity plans exist?	(-)
FEne10	**Co-ordination with other energy services in the city** Is there any coordination mechanism in place with other energy services/entities either at municipal or metropolitan level?	(-)
FEne11	**Learning from other energy services** Is there any knowledge exchange with other services?	(-)
Risk management		
FEne12	**Risk information related to the energy service** Do specific service plans include risk information (such as exposure and vulnerability, damage and loss quantification, etc.) related to the service and are regularly updated?	(-)
FEne13	**Damage and loss estimation** Does risk assessment include estimations of damage and loss for agreed climate change scenarios, based on current development and future urban and population growth?	(-)
FEne14	**Expected energy outage in the city area according to CC scenarios** Percentage of the city area expected to be affected by energy outage exceeding 6 h, according to climate change scenarios.	(% city area)
FEne15	**Expected energy outage for sensitive customers according to CC scenarios** % of sensitive customers expected to be affected by energy outage exceeding 6 h, according to climate change scenarios.	(% sensitive customers)

Table A7. *Cont.*

OBJECTIVE Criterion PI		PI Unit
Risk management		
FEne16	Expected energy outage for other services according to CC scenarios	(% customers other services)
	% of customers of other services expected to be affected by energy outage exceeding 6 h, according to climate change scenarios.	
FEne17	Expected energy outage for households according to CC scenarios	(% households)
	% of households expected to be affected by energy outage exceeding 6 h, according to climate change scenarios.	
FEne18	Expected total duration of energy outage period according to CC scenarios	(Days)
	Total duration (days) of expected energy outage, according to climate change scenarios.	
Reliable service		
FEne19	Energy outage in the city area last year	(% city area)
	Percentage of the city area affected by energy outage exceeding 6 h last year.	
FEne20	Energy outage for sensitive customers last year	(% sensitive customers)
	% of sensitive customers affected by energy outage exceeding 6 h last year.	
FEne21	Energy outage for other services last year	(% customers other services)
	% of customers of other services affected by energy outage exceeding 6 h last year.	
FEne22	Energy outage in households last year	(% households)
	% of households affected by energy outage exceeding 6 h last year.	
FEne23	Total duration of energy outage period last year	(Days)
	Total duration of energy outage periods last year (days).	
FEne24	Energy losses last year	(-)
	Energy losses last year (rate of electricity losses in distribution networks measured as the ratio between losses and supplies of electricity).	
Flexible service		
FEne25	Alternative energy sources	(% energy from renewable sources)
	% of energy coming from renewable sources.	
FEne26	Energy sources	(-)
	Which energy sources are used in the city?	
FEne27	Energy sources location	(-)
	Where are the city's energy source points located?	
FEne28	Service management	(-)
	Services are appropriately managed, i.e., technological tools are used, existing competences are adequate, and a command chain is at place?	

Table A7. *Cont.*

OBJECTIVE Criterion PI		PI Unit
AUTONOMOUS ENERGY SERVICE		
Service importance to the city		
FEne29	Stakeholders perception Is there a mechanism to provide service score, based on stakeholders' perception and is it applied? If yes, quantify the service score from stakeholder perception.	(-)
FEne30	Cascading impacts Is there an understanding of potentially cascading failures between different services, under different scenarios? (UNISDR Scorecard P2.4 (adapted))	(-)
Service inter-dependency with other services considering climate change		
FEne31	Critical services dependence on energy service according to CC scenarios To what extent are critical services (CS -RESCCUE services) dependent on the energy service, based on climate change scenarios?	(-)
FEne32	Energy services autonomy from other critical services according to CC scenarios To what extent is the energy service dependent on other critical services (CS -RESCCUE services), based on climate change scenarios?	(-)
ENERGY SERVICE PREPAREDNESS		
Service preparedness for disaster response		
FEne33	Energy service event management plans Is there a disaster management/preparedness/emergency response plan outlining service mitigation, preparedness and response to local emergencies? (UNISDR Scorecard P9.2 (adapted))	(-)
FEne34	Energy services interdepartmental collaboration for emergency Is there an emergency operations' centre, automating standard operating procedures specifically designed to deal with "most probable" and "most severe" scenarios? (UNISDR Scorecard P9.6 (adapted))	(-)
FEne35	Energy services early warning Does the service have a plan or standard operating procedure to act on early warnings and forecasts? Is the city warned by this system? (UNISDR Scorecard P9.1 (adapted))	(-)
FEne36	Energy service drills Are practices and drills carried out internally and periodically?	(-)
Service preparedness for climate change		
FEne37	Service commitment with mitigation of CC effects Is the service committed with an established mitigation target regarding reduction of GHG within its strategic planning?	(% reduction GHG)
FEne38	Existence of agreed CC scenarios and alignment with the city CC scenarios Are there agreed climate change scenarios, setting out service exposure and vulnerability, from each hazard level? Are they aligned with the city-wide climate change scenarios?	(-)
FEne39	Knowledge of exposure and service vulnerability for CC scenarios The analysis of exposure and service vulnerability for climate change scenarios addresses: a) People (…)	(-)

Table A7. *Cont.*

OBJECTIVE Criterion PI		PI Unit
Service preparedness for climate change		
FEne40	Service planning for adaptation to CC Is adaptation to climate change being considered in the service plans and enforced in new projects?	(-)
FEne41	Implemented measures to address CC mitigation and adaptation What type of measures has the service implemented to address climate change mitigation and adaptation?	(-)
FEne42	Planned measures to address CC mitigation and adaptation What type of measures is the service planning to implement to address climate change mitigation and adaptation?	(-)
FEne43	Equipment capacity of the service Has the service adequate equipment capacity, in normal and emergency circumstances?	(-)
FEne44	Staffing capacity of the service Has the service adequate staffing capacity, in normal and emergency circumstances?	(-)
Service preparedness for recovery and build back		
FEne45	Energy service CC recovery planning Is there a strategy or process in place for post-event service recovery and reconstruction? (UNISDR Scorecard P10.1)	(-)
FEne46	Energy service damage and loss post-event assessment Does the service have a system in place to provide Post-Disaster Needs Assessment?	(-)
FEne47	Current post-event assessment system If yes, has such system been defined, implemented, tested and historic data is registered?	(-)
FEne48	Energy outage in the city area in the last relevant climate-related event Percentage of city area affected by energy outage exceeding 6 h in the last climate-related event, with similar or harsher climate variables than the most probable scenario.	(% city area)
FEne49	Energy outage in sensitive customers in the last relevant climate-related event % of sensitive customers affected by energy outage exceeding 6 h in the last climate-related event, with similar or harsher climate variables than the most probable scenario.	(% sensitive customers)
FEne50	Energy outage in other services in the last relevant climate-related event % of customers of other services affected by energy outage exceeding 6 h in the last climate-related event, with similar or harsher climate variables than the most probable scenario.	(% customers other services)
FEne51	Energy outage in households in the last relevant climate-related event % of households affected by energy outage exceeding 6 h in the last climate-related event, with similar or harsher climate variables than the most probable scenario.	(% households)
FEne52	Total duration of energy outage in the last relevant climate-related event Days of energy outage in the last relevant climate-related event.	(Days)
FEne53	Energy service lessons learnt and learning loops Are service-specific processes in place for lessons learnt, including failure analysis? If yes, are service-specific plans informed by them?	(-)
FEne54	Insurance What level of insurance cover exists in the service?	(-)

Table A8. Functional dimension for the Mobility Service.

OBJECTIVE Criterion PI		PI Unit
MOBILITY SERVICE PLANNING AND RISK MANAGEMENT		
Strategic planning		
FMob01	Mobility service strategic plan making and implementation Existence and implementation of a strategic plan for the mobility in the city. (UNISDR Scorecard P1.1 (adapted))	(-)
FMob02	Characterization of mobility needs The plan includes the characterization of the following population mobility habits: a) Type of mobility solutions used (…)	(-)
FMob03	Mobility plan monitoring and review If existing, is the plan periodically monitored and reviewed, ensuring it remains relevant and operational?	(-)
FMob04	Routes hierarchy characterization The city established a hierarchy of its routes.	(-)
FMob05	Land use zoning compliance Do mobility-specific plans comply with up-to-date land use and zoning regulations?	(-)
Resilience engaged mobility		
FMob06	Resilience in Mobility service strategy Resilience's aspects are included in the mobility plan?	(-)
FMob07	Mobility plan for Climate Change The plan considers climate change (hazards, projections, scenarios, impacts, etc.)?	(-)
FMob08	Budget for resilience The mobility plan has dedicated allocations for resilience-building actions (including disaster risk reduction (DRR))?	(-)
FMob09	Co-ordination with other Mobility services in the city Is there any coordination mechanism in place between mobility services/entities either at municipal or metropolitan level?	(-)
FMob10	Learning from other Mobility services Is there any knowledge exchange with other services?	(-)
Risk management		
FMob11	Risk information related to the Mobility service Does the mobility plan include risk information (such as exposure and vulnerability, identification of higher flow routes, damage and loss quantification, etc.) and is it regularly updated?	(-)
FMob12	Damage and loss estimation Does risk assessment include estimations of damage and loss for agreed climate change scenarios, based on current development and future urban and population growth?	(-)
FMob13	Expected mobility interruption in the city area according to CC scenarios No city area at risk of mobility interruptions exceeding 2 h, due to the most probable scenario, for these services:	(-)
FMob14	Expected mobility interruption in the higher flow routes according to CC scenarios Expected mobility interruption exceeding 2 hours in the higher flow routes according to climate change scenarios.	(-)
FMob15	Expected mobility interruption for population according to CC scenarios No population living in the area expected to be affected by mobility interruption exceeding 2 h, due to the most probable scenario, for these services: a) Road based (…)	(-)
FMob16	Expected mobility interruption for long-distance passengers according to CC scenarios No long-distance passengers expected to be affected by mobility interruption exceeding 2 h, due to the most probable scenario, for these services: a) Road based (…)	(-)
FMob17	Expected mobility interruption period according to CC scenarios Less than 2 h of expected mobility interruption, due to the most probable scenario, for these services: a) Road based (…)	(-)

Table A8. *Cont.*

OBJECTIVE Criterion PI		PI Unit
Reliable mobility		
FMob18	Public transport spatial coverage	(% city area)
	Public transport is available and covers: a) More than or equal to 80% of the city area (…)	
FMob19	Public transport daily coverage	(Hours/day)
	Public transport is available.	
FMob20	Mobility interruption in the higher flow routes last year	(-)
	Mobility interruption exceeding 2 hours in the higher flow routes last year.	
FMob21	Mobility interruption in the city area last year	(-)
	Less than 2.5% of the city area with mobility interruptions exceeding 2 h, last year, for these services: a) Road based (…)	
FMob22	Mobility interruption for population last year	(-)
	Less than 2.5% of the population living in the area affected by mobility interruption exceeding 2 h, last year, for these services: a) Road based (…)	
FMob23	Mobility interruption for long-distance passengers last year	(-)
	Less than 2.5% of the long-distance passengers affected by mobility interruption exceeding 2 h, last year, for these services: a) Road based (…)	
FMob24	Total duration of mobility interruption period last year	(-)
	Less than 0.5 days of mobility interruption, last year, for these services: a) Road based (…)	
FMob25	Routes with restrictions to circulation of heavy vehicles	(-)
	The city has identified the routes with restriction to the circulation of heavy vehicles.	
FMob26	Routes with restrictions to circulation of medical or emergency vehicles	(-)
	The city has identified the routes with restriction to the circulation of medical or emergency vehicles.	
Flexible mobility		
FMob27	Alternative mobility	(% everyday cycling mobility)
	% of everyday cycling mobility.	
FMob28	City mobility solutions	(-)
	Which solutions for mobility are available in the city?	
FMob29	Modal split for city road-based solutions	(% share)
	% share of each road-based solution.	
FMob30	Long distance mobility solutions	(-)
	Which solutions for long distance mobility are available in the city?	
FMob31	Mobility passenger transference	(-)
	Where are the city's mobility central node points located?	
FMob32	Use of mobility management tools	(-)
	Mobility in the city recurs to the following management tools: a) Traffic lighting is managed in an integrated and automatic way (…)	
AUTONOMOUS MOBILITY		
Service importance to the city		
FMob33	Stakeholders perception of city mobility	(-)
	Is there a mechanism to provide service score, based on stakeholders' perception and is it applied? If yes, quantify the service score from stakeholder perception.	
FMob34	Cascading impacts	(-)
	Is there an understanding of potentially cascading failures between different mobility services, under different scenarios? (UNISDR Scorecard P2.4 (adapted))	
Service inter-dependency with other services considering climate change		
FMob35	Critical services dependence on mobility according to CC scenarios	(-)
	To what extent are critical services (CS -RESCCUE services) dependent on the mobility, based on climate change scenarios?	
FMob36	Mobility autonomy from other critical services according to CC scenarios	(-)
	To what extent is the mobility dependent on other critical services (CS -RESCCUE services), based on climate change scenarios?	

Table A8. *Cont.*

OBJECTIVE Criterion PI		PI Unit
MOBILITY PREPAREDNESS		
Mobility preparedness for climate change		
FMob37	Mobility commitment with mitigation of CC effects	(% reduction GHG)
	Is city mobility committed with an established mitigation target regarding reduction of GHG within its strategic planning?	
FMob38	Mobility interruption in the city area in the last relevant climate-related event	(% city area)
	Percentage of city area affected by mobility interruption exceeding 2 h, in the last climate-related event, with similar or harsher climate variables than the most probable scenario.	
FMob39	Mobility interruption in the higher flow routes in the last relevant climate-related event	(-)
	Mobility interruption exceeded 2 h in higher flow routes in the last climate-related event, with similar or harsher climate variables than the most probable scenario.	
FMob40	Mobility interruption for population in the last relevant climate-related event	(-)
	Less than 2.5% of population living in the area affected by mobility interruption exceeding 2 h, in the last climate-related event, with similar or harsher climate variables than the most probable scenario, for these services: a) Road based (…)	
FMob41	Mobility interruption for long-distance passengers in the last relevant climate-related event	(-)
	Less than 2.5% of long-distance passengers affected by mobility interruption exceeding 2 h, in the last climate-related event, with similar or harsher climate variables than the most probable scenario, for these services: a) Road based (…)	
FMob42	Mobility interruption period in the last relevant climate-related event	(-)
	Less than 2 h that mobility services suffered from interruption, in the last climate-related event, with similar or harsher climate variables than the most probable scenario, for these services: a) Road based (…)	

Table A9. Physical dimension for the water infrastructure.

OBJECTIVE Criterion PI		PI Unit
SAFE WATER INFRASTRUCTURE		
Infrastructure assets criticality and protection		
PWts01	Water infrastructure critical assets	(-)
	Are the critical infrastructure assets for service provision identified?	
PWts02	Component importance	(-)
	The identification of infrastructure critical assets is based in the following:	
PWts03	Water infrastructure critical assets mapping, review and update	(-)
	Are the infrastructure critical assets identified on hazard maps and included in data on risk?	
PWts04	Exchange of information	(-)
	Is there a regular exchange of information regarding infrastructure critical assets, hazard maps and data on risk with the city?	
PWts05	Protective buffers mapping and information to the city	(-)
	Have protective buffers to safeguard infrastructure assets been defined, are they clearly identified on hazard maps and data on risk and is the city informed?	
Infrastructure assets robustness		
PWts06	Codes and standards for infrastructure	(-)
	Do codes or standards for infrastructure design and construction exist and are these implemented?	
PWts07	Maintenance of infrastructure	(-)
	Is infrastructure maintained on a regular basis (according to a preventive maintenance plan), resources for corrective maintenance are assured and all maintenance information is continuously registered?	
PWts08	Water pump failures last year	(Days)
	Average number of days that system pumps were out of order last year.	
PWts09	Water mains bursts last year	(No./100 km)
	Relative number of water mains bursts last year (No./system length (km) × 100 km).	
PWts10	Water service connections bursts last year	(No./1000 connections)
	Number of water connections bursts last year (No./connections in the system × 1000 connections).	
PWts11	Hydrant failures last year	(No./1000 hydrants)
	Average number of hydrant failures last year (No./hydrants in the system × 1000 hydrants).	

Table A9. *Cont.*

OBJECTIVE Criterion PI		PI Unit
Infrastructure assets robustness		
PWts12	Power failures last year Average number of days pumping stations were out of service due to power supply interruptions last year.	(Days)
PWts13	Water quality last year Percentage of performed laboratory analysis that were in accordance to legal or regulatory requirements last year.	(%)
PWts14	Level of failure of critical infrastructure asset last year Percentage of critical infrastructure asset out of order last year.	(%)
PWts15	Coverage of expenditure in infrastructure last year Ratio between expenditure with rehabilitation, operation and management of infrastructure and annual operating budget of last year.	(-)
PWts16	Time for restoration last year Maximum out-of-service period for all failures in infrastructure, including recovery time, last year (days).	(Days)
PWts17	Real water losses Volume of real physical water losses, through any leaks, damaged pipes or overflows (m^3/(km.day)).	(m^3/(km.day))
PWts18	Energy efficiency in pumping stations Average normalized energy consumption in PS - pumping stations = (Total energy consumption for pumping/sum (Water volume in PS i × Manometric pressure head i/100).	(kWh/m^3.100m)
PWts19	Pollution prevention Percentage of sludge from water treatment with appropriate final disposal.	(% appropriate sludge disposal)
AUTONOMOUS AND FLEXIBLE WATER INFRASTRUCTURE		
Infrastructure assets importance to and dependency on other services		
PWts20	Cascading impacts There is knowledge concerning potentially cascading failures between the components of the infrastructure and the following infrastructure, under the agreed scenarios:	(-)
PWts21	Infrastructure of other services dependency on water infrastructure The infrastructure of the following services is dependent on water infrastructure: a) Infrastructure of the wastewater service (…)	(-)
PWts22	Dependency on infrastructures of other services The infrastructure of the water service directly depends on the infrastructure of the following services: a) Infrastructure of the wastewater service (…)	(-)
PWts23	Level of dependency Percentage of customers affected by infrastructure dependent on other services.	(% customers affected)
Infrastructure assets autonomy		
PWts24	Autonomy from infrastructures of other services Percentage of infrastructure directly dependent on other services that have an autonomy solution managed by the water service.	(% infrastructure)
PWts25	Level of autonomy Percentage of customers covered by infrastructure dependent on other services that benefit from autonomy solutions (i.e., customers that benefit/customers affected).	(% customers covered)
PWts26	Autonomy activation How is infrastructure autonomy activated? Specify the time required to activate it, if possible.	(-)
PWts27	Autonomy period Weighted average of autonomy period (Ti) of each dependent infrastructure (i) (i.e., Sum (Ti × level of autonomy i)).	(Days)
PWts28	Water storage autonomy Days of water supply autonomy provided by supply and distribution storage tanks = water inflow (m^3/year)/(water storage volume (m^3) × 365)	(Days)
PWts29	Energy self-production Percentage of energy consumption coming from self-production.	(%)
Infrastructure assets redundancy		
PWts30	Redundancy Is there an understanding of infrastructure redundancy, clearly identified on hazard maps and data on risk?	(-)
PWts31	Redundancy activation How is infrastructure redundancy activated? Specify the time required to activate it, if possible.	(-)
PWts32	Level of redundancy Percentage of customers covered by redundant infrastructure, i.e., with alternative infrastructure able to provide the service.	(% customers covered)

Table A9. *Cont.*

OBJECTIVE Criterion PI		PI Unit
WATER INFRASTRUCTURE PREPAREDNESS		
Contribution to city resilience		
PWts33	Use of design solutions to improve city resilience The design of the infrastructure incorporates the use of the following solutions to improve city resilience: a) Soakaways and porous pavement (…)	(-)
PWts34	Greenhouse gas emission target Contribution to greenhouse gas emission reduction.	(-)
PWts35	Other contributions to city resilience The water infrastructure and related services provide other contributions to city resilience in emergency situation, such as: a) Shelter (…)	(-)
Infrastructure assets exposure to climate change		
PWts36	Level of exposure of critical infrastructure assets to the most probable scenario Identify the critical infrastructure asset for which less than 10% is exposed to different hazards for climate change scenarios.	(-)
PWts37	Coverage of expenditure in infrastructure for most probable scenario Ratio between predicted expenditure on infrastructure affected by climate change scenarios and annual operating budget of last year.	(%)
PWts38	Time for restoration for most probable scenario Maximum out-of-service period predicted for all failures in infrastructure, including recovery time, due to different hazards for climate change scenarios.	(Days)
Preparedness for climate change		
PWts39	Implemented infrastructural measures to address CC mitigation and adaptation What type of measures were implemented in infrastructure design to address climate change mitigation and adaptation?	(-)
PWts40	Planned infrastructural measures to address CC mitigation and adaptation What type of measures are being planned in infrastructure design to address climate change mitigation and adaptation?	(-)
Preparedness for recovery and build back		
PWts41	Water pump failures in the last relevant event Number of days system pumps were out of order due to the last climate-related event, with similar or harsher climate variables than the most probable scenario.	(Days)
PWts42	Water service mains failures in the last relevant event Number of mains failures due to the last climate-related event, with similar or harsher climate variables than the most probable scenario (No./system length (km) × 100 km).	(No./100 km)
PWts43	Water service connection mains bursts in the last relevant event Number of water service connections mains bursts due to the last climate-related event, with similar or harsher climate variables than the most probable scenario (No./connections in the system × 1000 connections).	(No./1000 connections)
PWts44	Hydrant bursts in the last relevant event Number of hydrant bursts due to the last climate-related event, with similar or harsher climate variables than the most probable scenario (No./hydrants in the system × 1000 hydrants).	(No./1000 hydrants)
PWts45	Power failures in the last relevant event Number of days pumping stations were out of service by power supply interruptions due to the last climate-related event, with similar or harsher climate variables than the most probable scenario.	(Days)
PWts46	Water quality compliance in the last relevant event Percentage of laboratory analysis that were in accordance to legal or regulatory requirements due to the last climate-related event, with similar or harsher climate variables than the most probable scenario.	(%)
PWts47	Level of failure of critical assets in the last relevant event Percentage of critical infrastructure asset out of order due to the last climate-related event, with similar or harsher climate variables than the most probable scenario.	(%)
PWts48	Coverage of expenditure in infrastructure in the last relevant event Ratio between expenditure on infrastructure affected by the last climate-related event, with similar or harsher climate variables than the most probable scenario and annual operating budget of last year.	(%)
PWts49	Time for restoration in the last relevant event Maximum out-of-service period for all failures in infrastructure, including recovery time, due to the last climate-related event, with similar or harsher climate variables than the most probable scenario.	(Days)

Table A10. Physical dimension for the wastewater infrastructure.

OBJECTIVE Criterion PI		PI Unit
SAFE WASTEWATER INFRASTRUCTURE		
Infrastructure assets criticality and protection		
PWwt01	Wastewater infrastructure critical assets	(-)
	Are the critical infrastructure assets for service provision identified?	
PWwt02	Component importance	(-)
	The identification of infrastructure critical assets is based in the following: a) Population served (…)	
PWwt03	Wastewater infrastructure critical assets mapping, review and update	(-)
	Are the infrastructure critical assets identified on hazard maps and included in data on risk?	
PWwt04	Exchange of information	(-)
	Is there a regular exchange of information regarding infrastructure critical assets, hazard maps and data on risk with the city?	
PWwt05	Protective buffers mapping and information to the city	(-)
	Have protective buffers to safeguard infrastructure assets been defined, are they clearly identified on hazard maps and data on risk and is the city informed?	
Infrastructure assets robustness		
PWwt06	Codes and standards for infrastructure	(-)
	Do codes or standards for infrastructure design and construction exist and are these implemented?	
PWwt07	Maintenance of infrastructure	(-)
	Is infrastructure maintained on a regular basis (according to a preventive maintenance plan), resources for corrective maintenance are assured and all maintenance information is continuously registered?	
PWwt08	Wastewater pump failures last year	(Days)
	Average number of days that system pumps were out of order last year.	
PWwt09	Wastewater sewer pipe collapses last year	(No./100 km)
	Relative number of collapses in wastewater sewers last year (No./system length (km) $\times$ 100 km).	
PWwt10	Wastewater connection collapses last year	(No./1000 connections)
	Number of collapses in wastewater connections last year (No./connections in the system $\times$ 1000 connections).	
PWwt11	Power failures last year	(Days)
	Average number of days pumping stations were out of service due to power supply interruptions last year.	
PWwt12	Combined sewer overflow failures last year	(CSO discharges/total CSO devices)
	Average number of combined sewer overflows last year.	
PWwt13	Wastewater quality last year	(%)
	Percentage of performed laboratory analysis that were in accordance to legal or regulatory requirements last year.	
PWwt14	Level of failure of critical infrastructure assets last year	(%)
	Percentage of critical infrastructure asset out of order last year.	
PWwt15	Coverage of expenditure in infrastructure last year	(-)
	Ratio between expenditure with rehabilitation, operation and management of infrastructure and annual operating budget of last year.	

Table A10. *Cont.*

OBJECTIVE Criterion PI		PI Unit
Infrastructure assets robustness		
PWwt16	Time for restoration last year Maximum out-of-service period for all failures in infrastructure, including recovery time, last year.	(Days)
PWwt17	Real undue inflows into the wastewater infrastructure Volume of real physical undue inflows into the wastewater infrastructure, through joints, damaged pipes or wrong connections $(m^3/(km.day))$.	$(m^3/(km.day))$
PWwt18	Energy efficiency in pumping stations Average normalised energy consumption in PS – pumping stations = (Total energy consumption for pumping/sum (wastewater volume in PS i × Manometric pressure head i/100).	$(kWh/m^3.100m)$
PWwt19	Pollution prevention Percentage of sludge from wastewater treatment with appropriate final disposal.	(% appropriate sludge disposal)
AUTONOMOUS AND FLEXIBLE WASTEWATER INFRASTRUCTURE		
Infrastructure assets importance to and dependency on other services		
PWwt20	Cascading impacts There is knowledge concerning potentially cascading failures between the components of the infrastructure and the following infrastructure, under the agreed scenarios: a) Other infrastructure of the wastewater service (…)	(-)
PWwt21	Infrastructure of other services' dependency on wastewater infrastructure The infrastructure of the following services is dependent on wastewater infrastructure: a) Infrastructure of the water service (…)	(-)
PWwt22	Dependency on infrastructures of other services The infrastructure of the wastewater service directly depends on the infrastructure of the following services: a) Infrastructure of the water service (…)	(-)
PWwt23	Level of dependency Percentage of customers affected by infrastructure dependent on other services.	(% customers affected)
Infrastructure assets autonomy		
PWwt24	Autonomy from infrastructures of other services Percentage of infrastructure directly dependent on other services that have an autonomy solution managed by the wastewater service.	(% infrastructure)
PWwt25	Level of autonomy Percentage of customers covered by infrastructure dependent on other services that benefit from autonomy solutions (i.e., customers that benefit/customers affected).	(% customers covered)
PWwt26	Autonomy activation How is infrastructure autonomy activated? Specify the time required to activate it, if possible.	(-)
PWwt27	Autonomy period Weighted average of autonomy period (Ti) of each dependent infrastructure (i) (i.e., Sum (Ti × level of autonomy i)).	(Days)
PWwt28	Energy self-production Percentage of energy consumption coming from self-production.	(%)

Table A10. Cont.

OBJECTIVE Criterion PI		PI Unit
Infrastructure assets redundancy		
PWwt29	Redundancy Is there an understanding of infrastructure redundancy, clearly identified on hazard maps and data on risk?	(-)
PWwt30	Redundancy activation How is infrastructure redundancy activated? Specify the time required to activate it, if possible.	(-)
PWwt31	Level of redundancy Percentage of customers covered by redundant infrastructure, i.e., with alternative infrastructure able to provide the service.	(% customers covered)
WASTEWATER INFRASTRUCTURE PREPAREDNESS		
Contribution to city resilience		
PWwt32	Use of design solutions to improve city resilience The design of the infrastructure incorporates the use of the following solutions to improve city resilience: a) Soakaways and porous pavement (…)	(-)
PWwt33	Greenhouse gas emission target Contribution to greenhouse gas emission reduction.	(-)
PWwt34	Other contributions to city resilience The wastewater infrastructure and related services provide other contributions to city resilience in emergency situation, such as: a) Shelter (…)	(-)
Infrastructure assets exposure to climate change		
PWwt35	Level of exposure of critical infrastructure assets to the most probable scenario Identify the critical infrastructure asset for which less than 10% is exposed to different hazards for climate change scenarios.	(-)
PWwt36	Coverage of expenditure in infrastructure for most probable scenario Ratio between predicted expenditure with infrastructure affected by climate change scenarios and annual operating budget of last year.	(%)
PWwt37	Time for restoration for most probable scenario Maximum out-of-service period predicted for all failures in infrastructure, including recovery time, due to different hazards for climate change scenarios.	(Days)
Preparedness for climate change		
PWwt38	Implemented infrastructural measures to address CC mitigation and adaptation What type of measures were implemented in infrastructure design to address climate change mitigation and adaptation?	(-)
PWwt39	Planned infrastructural measures to address CC mitigation and adaptation What type of measures are being planned in infrastructure design to address climate change mitigation and adaptation?	(-)
Preparedness for recovery and build back		
PWwt40	Wastewater pump failures in the last relevant event Number of days system pumps were out of order due to the last climate-related event, with similar or harsher climate variables than the most probable scenario.	(Days)
PWwt41	Wastewater sewer pipe failures in the last relevant event Number of failures in wastewater sewers due to the last climate-related event, with similar or harsher climate variables than the most probable scenario (No./system length (km) × 100 km).	(No./100km)
PWwt42	Wastewater connection failures in the last relevant event Number of failures in wastewater connections due to the last climate-related event, with similar or harsher climate variables than the most probable scenario (No./connections in the system × 1000 connections).	(No./100km)
PWwt43	Combined sewer overflow failures in the last relevant event Number of combined sewer overflow failures due to the last climate-related event, with similar or harsher climate variables than the most probable scenario.	(CSO discharges/total CSO devices)

Table A10. *Cont.*

OBJECTIVE Criterion PI		PI Unit
Preparedness for recovery and build back		
PWwt44	Power failures in the last relevant event Number of days pumping stations were out of service by power supply interruptions due to the last climate-related event, with similar or harsher climate variables than the most probable scenario.	(Days)
PWwt45	Wastewater quality compliance in the last relevant event Percentage of laboratory analysis that were in accordance to legal or regulatory requirements due to the last climate-related event, with similar or harsher climate variables than the most probable scenario.	(%)
PWwt46	Level of failure of critical assets in the last relevant event Percentage of critical infrastructure asset out of order due to the last climate-related event, with similar or harsher climate variables than the most probable scenario.	(%)
PWwt47	Coverage of expenditure in infrastructure in the last relevant event Ratio between expenditure on infrastructure affected by the last climate-related event, with similar or harsher climate variables than the most probable scenario and annual operating budget of last year.	(%)
PWwt48	Time for restoration in the last relevant event Maximum out-of-service period for all failures in infrastructure, including recovery time, due to the last climate-related event, with similar or harsher climate variables than the most probable scenario.	(Days)

Table A11. Physical dimension for the stormwater infrastructure.

OBJECTIVE Criterion PI		PI Unit
SAFE STORMWATER INFRASTRUCTURE		
Infrastructure assets criticality and protection		
PSwt01	Stormwater infrastructure critical assets Are the critical infrastructure assets for service provision identified?	(-)
PSwt02	Component importance The identification of infrastructure critical assets is based in the following: a) Population served (…)	(-)
PSwt03	Stormwater infrastructure critical assets mapping, review and update Are the infrastructure critical assets identified on hazard maps and included in data on risk?	(-)
PSwt04	Exchange of information Is there a regular exchange of information regarding infrastructure critical assets, hazard maps and data on risk with the city?	(-)
PSwt05	Protective buffers mapping and information to the city Have protective buffers to safeguard infrastructure assets been defined, are they clearly identified on hazard maps and data on risk and is the city informed?	(-)
Infrastructure assets robustness		
PSwt06	Codes and standards for infrastructure Do codes or standards for infrastructure design and construction exist and are these implemented?	(-)
PSwt07	Maintenance of infrastructure Is infrastructure maintained on a regular basis (according to a preventive maintenance plan), resources for corrective maintenance are assured and all maintenance information is continuously registered?	(-)
PSwt08	Stormwater pump failures last year Average number of days that system pumps were out of order last year.	(Days)
PSwt09	Stormwater sewer pipe collapses last year Relative number of pipe collapses last year (No./system length (km) × 100 km).	(No./100 km)
PSwt10	Stormwater connection collapses last year Number of collapses in stormwater connections last year (No./connections in the system × 1000 connections).	(No./1000 connections)

Table A11. *Cont.*

OBJECTIVE Criterion PI		PI Unit
Infrastructure assets robustness		
PSwt11	Inlet failures last year Average number of inlet failures last year (No./inlets in the system × 1000 inlets).	(No./1000 inlets)
PSwt12	Power failures last year Average number of days pumping stations were out of service due to power supply interruptions last year.	(Days)
PSwt13	Stormwater quality last year Percentage of performed laboratory analysis that were in accordance to legal or regulatory requirements last year.	(%)
PSwt14	Level of failure of critical infrastructure assets last year Percentage of critical infrastructure asset out of order last year.	(%)
PSwt15	Coverage of expenditure in infrastructure last year Ratio between expenditure with rehabilitation, operation and management of infrastructure and annual operating budget of last year.	(-)
PSwt16	Time for restoration last year Maximum out-of-service period for all failures in infrastructure, including recovery time, last year.	(Days)
Infrastructure assets robustness		
PSwt17	Real undue inflows into the stormwater infrastructure Volume of real physical undue inflows into the stormwater infrastructure (e.g., soil, wastewater, industrial, saline, water supply inflows), through joints, damaged pipes or wrong connections (m^3/(km.day)).	(m^3/(km.day))
PSwt18	Energy efficiency in pumping stations Average normalized energy consumption in PS - pumping stations = (Total energy consumption for pumping/sum (stormwater volume in PS i × Manometric pressure head i/100).	(-)
PSwt19	Pollution prevention Percentage of sludge from stormwater treatment with appropriate final disposal.	(% appropriate sludge disposal)
AUTONOMOUS AND FLEXIBLE STORMWATER INFRASTRUCTURE		
Infrastructure assets importance to and dependency on other services		
PSwt20	Cascading impacts There is knowledge concerning potentially cascading failures between the components of the infrastructure and the following infrastructure, under the agreed scenarios: a) Other infrastructure of the stormwater service (…)	(-)
PSwt21	Infrastructure of other services' dependency on stormwater infrastructure The infrastructure of the following services is dependent on stormwater infrastructure: a) Infrastructure of the water service (…)	(-)
PSwt22	Dependency on infrastructures of other services The infrastructure of the stormwater service directly depends on the infrastructure of the following services: a) Infrastructure of the water service (…)	(-)
PSwt23	Level of dependency Percentage of customers affected by infrastructure dependent on other services.	(% customers affected)
Infrastructure assets autonomy		
PSwt24	Autonomy from infrastructures of other services Percentage of infrastructure directly dependent on other services that have an autonomy solution managed by the stormwater service.	(% infrastructure)
PSwt25	Level of autonomy Percentage of customers covered by infrastructure dependent on other services that benefit from autonomy solutions (i.e., customers that benefit/customers affected).	(% customers covered)

Table A11. *Cont.*

OBJECTIVE Criterion PI		PI Unit
Infrastructure assets autonomy		
PSwt26	**Autonomy activation** How is infrastructure autonomy activated? Specify the time required to activate it, if possible.	(-)
PSwt27	**Autonomy period** Weighted average of autonomy period (Ti) of each dependent infrastructure (i) (i.e., Sum (Ti × level of autonomy i)).	(Days)
PSwt28	**Capacity for zero floods** Based on the historical data, estimative of the maximum return period without city-wide flood ensured by the existing stormwater infrastructure.	(Years)
PSwt29	**Energy self-production** Percentage of energy consumption coming from self-production.	(%)
Infrastructure assets redundancy		
PSwt30	**Redundancy** Is there an understanding of infrastructure redundancy, clearly identified on hazard maps and data on risk?	(-)
PSwt31	**Redundancy activation** How is infrastructure redundancy activated? Specify the time required to activate it, if possible.	(-)
STORMWATER INFRASTRUCTURE PREPAREDNESS		
Contribution to city resilience		
PSwt32	**Use of design solutions to improve city resilience** The design of the infrastructure incorporates the use of the following solutions to improve city resilience: a) Soakaways and porous pavement (…)	(-)
PSwt33	**Greenhouse gas emission target** Contribution to greenhouse gas emission reduction.	(-)
PSwt34	**Other contributions to city resilience** The stormwater infrastructure and related services provide other contributions to city resilience in emergency situation, such as: a) Shelter (…)	(-)
Infrastructure assets exposure to climate change		
PSwt35	**Level of exposure of critical infrastructure assets to the most probable scenario** Identify the critical infrastructure asset for which less than 10% is exposed to different hazards for climate change scenarios.	(-)
PSwt36	**Coverage of expenditure in infrastructure for most probable scenario** Ratio between predicted expenditure with infrastructure affected by climate change scenarios and annual operating budget of last year.	(%)
PSwt37	**Time for restoration for most probable scenario** Maximum out-of-service period predicted for all failures in infrastructure, including recovery time, due to different hazards for climate change scenarios.	(Days)
Preparedness for climate change		
PSwt38	**Implemented infrastructural measures to address CC mitigation and adaptation** What type of measures were implemented in infrastructure design to address climate change mitigation and adaptation?	(-)
PSwt39	**Planned infrastructural measures to address CC mitigation and adaptation** What type of measures are being planned in infrastructure design to address climate change mitigation and adaptation?	(-)
Preparedness for recovery and build back		
PSwt40	**Stormwater pump failures in the last relevant event** Number of days system pumps were out of order due to the last climate-related event, with similar or harsher climate variables than the most probable scenario.	(Days)
PSwt41	**Stormwater sewer pipe failures in the last relevant event** Number of failures in stormwater sewers due to the last climate-related event, with similar or harsher climate variables than the most probable scenario (No./system length (km) × 100 km).	(No./100 km)
PSwt42	**Stormwater connection failures in the last relevant event** Number of failures in stormwater connections due to the last climate-related event, with similar or harsher climate variables than the most probable scenario (No./connections in the system × 1000 connections).	(No./1000 connections)

Table A11. *Cont.*

OBJECTIVE Criterion PI		PI Unit
Preparedness for recovery and build back		
PSwt43	Inlets failures in the last relevant event	(No./1000 inlets)
	Number of inlets failures due to the last climate-related event, with similar or harsher climate variables than the most probable scenario (No./inlets in the system × 1000 inlets).	
PSwt44	Power failures in the last relevant event	(Days)
	Number of days pumping stations were out of service by power supply interruptions due to the last climate-related event, with similar or harsher climate variables than the most probable scenario.	
PSwt45	Stormwater quality compliance in the last relevant event	(%)
	Percentage of laboratory analysis that were in accordance to legal or regulatory requirements due to the last climate-related event, with similar or harsher climate variables than the most probable scenario.	
PSwt46	Level of failure of critical assets in the last relevant event	(%)
	Percentage of critical infrastructure asset out of order due to the last climate-related event, with similar or harsher climate variables than the most probable scenario.	
PSwt47	Coverage of expenditure in infrastructure in the last relevant event	(%)
	Ratio between expenditure on infrastructure affected by the last climate-related event, with similar or harsher climate variables than the most probable scenario and annual operating budget of last year.	
PSwt48	Time for restoration in the last relevant event	(Days)
	Maximum out-of-service period for all failures in infrastructure, including recovery time, due to the last climate-related event, with similar or harsher climate variables than the most probable scenario.	

Table A12. Physical dimension for the waste infrastructure.

OBJECTIVE Criterion PI		PI Unit
SAFE WASTE INFRASTRUCTURE		
Infrastructure assets criticality and protection		
PSlw01	Solid waste infrastructure critical assets	(-)
	Are the critical infrastructure assets for service provision identified?	
PSlw02	Component importance	(-)
	The identification of infrastructure critical assets is based in the following: a) Population served (…)	
PSlw03	Solid waste infrastructure critical assets mapping, review and update	(-)
	Are the infrastructure critical assets identified on hazard maps and included in data on risk?	
PSlw04	Exchange of information	(-)
	Is there a regular exchange of information regarding infrastructure critical assets, hazard maps and data on risk with the city?	
PSlw05	Protective buffers mapping and information to the city	(-)
	Have protective buffers to safeguard infrastructure assets been defined, are they clearly identified on hazard maps and data on risk and is the city informed?	

Table A12. *Cont.*

OBJECTIVE Criterion PI		PI Unit
Infrastructure assets robustness		
PSlw06	**Codes and standards for infrastructure** Do codes or standards for infrastructure design and construction exist and are these implemented?	(-)
PSlw07	**Maintenance of infrastructure** Is infrastructure maintained on a regular basis (according to a preventive maintenance plan), resources for corrective maintenance are assured and all maintenance information is continuously registered?	(-)
PSlw08	**Waste collection infrastructure components failures last year** Average number of days with collection infrastructure components out of service last year.	(Days)
PSlw09	**Waste management service facilities unavailable last year** Relative number of waste management facilities unavailable for longer than 4 days, last year (facilities unavailable /total number of facilities).	(% facilities)
PSlw10	**Waste management fleet failures last year** Average number of days that at least 10% of the waste management fleet was out of service last year.	(-)
PSlw11	**Waste containers dumped or displaced last year** Relative number of waste containers dumped or displaced last year (number affected/total number of containers).	(% containers)
PSlw12	**Power failures interrupting service last year** Average number of days waste management were out of service due to power supply interruptions last year.	(Days)
PSlw13	**Laboratory analysis compliance** Percentage of laboratory analysis performed in disposal site that were in accordance to legal or regulatory requirements last year.	(%)
PSlw14	**Level of failure of critical infrastructure assets last year** Percentage of critical infrastructure asset out of order last year.	(%)
PSlw15	**Coverage of expenditure in infrastructure last year** Ratio between expenditure with rehabilitation, operation and management of infrastructure and annual operating budget of last year.	(-)
PSlw16	**Time for restoration last year** Maximum out-of-service period for all failures in infrastructure, including recovery time, last year.	(Days)
PSlw17	**Pollution prevention** Percentage of leachate from solid waste treatment with appropriate final disposal.	(% appropriate leachate disposal)
AUTONOMOUS AND FLEXIBLE WASTE INFRASTRUCTURE		
Infrastructure assets importance to and dependency on other services		
PSlw18	**Cascading impacts** There is knowledge concerning potentially cascading failures between the components of the infrastructure and the following infrastructure, under the agreed scenarios: a) Other infrastructure of the solid waste service (…)	(-)
PSlw19	**Infrastructure of other services' dependency on solid waste infrastructure** The infrastructure of the following services is dependent on waste infrastructure: a) Infrastructure of the water service (…)	(-)
PSlw20	**Dependency on infrastructures of other services** The infrastructure of the waste service directly depends on the infrastructure of the following services: a) Infrastructure of the water service (…)	(-)
PSlw21	**Level of dependency** Percentage of customers affected by infrastructure dependent on other services.	(% customers affected)

Table A12. *Cont.*

OBJECTIVE Criterion PI		PI Unit
Infrastructure assets autonomy		
PSlw22	Autonomy from infrastructures of other services	(% infrastructure)
	Percentage of infrastructure directly dependent on other services that have an autonomy solution managed by the solid waste service.	
PSlw23	Level of autonomy	(% customers covered)
	Percentage of customers covered by infrastructure dependent on other services that benefit from autonomy solutions (i.e., customers that benefit/customers affected).	
PSlw24	Autonomy activation	(-)
	How is infrastructure autonomy activated? Specify the time required to activate it, if possible.	
PSlw25	Autonomy period	(Days)
	Weighted average of autonomy period (Ti) of each dependent infrastructure (i) (i.e., Sum (Ti × level of autonomy i)).	
PSlw26	Waste storage autonomy	(Days)
	Days of waste storage autonomy provided by containers and transfer locations.	
PSlw27	Energy self-production	(%)
	Percentage of energy consumption coming from self-production.	
Infrastructure assets redundancy		
PSlw28	Redundancy	(-)
	Is there an understanding of infrastructure redundancy, clearly identified on hazard maps and data on risk?	
PSlw29	Redundancy activation	(-)
	How is infrastructure redundancy activated? Specify the time required to activate it, if possible.	
PSlw30	Level of redundancy	(% customers covered)
	Percentage of customers covered by redundant infrastructure, i.e., with alternative infrastructure able to provide the service.	
WASTE INFRASTRUCTURE PREPAREDNESS		
Contribution to city resilience		
PSlw31	Use of design solutions to improve city resilience	(-)
	The design of the infrastructure incorporates the use of the following solutions to improve city resilience: a) Soakaways and porous pavement (…)	
PSlw32	Recovered material from waste treatment	(% recovered material)
	% of recovered material from treatment per year (including composting, recycling and direct recovery).	
PSlw33	Greenhouse gas emission target	(-)
	Contribution to greenhouse gas emission reduction.	
PSlw34	Other contributions to city resilience	(-)
	The solid waste infrastructure and related services provide other contributions to city resilience in emergency situation, such as: a) Shelter (…)	
Infrastructure assets exposure to climate change		
PSlw35	Level of exposure of critical infrastructure assets to the most probable scenario	(-)
	Identify the critical infrastructure asset for which less than 10% is exposed to different hazards for climate change scenarios.	
PSlw36	Coverage of expenditure in infrastructure for most probable scenario	(%)
	Ratio between predicted expenditure with infrastructure affected by climate change scenarios and annual operating budget of last year.	
PSlw37	Time for restoration for most probable scenario	(Days)
	Maximum out-of-service period predicted for all failures in infrastructure, including recovery time, due to different hazards for climate change scenarios.	

Table A12. *Cont.*

OBJECTIVE Criterion PI		PI Unit
Preparedness for climate change		
PSlw38	Implemented infrastructural measures to address CC mitigation and adaptation What type of measures were implemented in infrastructure design to address climate change mitigation and adaptation?	(-)
PSlw39	Planned infrastructural measures to address CC mitigation and adaptation What type of measures are being planned in infrastructure design to address climate change mitigation and adaptation?	(-)
Preparedness for recovery and build back		
PSlw40	Waste collection infrastructure components failures last relevant event Number of days waste collection infrastructure components were out of service due to the last climate-related event, with similar or harsher climate variables than the most probable scenario.	(Days)
PSlw41	Waste management service facilities unavailable in the last relevant event Number of waste management service facilities unavailable in the last climate-related event, with similar or harsher climate variables than the most probable scenario.	(% facilities)
PSlw42	Waste management fleet failures in the last relevant event Number of waste management fleet failures due to the last climate-related event, with similar or harsher climate variables than the most probable scenario.	(-)
PSlw43	Waste containers dumped or displaced in the last relevant event Number of waste containers dumped or displaced due to the last climate-related event, with similar or harsher climate variables than the most probable scenario.	(% containers)
PSlw44	Power failures in the last relevant event Number of days waste management facilities were out of service by power supply interruptions due to the last climate-related event, with similar or harsher climate variables than the most probable scenario.	(Days)
PSlw45	Laboratory analysis compliance in the last relevant event Percentage of laboratory analysis performed in disposal site that were in accordance to legal or regulatory requirements in the last relevant event.	(%)
PSlw46	Level of failure of critical assets in the last relevant event Percentage of critical infrastructure asset out of order due to the last climate-related event, with similar or harsher climate variables than the most probable scenario.	(%)
PSlw47	Coverage of expenditure in infrastructure in the last relevant event Ratio between expenditure on infrastructure affected by the last climate-related event, with similar or harsher climate variables than the most probable scenario and annual operating budget of last year.	(%)
PSlw48	Time for restoration in the last relevant event Maximum out-of-service period for all failures in infrastructure, including recovery time, due to the last climate-related event, with similar or harsher climate variables than the most probable scenario.	(Days)

Table A13. Physical dimension for the energy infrastructure.

OBJECTIVE Criterion PI		PI Unit
SAFE ENERGY INFRASTRUCTURE		
Infrastructure assets criticality and protection		
PEne01	Energy infrastructure critical assets	(-)
	Are the critical infrastructure assets for service provision identified?	
PEne02	Component importance	(-)
	The identification of infrastructure critical assets is based in the following:	
PEne03	Energy infrastructure critical assets mapping, review and update	(-)
	Are the infrastructure critical assets identified on hazard maps and included in data on risk?	
PEne04	Exchange of information	(-)
	Is there a regular exchange of information regarding infrastructure critical assets, hazard maps and data on risk with the city?	
PEne05	Protective buffers mapping and information to the city	(-)
	Have protective buffers to safeguard infrastructure assets been defined, are they clearly identified on hazard maps and data on risk and is the city informed?	
Infrastructure assets robustness		
PEne06	Codes and standards for infrastructure	(-)
	Do codes or standards for infrastructure design and construction exist and are these implemented?	
PEne07	Maintenance of infrastructure	(-)
	Is infrastructure maintained on a regular basis (according to a preventive maintenance plan), resources for corrective maintenance are assured and all maintenance information is continuously registered?	
PEne08	Power station failure last year	(Days)
	Average number of days that power stations were out of service due to infrastructure problems last year.	
PEne09	Power substation failure last year	(Days)
	Average number of days that power substations were out of service due to infrastructure problems last year.	
PEne10	Power distribution network failures last year	(-)
	Number of failures in the distribution network last year.	
PEne11	Local power installations failures last year	(-)
	Number of sectional and transformation power stations and public lighting installations failures last year.	
PEne12	Level of failure of critical infrastructure assets last year	(%)
	Percentage of critical infrastructure assets out of order by failure last year.	
PEne13	Coverage of expenditure in infrastructure last year	(-)
	Ratio between expenditure with rehabilitation, operation and management of infrastructure and annual operating budget of last year.	
PEne14	Time for restoration last year	(Days)
	Maximum out-of-service period for all failures in infrastructure, including recovery time, last year.	
PEne15	Use of cooling waters	(l/kWh)
	Water use per year for cooling power stations.	
AUTONOMOUS AND FLEXIBLE ENERGY INFRASTRUCTURE		
Infrastructure assets importance to and dependency on other services		
PEne16	Cascading impacts	(-)
	There is knowledge concerning potentially cascading failures between the components of the infrastructure and the following infrastructure, under the agreed scenarios: a) Other infrastructure of the energy service (…)	
PEne17	Infrastructure of other services' dependency on energy infrastructure	(-)
	The infrastructure of the following services is dependent on energy infrastructure: a) Infrastructure of the wastewater service (…)	
PEne18	Dependency on infrastructures of other services	(-)
	The infrastructure of the energy service directly depends on the infrastructure of the following services: a) Infrastructure of the wastewater service (…)	
PEne19	Level of dependency	(% customers affected)
	Percentage of customers affected by infrastructure dependent on other services.	

Table A13. *Cont.*

OBJECTIVE Criterion PI		PI Unit
Infrastructure assets autonomy		
PEne20	Autonomy from infrastructures of other services	(% infrastructure)
	Percentage of infrastructure directly dependent on other services that have an autonomy solution managed by the energy service.	
PEne21	Level of autonomy	(% customers covered)
	Percentage of customers covered by infrastructure dependent on other services that benefit from autonomy solutions (i.e., customers that benefit/customers affected).	
PEne22	Autonomy activation	(-)
	How is infrastructure autonomy activated? Specify the time required to activate it, if possible.	
PEne23	Autonomy period	(Days)
	Weighted average of autonomy period (Ti) of each dependent infrastructure (i) (i.e., Sum (Ti × level of autonomy i)).	
Infrastructure assets redundancy		
PEne24	Redundancy	(-)
	Is there an understanding of infrastructure redundancy, clearly identified on hazard maps and data on risk?	
PEne25	Redundancy activation	(-)
	How is infrastructure redundancy activated? Specify the time required to activate it, if possible.	
PEne26	Level of redundancy	(% customers covered)
	Percentage of customers covered by redundant infrastructure, i.e., with alternative infrastructure able to provide the service.	
ENERGY INFRASTRUCTURE PREPAREDNESS		
Contribution to city resilience		
PEne27	Use of design solutions to improve city resilience	(-)
	The design of the infrastructure incorporates the use of the following solutions to improve city resilience: a) Soakaways and porous pavement (…)	
PEne28	Greenhouse gas emission target	(-)
	Contribution to greenhouse gas emission reduction.	
PEne29	Other contributions to city resilience	(-)
	The energy infrastructure and related services provide other contributions to city resilience in emergency situation, such as: a) Shelter (…)	
Infrastructure assets exposure to climate change		
PEne30	Level of exposure of critical infrastructure assets to the most probable scenario	(-)
	Identify the critical infrastructure asset for which less than 10% is exposed to different hazards for climate change scenarios.	
PEne31	Coverage of expenditure in infrastructure for most probable scenario	(%)
	Ratio between predicted expenditure with infrastructure affected by climate change scenarios and annual operating budget of last year.	
PEne32	Time for restoration for most probable scenario	(Days)
	Maximum out-of-service period predicted for all failures in infrastructure, including recovery time, due to different hazards for climate change scenarios.	

Table A13. Cont.

OBJECTIVE Criterion PI		PI Unit
Preparedness for climate change		
PEne33	Implemented infrastructural measures to address CC mitigation and adaptation What type of measures were implemented in infrastructure design to address climate change mitigation and adaptation?	(-)
PEne34	Planned infrastructural measures to address CC mitigation and adaptation What type of measures are being planned in infrastructure design to address climate change mitigation and adaptation?	(-)
Preparedness for recovery and build back		
PEne35	Power stations failure in the last relevant event Average number of days that power stations were out of service due to infrastructure problems due to the last climate-related event, with similar or harsher climate variables than the most probable scenario.	(Days)
PEne36	Power substation failure in the last relevant event Average number of days that power substations were out of service due to infrastructure problems due to the last climate-related event, with similar or harsher climate variables than the most probable scenario.	(Days)
PEne37	Power distribution network failures in the last relevant event Number of failures in the distribution network due to the last climate-related event, with similar or harsher climate variables than the most probable scenario.	(-)
PEne38	Local power installation failures in the last relevant event Number of sectional and transformation power stations and public lighting installation failures due to the last climate-related event, with similar or harsher climate variables than the most probable scenario.	(-)
PEne39	Level of failure of critical assets in the last relevant event Percentage of critical infrastructure asset out of order by failure due to the last climate-related event, with similar or harsher climate variables than the most probable scenario.	(%)
PEne40	Coverage of expenditure in infrastructure in the last relevant event Ratio between expenditure on infrastructure affected by the last climate-related event, with similar or harsher climate variables than the most probable scenario and annual operating budget of last year.	(-)
PEne41	Time for restoration in the last relevant event Maximum out-of-service period for all failures in infrastructure, including recovery time, due to the last climate-related event, with similar or harsher climate variables than the most probable scenario.	(Days)

Table A14. Physical dimension for the mobility infrastructure.

OBJECTIVE Criterion PI		PI Unit
SAFE MOBILITY INFRASTRUCTURE		
Infrastructure assets criticality and protection		
PMob01	Mobility infrastructure critical assets Are the critical infrastructure assets for mobility identified?	(-)
PMob02	Component importance for city mobility The identification of infrastructure critical assets for city mobility is based in the following: a) Population served (…)	(-)
PMob03	Mobility infrastructure critical assets mapping, review and update Are the infrastructure critical assets identified on hazard maps and included in data on risk?	(-)
PMob04	Protective buffers mapping and information to the city Have protective buffers to safeguard infrastructure assets been defined and are they clearly identified on hazard maps and data on risk?	(-)

Table A14. *Cont.*

OBJECTIVE Criterion PI		PI Unit
Infrastructure assets robustness		
PMob05	Codes and standards for infrastructure Do codes or standards for infrastructure design and construction exist and are these implemented?	(-)
PMob06	Maintenance of infrastructure Is infrastructure maintained on a regular basis (according to a preventive maintenance plan), resources for corrective maintenance are assured and all maintenance information is continuously registered?	(-)
PMob07	Road and rail routes failures last year Critical routes were out of order for less than 2 h on average last year, for these infrastructures: a) Road based (…)	(-)
PMob08	Transport interfaces failures last year Average number of hours that critical transport interfaces were out of order due to infrastructural failures last year.	(Hours)
PMob09	Power-related failures in road and rail routes last year Critical routes were out of order for less than 2 h on average, due to power-related failures, last year.	(-)
PMob10	Power-related failures in transport interfaces last year Average number of hours that critical transport interfaces were out of order due to power-related failures, last year.	(Hours)
PMob11	Flooding-related failures in road and rail routes last year Critical routes were out of order for less than 2 h on average, due to flooding, last year.	(-)
PMob12	Flooding-related failures in transport interfaces last year Average number of hours that critical transport interfaces were out of order due to flooding-related failures on average, last year.	(Hours)
PMob13	Coverage of expenditure in infrastructure last year Ratio of expenditure with rehabilitation, operation and management of infrastructure (routes and interfaces) and annual operating budget of last year between 0.9 and 1.0 or between 1.1 and 1.2, for these infrastructures: a) Road based (…)	(-)
PMob14	Time for restoration last year Mobility critical infrastructure (routes and interfaces) with a maximum out-of-service period for all failures in infrastructure, including recovery time, less than or equal to 7 h last year, for these infrastructures: a) Road based (…)	(-)
PMob15	Clean fuel public transport Existence of alternative clean fuel public transport in the city.	(-)
AUTONOMOUS AND FLEXIBLE MOBILITY INFRASTRUCTURE		
Infrastructure assets importance to and dependency on other services		
PMob16	Cascading impacts There is knowledge concerning potentially cascading failures between the components of the mobility infrastructure (road, train, air and water-based transport that applies) and the following infrastructure, under the agreed scenarios: a) Full knowledge between the components of the mobility infrastructure (…)	(-)
PMob17	Infrastructure of other services' dependency on mobility infrastructure The infrastructure of the following services is dependent on mobility infrastructure: a) Infrastructure of the water service (…)	(-)
PMob18	Dependency on infrastructures of other services The infrastructure of the mobility service directly depends on the infrastructure of the following services: a) Infrastructure of the water service.	(-)

Table A14. *Cont.*

OBJECTIVE Criterion PI		PI Unit
Infrastructure assets autonomy and redundancy		
PMob19	Energy self-production	(%)
	Percentage of energy consumption coming from self-production.	
PMob20	Redundancy	(-)
	Is there an understanding of infrastructure redundancy, clearly identified on hazard maps and data on risk?	
MOBILITY INFRASTRUCTURE PREPAREDNESS		
Contribution to city resilience		
PMob21	Use of design solutions to improve city resilience	(-)
	The design of the infrastructure incorporates the use of solutions to improve city resilience: a) Renewable energy generation (…)	
PMob22	Greenhouse gas emission target	(-)
	There is a prediction of GHG emissions reduction, aiming at the targets defined at the strategic planning level, from the following components of assets: a) Infrastructure operation (…)	
PMob23	Other contributions to city resilience	(-)
	The mobility infrastructure and related services provide other contributions to city resilience in emergency situation, such as: a) Shelter (…)	
Infrastructure assets exposure to climate change		
PMob24	Level of exposure of mobility infrastructure to the most probable scenario	(-)
	Identify the critical assets for which less than 10% is exposed to different hazards for climate change scenarios.	
PMob25	Coverage of expenditure in infrastructure for most probable scenario	(-)
	Ratio between predicted expenditure with infrastructure (routes and interfaces) affected by climate change scenarios and annual operating budget of last year between 0.9 and 1.0 or 1.1 and 1.2, for these infrastructures: a) Road based (…)	
PMob26	Time for restoration for most probable scenario	(-)
	Transport networks with maximum out-of-service period for all failures in infrastructure (routes and interfaces), including recovery time, for less than 7 h, due to different hazards for climate change scenarios, for these infrastructures: a) Road based (…)	
Preparedness for climate change		
PMob27	Implemented infrastructural measures to address CC mitigation and adaptation	(-)
	What type of measures were implemented in infrastructure design to address climate change mitigation and adaptation?	
PMob28	Planned infrastructural measures to address CC mitigation and adaptation	(-)
	What type of measures are being planned in infrastructure design to address climate change mitigation and adaptation?	
Preparedness for recovery and build back		
PMob29	Road and rail routes failures in the last relevant event	(-)
	Critical routes were out of order for less than 2 h on average due to the last climate-related event, with similar or harsher climate variables than the most probable scenario, for these infrastructures: a) Road based (…)	
PMob30	Transport interfaces failures in the last relevant event	(Hours)
	Average number of hours that critical transport interfaces were out of order due to infrastructural failures due to the last climate-related event, with similar or harsher climate variables than the most probable scenario.	
PMob31	Power-related failures in road and rail routes in the last relevant event	(-)
	Critical routes were out of order for less than 2 h on average, by power-related failures, due to the last climate-related event, with similar or harsher climate variables than the most probable scenario.	
PMob32	Power-related failures in transport interfaces in the last relevant event	(-)
	Critical routes were out of order for less than 2 h due to flooding on average, due to the last climate-related event, with similar or harsher climate variables than the most probable scenario.	
PMob33	Flooding-related failures in road and rail routes in the last relevant event	(Hours)
	Average number of hours that critical transport interfaces were out of order due to flooding-related failures on average, due to the last climate-related event, with similar or harsher climate variables than the most probable scenario.	

Table A14. Cont.

OBJECTIVE Criterion PI		PI Unit
Preparedness for recovery and build back		
PMob34	Flooding-related failures in transport interfaces in the last relevant event Average number of hours that critical transport interfaces were out of order due to power-related failures, due to the last climate-related event, with similar or harsher climate variables than the most probable scenario.	(Hours)
PMob35	Coverage of expenditure in infrastructure in the last relevant event Ratio of expenditure on rehabilitation, operation and management of infrastructure (routes and interfaces) affected by the last climate-related event, with similar or harsher climate variables than the most probable scenario, and annual operating budget of last year, is between 0.9 and 1.0 or 1.1 and 1.2, for these infrastructures: a) Road based (…)	(-)
PMob36	Time for restoration in the last relevant event Mobility critical infrastructure (routes and interfaces) with a maximum out-of-service period for all failures in infrastructure, including recovery time, less than or equal to 7 h due to the last climate-related event, with similar or harsher climate variables than the most probable scenario, for these infrastructures: a) Road based (…)	(-)

References

1. Revi, A.; Satterthwaite, D.E.; Aragón-Durand, F.; Corfee-Morlot, J.; Kiunsi, R.B.R.; Pelling, M.; Roberts, D.C.; Solecki, W. Urban areas. In *Climate Change 2014: Impacts, Adaptation, and Vulnerability. Part A: Global and Sectoral Aspects. Contribution of Working Group II to the Fifth Assessment Report of the Intergovernmental Panel on Climate Change*; Field, C.B., Barros, V.R., Dokken, D.J., Mach, K.J., Mastrandrea, M.D., Bilir, T.E., Chatterjee, M., Ebi, K.L., Estrada, Y.O., Genova, R.C., Eds.; Cambridge University Press: Cambridge, UK; New York, NY, USA, 2014; pp. 535–612.

2. Walloth, C.; Gurr, J.M.; Schmidt, J.A. (Eds.) *Understanding Complex Urban Systems: Multidisciplinary Approaches to Modelling*; Springer International Publishing: Cham, Switzerland, 2014.

3. Panda, A. Foreward. In Proceedings of the 8th International Conference on Building Resilience, ICBR, Lisbon, Portugal, 14–16 November 2018.

4. UN-GA. Sustainable development: Disaster risk reduction. Report of the open-ended intergovernmental expert working group on indicators and terminology relating to disaster risk reduction. In Proceedings of the United Nations—General Assembly, Seventy-First Session, Agenda Item, New York, NY, USA, 1 December 2016.

5. Patel, R.; Nosal, L. Defining the Resilient City. In *Working Paper 6*; United Nations University Centre for Policy Research: New York, NY, USA, 2016.

6. Sharifi, A. A critical review of selected tools for assessing community resilience. *Ecol. Indic.* **2016**, *69*, 629–647. [CrossRef]

7. UN-Habitat City Resilience Profiling Programme. Guide to the City Resilience Profiling Tool. United Nations Human Settlements Programme (UN-Habitat). 2018. Available online: http://urbanresiliencehub.org/wp-content/uploads/2018/10/CRPT-Guide-Pages-Online.pdf (accessed on 24 September 2018).

8. UNDRR. Disaster resilience scorecard for cities. Preliminary level assessment. In *United Nations International Strategy for Disaster Reduction United*; Nations Office for Disaster Reduction: Geneva, Switzerland, 2017.

9. UNDRR. Disaster resilience scorecard for cities. Detailed level assessment. In *United Nations International Strategy for Disaster Reduction United*; Nations Office for Disaster Reduction: Geneva, Switzerland, 2017.

10. ARUP. *City Resilience Framework. 100 Resilient Cities*; The Rockefeller Foundation, ARUP: New York, NY, USA, 2015.

11. ISO. *ISO 9001:2015—Quality Management Systems*; International Organisation for Standardisation: Geneva, Switzerland, 2015.

12. ICLEI 2010. Changing Climate, Changing Communities: Guide and Workbook for Municipal Climate Adaptation, ICLEI—Local Governments for Sustainability. 2010. Available online: https://icleicanada.org/wp-content/uploads/2019/07/Guide.pdf (accessed on 10 September 2018).

13. Rockefeller Foundation and Arup Group. City Resilience Framework. 2014. Available online: https://assets.rockefellerfoundation.org/app/uploads/20140410162455/City-Resilience-Framework-2015.pdf (accessed on 24 September 2018).

14. World Bank. City Strength: Resilient Cities Program. World Bank Group. 2015. Available online: https://openknowledge.worldbank.org/handle/10986/22470 (accessed on 10 September 2018).

15. EPA. *Evaluating Urban Resilience to Climate Change: A Multi-Sector Approach*; U.S. Environmental Protection Agency: Washington, DC, USA, 2017; EPA/600/R-16/365F.

16. Summers, J.K.; Smith, L.M.; Harwell, L.C.; Buck, K.D. Conceptualizing holistic community resilience to climate events: Foundation for a climate resilience screening index. *GeoHealth* **2017**, *1*, 151–164. [CrossRef] [PubMed]

17. Schipper, E.L.; Langston, L. *A Comparative Overview of Resilience Measurement Frameworks—Analysing Indicators and Approaches*; Working paper 422; Overseas Development Institute: London, UK, 2015.

18. Abdrabo, M.; Hassaan, M. Assessing Resilience of the Nile Delta Urban Centers to Sea Level Rise Impacts. In Proceedings of the 5th Global Forum on Urban Resilience & Adaptation, Bonn, Germany, 29–31 May 2014.

19. Joerin, J.; Shaw, R. Chapter 3 Mapping Climate and Disaster Resilience in Cities. In *Climate and Disaster Resilience in Cities (Community, Environment and Disaster Risk Management)*; Shaw, R., Sharma, A., Eds.; Emerald Group Publishing Limited: Bingley, UK, 2011; Volume 6, pp. 47–61. [CrossRef]

20. Peacock, W.G.; Brody, S.; Seitz, W.; Merrell, W.; Vedlitz, A.; Zahran, S.; Harriss, R.; Stickney, R. *Advancing Resilience of Coastal Localities: Developing, Implementing, and Sustaining the Use of Coastal Resilience Indicators: A Final Report*; Hazard Reduction and Recovery Center, Texas A&M University: College Station, TX, USA, 2010.

21. Batica, J. Methodology for Flood Resilience Assessment in Urban Environments and Mitigation Strategy Development. Ph.D. Thesis, Universite Nice Sophia Antipolis, Nice, France, 2015. Available online: https://tel.archives-ouvertes.fr/tel-01159935/document (accessed on 25 July 2019).

22. Ainuddin, S.; Routray, J.K. Earthquake hazards and community resilience in Baluchistan. *Nat. Hazards* **2012**, *63*, 909–937. [CrossRef]

23. Yoon, D.K.; Kang, J.E.; Brody, S.D. A measurement of community disaster resilience in Korea. *Environ. Plan. Manag.* **2016**, *59*, 436–460. [CrossRef]

24. Kwasinski, A.; Trainor, J.; Wolshon, B.; Lavelle, F.M. *A Conceptual Framework for Assessing Resilience at the Community Scale*; National Institute of Standards and Technology: Gaithersburg, MD, USA, 2016; NIST GCR 16-001.

25. UKWIR. *Resilience—Performance Measures, Costs and Stakeholder Communication*; UK Water Industry Research, Report Ref. No. 17/RG/06/4; UKWIR: London, UK, 2017.

26. Cox, R.S.; Hamlen, M. Community disaster resilience and the rural resilience index. *Am. Behav. Sci.* **2014**, *59*, 220–237. [CrossRef]

27. Brugmann, J. Financing the resilient city. *Environ. Urban.* **2012**, *24*, 215–232. [CrossRef]

28. Vallejo, L.; Mullan, M. *Climate-Resilient Infrastructure: Getting the Policies Right*; OECD Environment Working Papers, No. 121; OECD Publishing: Paris, France, 2017. [CrossRef]

29. Pagani, G.; Fournière, H.; Cardoso, M.A.; Brito, R.S. *Report with the Resilience Diagnosis for each City*; RESCCUE Project: Barcelona, Spain, 2018.

30. Brito, R.S.; Pereira, C.L.; Lopes, P.; Cardoso, M.A. RESCCUE RAF App—Climate change Resilience Assessment Framework tool for urban areas. In Proceedings of the ECCA 2019, European Climate Change Adaptation Conference, Lisbon, Portugal, 28–31 May 2019.

MDPI
St. Alban-Anlage 66
4052 Basel
Switzerland
Tel. +41 61 683 77 34
Fax +41 61 302 89 18
www.mdpi.com

Sustainability Editorial Office
E-mail: sustainability@mdpi.com
www.mdpi.com/journal/sustainability